PHYSICS OF METALS
1. ELECTRONS

THE
PHYSICS OF METALS

1. ELECTRONS

EDITED BY

J. M. ZIMAN

CAMBRIDGE
AT THE UNIVERSITY PRESS
1971

CAMBRIDGE UNIVERSITY PRESS
Cambridge, New York, Melbourne, Madrid, Cape Town, Singapore,
São Paulo, Delhi, Dubai, Tokyo, Mexico City

Cambridge University Press
The Edinburgh Building, Cambridge CB2 8RU, UK

Published in the United States of America by Cambridge University Press, New York

www.cambridge.org
Information on this title: www.cambridge.org/9780521180795

© Cambridge University Press 1969

First published 1969
Reprinted 1971
First paperback edition 2010

A catalogue record for this publication is available from the British Library

Library of Congress Catalogue Card Number: 69–10436

ISBN 978-0-521-07106-2 Hardback
ISBN 978-0-521-18079-5 Paperback

PREFACE

By convention a *Festschrift* is a collection of pretty 'essays' by a miscellany of scholars, presented to an older and greater scholar at some significant milestone of his life. The present work conforms to this convention only in that it was conceived as a tribute to Professor Sir Nevill Mott upon the occasion of his 60th birthday.

The prime intention was to write a modern version of 'Mott and Jones'—a comprehensive and up-to-date account of the Theory of Metals and Alloys. But much has been discovered in 30 years, and this would have stretched to many thousands of pages. We therefore decided to concentrate on the two major topics where our interests mainly lie—the electronic structure and electrical properties of metals, on the one hand, and the mechanical properties of solids on the other. In the end, each topic grew and diverged into a separate book.

The present volume is, therefore, an account of current understanding of the electron theory of metals, with particular reference to band structure, Fermi surfaces and transport properties. Although each chapter is quite long, and deals explicitly and in detail with a separate topic, considerable unity of approach has been achieved by interchange of manuscripts and many cross-references between chapters. For the authors are by no means a miscellaneous lot; at one time or another they have all been professional colleagues of Mott at Bristol or at Cambridge, and therefore share with him, and with one another, many scientific interests and points of view. We had no difficulty in agreeing upon a general plan for the book and writing our separate contributions in a co-operative spirit.

Since there exist many excellent elementary books, there seemed little point in trudging once more over the familiar ground of Bloch states, Brillouin zones, Debye temperatures, Burgers vectors, etc., before reaching the line of battle. This book is intended for research workers and advanced students who have already mastered Mott & Jones (1936), or more recent works such as those of Peierls (1955), Wannier (1959) or Ziman (1964a). On the other hand, the emphasis (as in so much of Mott's own work), is on the *physics* of situations rather than on elaborate mathematical devices, so that no great knowledge of the subtleties of advanced quantum theory is required of the reader.

In a genuine 'monograph' the author always has difficulty in deciding what topics to include; in a collaborative work this problem often settles itself, not very satisfactorily, by default. We tried, for example, to write on all aspects of the study of the Fermi surface, but in the end have said very little about magneto-acoustic phenomena; again, on the vast topic of superconductivity, which ought to come within our title, we are all silent.

But rather than attempting a vague review of the 'state of the art' over a wide field, perhaps it is more useful to give a coherent, detailed account of a limited range of topics, building solid foundations of well-established knowledge as a basis for further progress. In the end, each contributor apologized for the length of his chapter, as if surprised at the amount of learning which he had acquired and was able to impart, on his chosen topic.

The length and solidity of the contributions helps explain the long delay between the conception of the work and its final publication. It is unfortunate in a ceremonial sense that the book will appear only just after Mott's 63rd birthday; but as a scholarly treatise it may well have benefited by this leisurely growth. It is regrettable that several chapters, being completed earlier than others, may not be totally contemporary in their references to recent work—but this is a trivial defect in such authoritative writings.

Despite their labours, the authors of these two volumes will not receive any royalties, which will all be paid over into a trust fund to encourage study and expertise in foreign languages amongst professional scientists. This arrangement, which was suggested by Mott himself, corresponds most felicitously with his own interest in languages and with the wide international circle of his friendships.

It is our duty to record grateful acknowledgement to authors and publishers who have given permission for the reproduction of various diagrams in this book. It is an especial pleasure to say how much we are indebted to Dr Michael Berry, of the University of Bristol, who has undertaken the considerable and responsible task of collating the manuscripts, inserting cross-references, preparing the bibliography and subject index, and generally seeing the book through the press.

J. M. Z.

CONTENTS

[vii]

SIR NEVILL MOTT

MOTT, *Sir Nevill (Francis)*, Kt. 1962; F.R.S. 1936; M.A. Cantab.; Cavendish Professor of Experimental Physics, since 1954; b. 30 Sept. 1905; s. of C. F. Mott, late Dir. of Educn., Liverpool, and Lilian Mary Reynolds; m. 1930, Ruth Horder; two d. Educ.: Clifton College; St John's College Cambridge. Lecturer at Manchester University, 1929–30; Fellow and Lecturer, Gonville and Caius College, Cambridge, 1930–33; Melville Wills Professor of Theoretical Physics in the University of Bristol, 1933–48; Henry Overton Wills Professor and Director of the Henry Herbert Wills Physical Laboratories, University of Bristol, 1948–54. Master of Gonville and Caius College, University of Cambridge, 1959–66. Hughes Medal of Roy. Soc., 1941, Royal Medal, 1953; corr. mem., Amer. Acad. of Arts and Sciences, 1954: President, International Union of Physics, 1951–57; President, Mod. Languages Assoc., 1955; Pres., Physical Soc., 1956–58; Mem. Governing Bd. of Nat. Inst. for Research in Nuclear Science, 1957–60; Mem. Central Advisory Council for Education for England, 1956–59; Mem. Academic Planning Committee and Council of University of Sussex; Chm. Ministry of Education's Standing Cttee. on Supply of Teachers, 1959–62; Mem., Inst. of Strategic Studies; Chairman: Nuffield Foundation's Cttee. on Physics Education 1961; Physics Education Cttee. (Royal Society and Institute of Physics) 1965. Foreign Associate, Nat. Acad. of Sciences of U.S.A., 1957; Hon Mem. Akademie der Naturforscher Leopoldina, 1964. Hon D.Sc. (Louvain, Grenoble, Paris, Poitiers, Bristol, Ottawa, Liverpool, Reading, Sheffield, London). Publications: *An Outline of Wave Mechanics*, 1930; *The Theory of Atomic Collisions* (with H. S. W. Massey), 1933; *The Theory of the Properties of Metals and Alloys* (with H. Jones), 1936; *Electronic Processes in Ionic Crystals* (with R. W. Gurney), 1940; *Wave Mechanics and its Applications* (with I. N. Sneddon), 1938; *Elements of Wave Mechanics*, 1952; *Atomic Structure and the Strength of Metals*, 1956; various contribs. to scientific periodicals about Atomic Physics, Metals, Semiconductors and Photographic Emulsions. Address: The Cavendish Laboratory, Cambridge. Club: Athenaeum.†

A candidate for an academic appointment is expected to provide his *curriculum vitae* and a list of publications: but what really counts are the reports of his referees. It would not be proper to attempt to assess in public the scientific work of a scholar who is still so active, nor to say anything about his style of thought and technical skill. But the list of published books and papers which is included with this volume would not explain the influence that Mott has had over the thinking of a whole generation of physicists and metallurgists. The historians of science will never know of those timely meetings, arranged informally to discuss some current scientific problem, out of which sprouted new subjects and new ideas. They cannot catalogue those

† Reproduced with thanks from *Who's Who*, 1968.

long sessions of gentle quizzing, where many cherished opinions melt away and new theories formulate themselves. They cannot record the strong questioning voice, teasing simple clarity out of high flown complexity at the conference session, nor can they document the enormous correspondence, hand or typewritten, on all manner of scientific, cultural and educational topics, with all manner of friends, colleagues and acquaintances. They cannot measure the influence of his books and scientific writings, nor the impetus he has given to the writings of others, as an editor of journals and of monograph series.

The contributors to these volumes wish to express, on behalf of all those, from all parts of the world, who have enjoyed his companionship and leadership, as colleagues and as visitors at Bristol and at Cambridge, how much they owe to Nevill Mott—to his friendship, his kindness, and his understanding both in scientific and human affairs, and his encouragement and stimulation in their scientific endeavours.

PUBLICATIONS OF PROFESSOR N. F. MOTT

BOOKS

An Outline of Wave Mechanics, pp. 156. Cambridge University Press, 1930.
The Theory of Atomic Collisions (with H. S. W. Massey) pp. 858. third edition, 1965, Clarendon Press, Oxford, 1934.
The Theory of the Properties of Metals and Alloys (with H. Jones), pp. 326. Clarendon Press, Oxford, 1936.
Electronic Processes in Ionic Crystals (with R. W. Gurney), pp. 275. Clarendon Press, Oxford, 1940.
Wave Mechanics and its Applications (with I. N. Sneddon), pp. 393. Clarendon Press, Oxford, 1948.
Elements of Wave Mechanics, pp. 156. Cambridge University Press, 1952.
Atomic Structure and the Strength of Metals, pp. 64. Pergamon Press, 1956.

PAPERS

Classical limit of the space distribution law of a gas in a field of force. *Proc. Camb. Phil. Soc.* **24**, 76–9, 1928.
Solution of the wave equation for the scattering of particles by a Coulombian centre of force. *Proc. Roy. Soc.* A, **118**, 542–9, 1928.
Scattering of fast electrons by atomic nuclei. *Proc. Roy. Soc.* A, **124**, 425–42, 1929.
Interpretion of the relativity wave equation for two electrons. *Proc. Roy. Soc.* **124**, 422–5, 1929.
Wave mechanics of α-ray tracks. *Proc. Roy. Soc.* A, **126**, 79–84, 1929.
Quantum theory of electronic scattering by helium. *Proc. Camb. Phil. Soc.* **25**, 304–9, 1929.
Exclusion principle and aperiodic systems. *Proc. Roy. Soc.* A, **125**, 222–30, 1929.
Collision between two electrons. *Proc. Roy. Soc.* A, **126**, 259–67, 1930.
Scattering of electrons by atoms. *Proc. Roy. Soc.* A, **127**, 658–65, 1930.
Influence of radiative forces on the scattering of electrons. *Proc. Camb. Phil. Soc.* **27**, 255–67, 1931.
Theory of effect of resonance levels on artificial disintegration. *Proc. Roy. Soc.* A, **133**, 228–40, 1931.
Theory of excitation by collision with heavy particles. *Proc. Camb. Phil. Soc.* **27**, 553–60, 1931.
Polarization of electrons by double scattering. *Proc. Roy. Soc.* A, **135**, 429–58, 1932.
Contribution (on the anomalous scattering of fast particles in hydrogen and helium) to the discussion which followed Rutherford's address 'Structure of Atomic Nuclei', *Proc. Roy. Soc.* A, **136**, 735–62, 1932.
(With J. M. Jackson.) Energy exchange between inert gas atoms and a solid surface. *Proc. Roy. Soc.* A, **137**, 703–17, 1932.
(With H. M. Taylor.) Theory of internal conversion of γ-rays. *Proc. Roy. Soc.* A, **138**, 665–95, 1932.

(With C. D. Ellis.) Internal conversion of γ-rays and nuclear level systems of thorium B and C. *Proc. Roy. Soc.* A, **139**, 369–79, 1933.

(With C. D. Ellis.) Energy relations in the β-ray type of radioactive disintegration. *Proc. Roy. Soc.* A, **141**, 502–11, 1933.

(With H. M. Taylor). Internal conversion of γ-rays, Part II. *Proc. Roy. Soc.* A, **142**, 215–36, 1933.

(With C. Zener.) Optical properties of metals. *Proc. Camb. Phil. Soc.* **30**, 249–70, 1934.

(With H. Jones and H. W. B. Skinner.) Theory of the form of the X-ray emission bands of metals. *Phys. Rev.* **45**, 379–84. 1934.

Electrical conductivity of metals. *Proc. Phys. Soc. Lond.* **46**, 680–92, 1934.

Resistance of liquid metals. *Proc. Roy. Soc.* A, **146**, 465–72, 1934.

Discussion of the transition metals on the basis of quantum mechanics, *Proc. Phys. Soc. Lond.* **47**, 571–88, 1935.

Contribution to 'Discussion on supraconductivity and other low temperature phenomena'. *Proc. Roy. Soc.* A, **152**, 1–46, 1935.

Electrical conductivity of transition metals. *Proc. Roy. Soc.* A, **153**, 699–717, 1936.

Thermal properties of an incompletely degenerate Fermi gas. *Proc. Camb. Phil. Soc.* **32**, 108–11, 1936.

Resistivity of dilute solid solutions. *Proc. Camb. Phil. Soc.* **32**, 281–90, 1936.

(With H. R. Hulme, F. Oppenheimer and H. M. Taylor.) Internal conversion coefficient for γ-rays. *Proc. Roy. Soc.* A, **155**, 315–30, 1936.

Optical constants of copper–nickel alloys. *Phil. Mag.* **22**, 287–90, 1936.

Resistance and thermoelectric properties of the transition metals, *Proc. Roy. Soc.* A, **156**, 368–82, 1936.

Energy of the superlattice in β brass. *Proc. Phys. Soc. Lond.* **49**, 258–62, Disc. 263, 1937.

Theory of optical constants of Cu–Zn alloys. *Proc. Phys. Soc. Lond.* **49**, 354–6, 1937.

(With H. H. Potter). Sharpness of the magnetic Curie point, *Nature, Lond.* **139**, 411, 1937.

Cohesive forces in metals. *Sci. Prog. Lond.* **31**, 414–24, 1937.

(With H. Jones.) Electronic specific heat and X-ray absorption of metals and other properties related to electron bands. *Proc. Roy. Soc.* A, **162**, 49–62, 1937.

Conduction of electricity in solids, Bristol Conference, Introduction. Also R. W. Gurney and N. F. Mott, Trapped electrons in polar crystals. *Proc. Phys. Soc. Lond.* **49**, 1–2, 32–5, 1937.

(With R. W. Gurney.) Theory of photolysis of silver bromide and photographic latent image. *Proc. Roy. Soc.* A, **164**, 151–67, 1938.

Theory of photoconductivity. *Proc. Phys. Soc. Lond.* **50**, 196–200, 1938.

(With M. J. Littleton.) Conduction in polar crystals, Part I. Conduction in polar crystals, Part II. (With R. W. Gurney.) Conduction in polar crystals, Part III. *Trans. Faraday Soc.* **34**, 485–511, 1938.

Action of light on photographic emulsions. *Phot. J.* **78**, 286–92, 1938.

Magnetism and the electron theory of metals. *Inst. Physics, Physics in Industry: Magnetism*, 1–15, 1938.

Contribution to discussion of 'Photochemical processes in crystals', by R. Hilsch and R. W. Pohl. *Trans. Faraday Soc.* **34**, 883–8; Disc. 888–92.

Energy levels in real and ideal crystals. *Trans. Faraday Soc.* **34**, 822–7; Disc. 832–4.

Absorption of light by crystals. *Proc. Roy. Soc.* A, **167**, 384–91, 1938.

Contribution to discussion on 'Plastic flow in metals'. *Proc. Roy. Soc.* A, **168**, 302–17, 1938.

Contact between metal and insulator or semiconductor, *Proc. Camb. Phil. Soc.* **34**, 568–72, 1938.

(With R. W. Gurney.) Recent theories of the liquid state. *Reports on Progress in Physics* **5**, 46–63, 1938.

(With R. W. Gurney.) Theory of liquids. *Trans. Faraday Soc.* **35**, 364–8, 1939.

Faraday Society Discussion on 'Luminescence,' *Trans. Faraday Soc.* **35**, 1–240, 1939. R. W. Gurney and N. F. Mott 'Luminescence in solids', 69–73.

Theory of crystal rectifiers. *Proc. Roy. Soc.* A, **171**, 27–38, 1939.

Cu–Cu$_2$O Photo-cells. *Proc. Roy. Soc.* A, **171**, 281–5, 1939.

(With H. Fröhlich.) Mean free path of electrons in polar crystals. *Proc. Roy. Soc.* A, **171**, 496–504. 1939.

Decomposition of metallic azides. *Proc. Roy. Soc.* A, **172**, 325–34, 1939.

Theory of formation of protective oxide films on metals. *Trans. Faraday Soc.* **35**, 1175–7, 1939.

Reactions in solids. *Rep. prog. Phys.* **6**, 186–211, 1939.

Conference on Internal Strains in Solids. *Proc. Phys. Soc. Lond.* **52**, 1–178, 1940. N. F. Mott and F. R. N. Nabarro, Estimation of degree of precipitation hardening, pp. 86–9.

Theory of formation of protective oxide films on metals, Part II. *Trans. Faraday Soc.* **36**, 472–83, 1940.

Photographic latent image. *Photogr. J.* **81**, 63–9, 1941.

Theorie de l'image latent photographique. *J. Phys. Radium* (8), **7**, 249–52, 1946.

Atomic physics and the strength of metals. *J. Inst. Metals* **72**, 367–80, 1946.

The present position of theoretical physics. *Endeavour* **5**, 107–9, 1946.

(With T. B. Grimley.) The contact between a solid and a liquid electrolyte. *Disc. Faraday Soc.* **1**, 3–11, Disc. 43–50, 1947.

Fragmentation of shell cases. *Proc. Roy. Soc.* A, **189**, 300–8, 1947.

The theory of the formation of protective oxide films on metals, III. *Trans. Faraday Soc.* **43**, 429–34, 1947.

L'oxydation des metaux, *J. de Chimie Physique* **44**, 172, 1947.

Slip at grain boundaries and grain growth in metals. *Proc. Phys. Soc.* **60**, 391–4, 1948.

Notes on latent image theory. *Photogr. J.* **88** B, 119–22, 1948.

(With F. R. N. Nabarro.) Dislocation theory and transient creep, Report of conference on 'Strength of solids' (1947). (*Phys. Soc. Lond.*) 1–19, 1948.

A contribution to the theory of liquid helium, II. *Phil. Mag.* **40**, 61–71, 1949.

Theories of the mechanical properties of metals. *Research*, **2**, 162–9, 1949.

The basis of the electron theory of metals, with special reference to the transition metals. *Proc. Phys. Soc. Lond.* A, **62**, 416–22, 1949.

Note on the slowing down of mesons, *Proc. Phys Soc. Lond.* A, **62**, 196, 1949.

Semiconductors and rectifiers, The 40th Kelvin Lecture. *Proc. Inst. Elec. Eng.* **96**, 253, 1949.

(With N. Cabrera.) Theory of the oxidation of metals. *Rep. Prog. Phys.* **12**, 163–84, 1948–49.

Mechanical properties of metals, *Physica* **15**, 119–34, 1949.

(With Y. Cauchois.) The interpretation of X-ray absorption spectra of solids. *Phil. Mag.* **40**, 1260–9.

B

Notes on the transistor and surface states in semiconductors. *Rep. Br. Elec. Res. Ass.* (Ref. L/T 216, 6 pp.), 1949.

(With J. K. Mackenzie.) A note on the theory of melting. *Proc. Phys. Soc. Lond.* A, **63**, 411–12, 1950.

Theory of crystal growth. *Nature, Lond.* **165**, 295–7, 1950.

The mechanical properties of metals. *Proc. Phys. Soc. Lond.* B **64**, 729–41.

Diffusion, work-hardening, recovery and creep. *Solvay Conference, Report,* 1951.

Recent advances in the electron theory of metals. *Prog. Metal Phys.* **3**, 76–114, 1952.

A theory of work-hardening of metal crystals. *Phil. Mag.* **43**, 1151–78, 1952.

The mechanism of work hardening of metals. *39th Thomas Hawksley Lecture; Institution of Mechanical Engineers,* 1952, p. 3.

Dislocations and the theory of solids. *Nature, Lond.* **171**, 234–7, 1953.

Note on the electronic structure of the transition metals. *Phil. Mag.* **44**, 187–90, 1953.

A theory of work-hardening of metals, II. Flow without slip-lines, recovery and creep. *Phil. Mag.* **44**, 742–65, 1953.

Ricerche recenti in teoria dei solidi. Suppl. *Nuovo Cim.* **10**, 212–24, 1953.

Dislocations, plastic flow and creep (Bakerian Lecture). *Proc. Roy. Soc.* A, **220**, 1–14, 1953.

Difficulties in the theory of dislocations. *Proc. Internat. Conf. Theoretical Phys., Kyoto and Tokyo,* Sept. 1953, pp. 565–70.

Creep in metals: the rate determining process, Paper No. 2 of *Symposium on Creep and Fracture of Metals at High Temperatures,* 31 May, 1 and 2 June, 1954, National Physical Laboratory publication, pp. 1–4.

Science and modern languages. *Mod. Lang.* **36**, 45–50, 1955.

A theory of fracture and fatigue. *J. Phys. Soc. Jap.* **10**, 650–6, 1955.

Physics of the solid state. *Advancement of Science,* no. 46, Sept. 1955, pp. 1–9.

Science and education (Thirty-sixth Earl Grey Memorial Lecture). Delivered at King's College, Newcastle upon Tyne, 11 May 1956, printed by Andrew Reid and Co., Newcastle upon Tyne, pp. 1–15.

Creep in metal crystals at very low temperatures. *Phil. Mag.* **1**, 568–72, 1956.

On the transition to metallic conduction in semiconductors. *Can. J. Phys.* **34**, 1356–68, 1956.

Fracture in metals. *J. Iron Steel Inst.* **183**, 233–43, 1956.

Theoretical chemistry of metals. *Nature, Lond.* **178**, 1205–7, 1956.

A discussion on work-hardening and fatigue in metals. *Proc. Roy. Soc.* A, **242**, 145–7, 1957.

(With J. W. Mitchell,) The nature and formation of the photographic latent image. *Phil. Mag.* **2**, 1149–70, 1957.

(With K. W. H. Stevens.) The band structure of the transition metals. *Phil. Mag.* **2**, 1364–86, 1957.

The physics and chemistry of metals. *Year Book of the Physical Society* 1957, pp. 1–13.

A theory of the origin of fatigue cracks. *Acta Met.* **6**, 195–7, 1958.

The transition from the metallic to the non-metallic state. Suppl. *Nuovo Cim.* **7**, 312–28, 1958.

(With T. P. Hoar.) A mechanism for the formation of porous anodic oxide films on aluminium. *J. Phys. Chem. Sol.* **9**, 97–9, 1959.

La rupture des metaux (*J. Iron Steel Inst.* **183**, 233–43, 1956). Paris: Imprimerie Nationale, 1959.

The work-hardening of metals. *Trans. Met. Soc. of AIME* **218**, 962–8, 1960.

(With R. J. Watts-Tobin.) The interface between a metal and an electrolyte. *Electrochim. Acta*, **3**, 79–107, 1961.

The transition to the metallic state. *Phil. Mag.* **6**, 287–309, 1961.

(With W. D. Twose.) The theory of impurity conduction. *Phil. Mag. Suppl.* (*Adv. Phys.*), **10**, 107–63, 1961.

(With R. Parsons and R. J. Watts-Tobin.) The capacity of a mercury electrode in electrolytic solution. *Phil. Mag.* **7**, 483–93, 1962.

The cohesive forces in metals and alloys. *Rep. Prog. Phys.* **25**, 218–43, 1962.

Opening address to the International Conference on the Physics of Semi conductors held at Exeter, July 1962, The Institute of Physics and The Physical Society, 1–4, 1962.

The structure of Metallic Solid Solutions. *J. Phys. Radium*, **23**, 594–6, 1962.

Atomic physics and the strength of metals. The Rutherford Memorial Lecture, 1962, *Proc. Roy. Soc.* A, **275**, 149–60, 1963.

Physics in the Modern World. *Proc. Ghana Acad. Sci.* **2**, 80–4, 1964.

Electrons in transition metals. *Adv. Phys.* **13**, 325–422, 1964.

On teaching quantum phenomena. *Contemp. Phys.* **5**, 401–18, 1964.

The theory of magnetism in transition metals. *Proceedings of the International Conference on Magnetism, Nottingham*, 7–11 September 1964, pp. 67–8.

An outline of the theory of transport properties. '*Liquids: Structure, Properties, Solid Interactions*, pp. 152–71. edited by T. J. Hughel, Elsevier Publishing Company, Amsterdam, 1965.

The electrical properties of liquid mercury. *Phil. Mag.* **13**, 989–1014, 1966.

Electrons in disordered structures. *Adv. Phys.* **16**, 49–144, 1967.

(With R. S. Allgaier.) Localized states in disordered lattices. *Phys. Stat. Sol.* **21**, 343, 1967.

The Solid State. *Scientific Am.* **217**, 80, 1967.

The transition from metal to insulator. *Endeavour*, **26**, 155–8, 1967.

CHAPTER 1

ELECTRONIC STRUCTURE OF METALS†

by VOLKER HEINE‡

1.1 PSEUDISM

'Electronic structure', interpreted widely, covers all that the outer conduction electrons in metals do, and with it practically all solid state properties, in the sense that the energy of a vacancy, for example, is given in terms of the energy of the whole electronic system. The present chapter is concerned with electronic structure that can be treated theoretically from a 'fundamental' point of view, i.e. based on the solution of the Schrödinger equation with more or less well-defined and justifiable approximations. The theories of magnetism and transport properties come within this definition, and form separate chapters. Otherwise, until recent years, it was only the band structure $\mathscr{E}(\mathbf{k})$ of an electron with wave-vector $\mathbf{k}$ travelling through the periodic potential that could be discussed from fundamentals, together with a few immediately related properties such as the electronic specific heat. A phonon spectrum had to be analysed in terms of ad hoc force constants. Now, however, it can be calculated in favourable cases from the same basic potential set up for computing $\mathscr{E}(\mathbf{k})$. Other examples are stacking fault energies, phase transitions under pressure, and the resistivities of liquid metals. For simple metals, the area that can be treated 'fundamentally' is still centred on the band structure, but has begun to expand.

The concept unifying much of what we have to say is that of the *pseudo-potential*. While the term is relatively new in the present context, some of the ideas it draws together predate it qualitatively by twenty years. Recent developments sharpen and exploit them.

Figs. 1.1 and 1.2 show the results of measurements on the Fermi surface of lead (Anderson & Gold, 1965). The arcs of circles are what

† This work was supported in part by the National Aeronautics and Space Administration, U.S.A., while the author was a Visiting Professor at the Physics Department and Institute for the Study of Metals of the University of Chicago, 1965–6.
‡ Dr Heine is University Lecturer in Physics at the Cavendish Laboratory, Free School Lane, Cambridge, and Fellow of Clare College, Cambridge.

the Fermi surface would be for perfectly free electrons, namely, the Fermi sphere cut up by Brillouin zone planes (Fig. 1.1), the pieces being translated by appropriate reciprocal lattice vectors $\mathbf{g}$ and re-assembled in successive bands $\mathscr{E}_n(\mathbf{k})$ inside the fundamental Brillouin zone (Fig. 1.2). The observed Fermi surface can be recognized as a modest distortion from the free-electron model, and the same is true of all other metals studied except the transition, rare earth and actinide

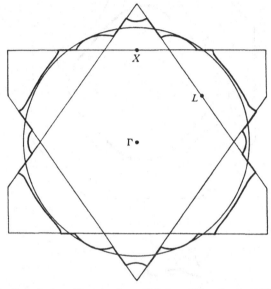

Fig. 1.1. The free electron Fermi sphere (light) and the observed Fermi surface of lead (heavy), shown with Brillouin zone planes in extended **k**-space.

metals with incomplete inner d or/and f shells (see Chapter 2). More-over, the distortions conform qualitatively, and sometimes quanti-tatively, to what would be expected on the basis of the nearly free electron (N.F.E.) approximation.[†] The band structures of the group IV semiconductors diamond, Si, Ge, gray Sn and the III–IV com-pounds have been probed by optical interband transitions, and the band structures inferred from the measurements also interpreted in N.F.E. terms (see, for example, Brust, 1964; Cohen & Bergstresser, 1966). Fermi surface measurements coupled with band structure calculations on the semimetals As, Sb, Bi indicate a N.F.E. situation there, too (Cohen, Falicov & Golin, 1964; Priestley et al. 1967; Lin & Falicov, 1966; Falicov & Lin, 1966). While the Fermi surface studies

[†] Throughout this chapter we shall not define terms that may easily be tracked down through the index of Ziman (1964a).

and optical properties provide detailed information about a part of the band structure, the soft X-ray emission spectra give a rough over-all picture which in bandwidth and shape conforms approximately to free electrons. Although most of the detailed evidence for the N.F.E. picture has been built up in the last ten years, the beginnings

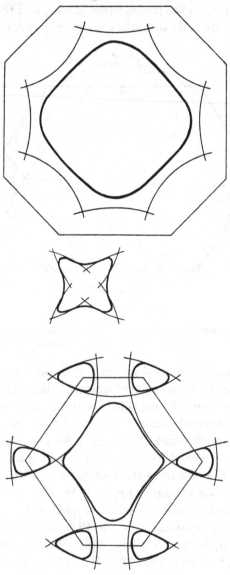

Fig. 1.2. Similar to Fig. 1.1, but with the various parts of the Fermi surface joined together inside the fundamental Brillouin zone.

can already be seen in Mott & Jones' (1936) treatment of diamond and Bi, for example, in N.F.E. terms.

The success of the N.F.E. model for the band structure $\mathscr{E}(\mathbf{k})$ does not imply, however, that the potential $V(\mathbf{r})$ in the solid is weak or can be treated by perturbation theory, as assumed in most textbook presentations of the N.F.E. method. $V(\mathbf{r})$ becomes very strong near the atomic nuclei, much larger than the bandwidth of the conduction electrons and far too strong to be treated as a perturbation. Inside the ion core of the metal atom $V(\mathbf{r})$ is a sufficiently deep potential well to produce several atomic-like oscillations in the wave-function (Fig. 1.3).

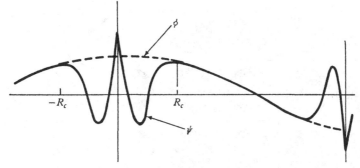

Fig. 1.3. Real wave-function ψ and pseudo-wave-function ϕ of an electron in a N.F.E. metal. R_c is the radius of the ion core.

In order to understand the success of the N.F.E. model, we have to take a broad view of how the electronic structure of a solid is formulated. First, there is the interaction of the electron wave with one atomic centre, and, secondly, there is the multiple scattering from the array of atoms taken together. This separation is present explicitly, for example, in the augmented plane wave (A.P.W.) and Korringa–Kohn–Rostoker (KKR) methods for calculating band structures, where each ion core is surrounded by a sphere of radius R (Fig. 1.4). $\mathscr{E}(\mathbf{k})$ is determined by solving the Schrödinger equation in the Swiss-cheese-like interstitial region, subject to a boundary condition on the spheres given by the radial derivative of the wave-function at R or by the phase shift (Ziman, 1964a, pp. 87–97). We may picture the atom as a black box (Fig. 1.5), and *the electronic structure is completely determined once we specify how strongly any incident wave from outside R, which we may decompose into plane waves $|\mathbf{k}\rangle$, is scattered into the direction $\mathbf{k}+\mathbf{q}$.*

The applicability of the N.F.E. model means that the *net scattering* by an atom can be *weak*, even though the potential is strong. How this is possible follows from considering the scattering expressed in terms of phase shifts η_l. These may be written

$$\eta_l = p_l \pi + \delta_l. \tag{1.1}$$

The integer p_l, chosen so that $|\delta_l| < \tfrac{1}{2}\pi$, counts the number of internal radial nodes. Since the usual phase shift formula for the scattering (Schiff, 1955, p. 105) only involves $\exp(2i\eta_l)$, any multiple of π in (1.1) does not contribute and the scattering is determined by δ_l, which is relatively small in those metals and semiconductors where the N.F.E. model holds.

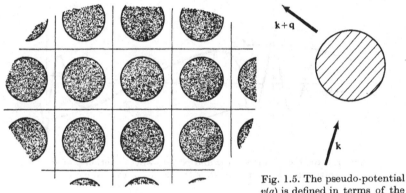

Fig. 1.4.

Fig. 1.5. The pseudo-potential $v(q)$ is defined in terms of the scattering amplitude of an atom, pictured as a black box, for scattering of electrons from $\mathbf{k}$ to $\mathbf{k}+\mathbf{q}$.

In this case it is convenient to define a weak pseudo-potential V_{ps} with small phase shifts equal to δ_l (*not* to the η_l of the real potential) so that it gives just the correct amount of scattering. Alternatively we may demand that the pseudo Schrödinger equation†

$$(-\nabla^2 + V_{ps})\phi = \mathscr{E}\phi \tag{1.2}$$

has the same eigenvalues as the real Schrödinger equation. The V_{ps} is weak, the scattering amplitude of V_{ps} from $\mathbf{k}$ to $\mathbf{k}+\mathbf{q}$ may then be expressed as a highly convergent perturbation series. Formally it is given by the element $\langle \mathbf{k}+\mathbf{q}|t|\mathbf{k}\rangle$ of the t-matrix (Messiah, 1961, pp. 807, 849)

$$t = V_{ps} + V_{ps} G_0 V_{ps} + V_{ps} G_0 V_{ps} G_0 V_{ps} + \ldots \tag{1.3}$$

† We use units with $2m = \hbar = e = 1$, except that energies will normally be given in Rydbergs where $1\ \mathrm{Ry} = 13\cdot6\ \mathrm{eV}$.

where G_0 is the free propagating Green function. The scattering and hence the t-matrix derived from a given potential is of course unique, but the converse is certainly not true: there are many different V_{ps} which give the same scattering, at least over the limited energy range of interest for electronic structure. Indeed the whole possibility of defining a pseudo-potential rests on the fact that another potential can be found with identical scattering to the real potential V. From this point of view the transition from the real potential V to the pseudo-potential represents a partial transformation towards the t-matrix in which the strong inner core part of V giving the radial oscillations in ψ has been t-matrixed.

One explicit V_{ps} demonstrates the weakness of the pseudo-potential very clearly. It is (Austin *et al.* 1962)

$$V_{ps}\phi = V\phi - \sum_c \langle \psi_c, V\phi \rangle \psi_c, \qquad (1.4a)$$

where the ψ_c are the $1s$, $2s$, $2p$, etc., orbitals in the ion core. (Strictly they are the $1s$, etc., solutions of the same Hamiltonian that operates on the valence electrons and so may differ slightly from the actual orbitals of the core.) It is not difficult to verify (Ziman, 1964 a, p. 97) that $(1.4a)$ gives the same valence eigenvalues in (1.2) as V itself does. Since the ψ_c have definite angular momenta l, the second term of (1.2) picks out and operates differently on the different l components of the pseudo-wave-function ϕ. If we consider for a moment only the $l = 0$ component, ϕ is approximately a constant inside the core because radial oscillations have been eliminated and an s-state has no angular nodes, and so ϕ may be taken outside the matrix element in (1.4)

$$V_{ps}\phi \approx [V - \sum_c \langle \psi_c, V \rangle \psi_c]\phi. \qquad (1.4b)$$

The $\langle \psi_c, V \rangle$ are the expansion coefficients of V in terms of the set ψ_c. If we had a complete set, the bracket in $(1.4b)$ would vanish identically. As it is, the ψ_c are a finite set of core orbitals which form quite a good expansion set inside the core. Thus the second term in $(1.4b)$ cancels most of the strong potential V inside the core, as illustrated for a free Si^{4+} ion in Fig. 1.6. In fact, (1.4) is a special case of a more general cancellation theorem developed by Phillips & Kleinman (1959), Antoncik (1959), Cohen & Heine (1961), and Austin, Heine & Sham (1962).

The whole success of the pseudo-potential method, where it is applicable, depends on the perturbation series (1.3) being reasonably rapidly convergent. For many purposes the first term

$$v(q) \equiv \langle \mathbf{k} + \mathbf{q} | V_{ps} | \mathbf{k} \rangle \qquad (1.5a)$$

may suffice, and we abbreviate it to $v(q)$. We may think of it as approximately the scattering amplitude, which is given precisely by (1.3), of the black box representing the atom. Part of the art of choosing pseudo-potentials is to reduce the corrections in (1.3) by making the high (large q) Fourier components $v(q)$ as small as possible. For

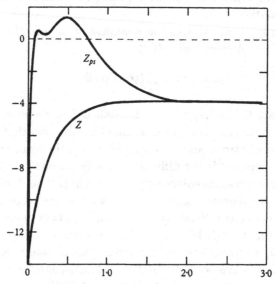

Fig. 1.6. The potential and pseudo-potential (for $l = 0$ states) of a Si^{4+} ion. The potential V is expressed in the form $V(r) = Z(r)/r$ and V_{ps} similarly in terms of $Z_{ps}(r)$. Note V and V_{ps} both become equal to the Coulomb potential $-4/r$ outside the core which has a radius of about one Bohr unit.

calculating band structures this is not too important: modern computers can handle large secular equations. But for calculating phonon spectra and electrical properties, for instance, the use of lowest-order Born approximation $v(q)$ can be a great help. In other problems, such as calculating the formation and migration energies of vacancies and interstitials in diamond-type semiconductors, Bennemann (1964, 1965) found the use of the full t-matrix essential. The degree of complication this causes is obvious from one glance at the papers! Finally, it should be emphasized that the scattering amplitude from $\mathbf{k}$ to $\mathbf{k} + \mathbf{q}$ depends in general not only on q but also on k, $|\mathbf{k} + \mathbf{q}|$ and the energy $\mathscr{E}$.

The same applies to the pseudo-potential, for example, the non-local operator (1.4a), and what we just wrote as $v(q)$ should be written

$$v(q;\, k,\, |\mathbf{k}+\mathbf{q}|,\, \mathscr{E}). \qquad (1.5b)$$

Actually (1.4) does not have an explicit energy dependence but some forms of V_{ps} do, in particular those involving the logarithmic derivative or phase shift which varies with energy.

We have been at pains to stress that electronic structure is a much wider field than calculating $\mathscr{E}(\mathbf{k})$. *Progress in science means unifying more and more experience through well-defined concepts which may be expressed in definite numbers. The scattering power of the atomic black box, or the pseudo-potential for* N.F.E. *solids, serves that purpose for electronic structure.* It remains the same, or nearly the same, whether we are calculating $\mathscr{E}(\mathbf{k})$ for a perfect metal or the frequency of a phonon in it. To a first approximation only the positions of the atoms change and the types of quantities one wishes to calculate depend largely on just these. This has one important consequence: it is not always necessary to calculate from first principles. Once the scattering power or pseudo-potential of an atom has been inferred from one set of experimental data, it may be applied to a study of other measurable properties. Although pseudo-potentials can be calculated with a gratifying accuracy, the precision needed for some applications goes beyond what can currently be achieved, and a pseudo-potential $v(q)$ matched to experiment, even at one or two values of q, is clearly superior. The pseudo-potential involves cancellation between large quantities, as seen in (1.4), and there are in addition many small effects of correlation and exchange. There may therefore be a natural limit, of the order of 0·01 Ry, to the accuracy with which pseudo-potentials will ever be calculated from first principles. After discussing band-structure calculations in §1.2, we therefore indicate in §1.3 how experimentally determined N.F.E. band structures may be analysed in terms of $v(q)$. In §1.5 we indicate how to calculate other properties of N.F.E. metals in terms of $v(q)$, applying this to a discussion of their crystal structures in §1.6.

The *transition metals* cannot of course be described in N.F.E. terms because of the narrow d-bands which have more of a tight binding character, but the general approach in terms of atomic scattering properties remains valid. We expect to be able to write the scattering in terms of a relatively weak pseudo-potential as far as the $l = 0$ and 1 components of the wave-function is concerned, but the $l = 2$

component reacts very strongly with the atomic potential as evidenced by the narrow bandwidth and its scattering has to be treated differently. The electronic structure of transition metals will be formulated in these terms in § 1.4. The noble metals form a borderline case. If the whole band structure is being considered, then they must be treated as transition metals with their d-bands. If, on the other hand, only states near the Fermi level above the d-band are significant, then a relatively simple pseudo-potential including the indirect effects of the d-bands may suffice.

Section 1.7, which logically should come at the beginning of the whole chapter, discusses what meaning can be attached to the 'one-electron' band structure from the many-body point of view, taking into account all the interactions between the electrons.

1.2 BAND-STRUCTURE CALCULATIONS

The approach developed in § 1.1, based on 'black box' atoms, depends partially on the fact that the potential in a metal or semiconductor is relatively flat in the interstitial region between the spheres (Fig. 4), at least when these are made large enough to touch the sides of the atomic polyhedron. Indeed, the potential is normally assumed to be a constant in this region in the Korringa–Kohn–Rostocker (KKR) and augmented plane wave (A.P.W.) methods for calculating the band structure $\mathscr{E}(\mathbf{k})$, though this simplification is not absolutely necessary for the latter. The electrons may be regarded as propagating approximately as plane waves between the atoms, and we expect $\mathscr{E}(\mathbf{k})$ to be defined by a secular equation of the form

$$\det \|\{(\mathbf{k} - \mathbf{g})^2 - \mathscr{E}\} \delta_{\mathbf{g}\mathbf{g}'} + \Gamma_{\mathbf{g}\mathbf{g}'}\| = 0, \qquad (1\cdot6)$$

where the $\mathbf{g}$'s are the reciprocal lattice vectors. $\Gamma_{\mathbf{g}\mathbf{g}'}$ *represents the atomic scattering*, plus that of the interstitial potential if any.

It is well known that the secular equation of the *orthogonalized plane wave* (O.P.W.) method is of the form (1.6), with $\Gamma_{\mathbf{g}\mathbf{g}'}$ (O.P.W.) the matrix element of the original O.P.W. form of the pseudo-potential

$$\Gamma_{\mathbf{g}\mathbf{g}'}(\text{O.P.W.}) = V_{\mathbf{g}-\mathbf{g}'} + \sum_c (\mathscr{E} - \mathscr{E}_c)\langle \mathbf{k} - \mathbf{g}|c\rangle\langle c|\mathbf{k} - \mathbf{g}'\rangle. \qquad (1.7)$$

The details of this and other band-structure methods have been described well in several texts (see, e.g. Ziman, 1964a, chapter 3, and Callaway, 1964), and we will restrict ourselves to a few comments on the practical state of the art. In the orthogonalized plane wave (O.P.W.)

method, convergence becomes very slow beyond about 0·01 Ry, because the representation of the inner oscillations of ψ in terms of plane waves and core functions ψ_c (Ziman, 1964a, p. 94) is only an *ad hoc* 'Ansatz' and it requires plane waves of very short wavelength to make the final corrections to ψ (Abarenkov & Heine, 1965). The success of the method depends on the fact that the Ansatz is remarkably good for atomic potentials which become progressively steeper near the origin. It does not work for a square well with infinitely high walls, for example. In a typical metal such as Al, the bandwidth is of the order of 1 Ry, the band gaps of order 0·1 Ry. An accuracy of 0·01 Ry in calculation is therefore often adequate. The virtue of the o.p.w. method is its relative simplicity, still falling as it does within the competence of an enthusiastic amateur. Its short-coming, besides the question of convergence, is its unsuitability for transition metals and anything approaching an ionic compound. However, Deegan & Twose (1967) have successfully adapted the o.p.w. method to d-bands.

The *augmented plane wave* (A.P.W) method overcomes the convergence problem of o.p.w.s by an exact integration of the wave equation inside the atomic sphere, and gives eigenvalues to 10^{-3} Ry without difficulty. It also copes with d-bands and has been applied to a few compounds and stoichiometric alloys, e.g. TiC, V_3Si (Earn & Switendick, 1965; Mattheiss, 1965). Its secular equation has

$$\Gamma_{gg'}(\text{A.P.W.}) = (4\pi R^2/\Omega)\left\{-[(\mathbf{k}-\mathbf{g}).(\mathbf{k}-\mathbf{g}')-\mathscr{E}]\frac{j_1(|\mathbf{g}-\mathbf{g}'|R')}{|\mathbf{g}-\mathbf{g}'|}\right.$$
$$\left.+\sum_{l=0}^{\infty}(2l+1)P_l(\cos\theta_{gg'})j_l(|\mathbf{k}-\mathbf{g}|R)j_l(|\mathbf{k}-\mathbf{g}'|R)\frac{\mathscr{R}_l'(R,\mathscr{E})}{\mathscr{R}_l(R,\mathscr{E})}\right\}$$

(1.8)

(where $\theta_{gg'}$ is the angle between $\mathbf{k}-\mathbf{g}$ and $\mathbf{k}-\mathbf{g}'$, Ω the volume of the unit cell, and $\mathscr{R}'$ the derivative of the radial wave-function $\mathscr{R}$). An A.P.W. calculation may take two years to develop from scratch. However, at least one experimentalist has learnt to use existing programmes to calculate a band structure as an aid in interpreting his data. The The o.p.w. and A.P.W. methods have been extended to include spin-orbit coupling, and in the case of A.P.W. all other relativistic effects (Weisz, 1966; Loucks, 1965a).

In the *Korringa–Kohn–Rostocker* (KKR) method, the propagation of the electron wave between the inscribed spheres of Fig. 1.4 is taken into account exactly through the KKR structure constants related to the free particle Green function. As in the A.P.W. method, the wave equation is integrated exactly inside the sphere, and the inside and outside solutions matched at the radius R. In this case a secular equation in angular momenta is obtained, instead of in reciprocal lattice vectors $\mathbf{g}$ as with A.P.W.s. Compared with O.P.W.s or A.P.W.s, the method is very easy to use, *if* the structure is *bcc* or *fcc* for which the required structure constants have been tabulated. Even for these the constants are only available along symmetry lines in the Brillouin zone, but these may serve to give quite a good picture of the band structure. Such calculations can be very useful as a guide in interpreting, for example, de Haas–van Alphen measurements (see §2.2.1.) on the Fermi surface. There is no problem about d-bands, and the method has been applied to the alloy CuZn (Johnson & Amar, 1965). It has also been extended to relativistic effects (Onodera *et al.* 1966; Treusch, 1967). A numerical accuracy of 10^{-3} Ry is achieved without difficulty, and the A.P.W. and KKR methods have been tested against each other by applying both to the same potential (Segall, 1961; Burdick, 1963), as have the O.P.W. and KKR (Segall, 1961). Incidentally, the cellular method has in the past suffered from some severe difficulties which Altmann and co-workers report to have overcome (Altmann, 1958). So far there are no results which can be checked against existing KKR or A.P.W. calculations on exactly the same potential.

From what has already been said, the KKR method is evidently closely related to the A.P.W. This becomes even clearer on carrying out a mathematical transformation on the KKR secular equation, which brings it to the form (1.6) with (Ziman, 1965)

$$\Gamma_{\mathbf{gg'}}(\text{KKRZ})$$
$$= -\frac{4\pi}{\kappa\Omega}\sum_{l}(2l+1)\tan\eta'_{l}\frac{j_{l}(|\mathbf{k}-\mathbf{g}|R)\,j_{l}(|\mathbf{k}-\mathbf{g'}|R)}{j_{l}(\kappa R)\,j_{l}(\kappa R)}\,P_{l}(\cos\theta_{\mathbf{gg'}}),\quad(1.9)$$

where
$$\cot\eta'_{l} = \cot\eta_{l} - n_{l}(\kappa R)/j_{l}(\kappa R),\quad(1.10)$$

and $\kappa^2 = \mathscr{E}$. Here $\Gamma(\text{KKRZ})$ is not identical with $\Gamma(\text{A.P.W.})$, but Morgan (1966) has shown that

$$\Gamma(\text{A.P.W.}) = \Gamma(\text{KKRZ}) + \Gamma^{0},\quad(1.11)$$

where the difference Γ^{0} between them is an 'empty lattice' term, which however is not identically zero. The secular equation based on (1.9) presumably has excellent convergence in $\mathbf{g}, \mathbf{g'}$ similar to the A.P.W.

method because it is based in the same way on exact solution of the
wave equation inside R. The summation over l in $\Gamma(\mathrm{KKRZ})$ converges much more rapidly than in $\Gamma(\mathrm{A.P.W.})$. The latter involves $\mathscr{R}_l$
directly, which tends to the free spherical wave j_l at large l because
the radial wave equation is dominated by the 'centrifugal term'
$l(l+1)/r^2$, whereas $\Gamma(\mathrm{KKRZ})$ depends through the phase shift only
on the deviation of $\mathscr{R}_l$ from j_l. The large l components of $\Gamma(\mathrm{A.P.W.})$
contribute purely to Γ^0 in (1.11).

It was emphasized in § 1.1 that *many* pseudo-potentials can be written
down, especially if we allow them to be energy dependent, which give
identical scattering. Thus the matrix elements $\Gamma(\mathrm{O.P.W.})$, $\Gamma(\mathrm{A.P.W.})$ and
$\Gamma(\mathrm{KKRZ})$ all define pseudo-potentials† which can be written down
in $\mathbf{r}$ (real) space (Lloyd, 1965), and it would have been consistent with
the notation (1.5) of §1.1 to have written v instead of Γ in (1.6) to
(1.9). If the phase shifts are near $p_l\pi$ in the sense of (1.1) and the scattering weak, it follows by inspection in the KKRZ case, for example,
that the corresponding pseudo-potential is small. Having identified
Γ in (1.6) with the pseudo-potential, we note that (1.6) is just the
secular equation of the pseudo-Schrödinger equation (1.2), and the
eigenvectors of (1.6) must be the Fourier expansions of the pseudo-
wave-functions ϕ. In N.F.E. cases where V_{ps} has eliminated the inner
radial nodes of the wave-function and V_{ps} is weak, ϕ will be approximately a plane wave (or simple linear combination of plane waves) such
as pictured in Fig. 1.3. Outside the core of the atom or the sphere of
radius R (Fig. 1.4), the ϕ must still correctly describe the electron
motion between the atoms. In other words $\phi_\mathbf{k}$ must be equal to the
real wave-function $\psi_\mathbf{k}$ there, apart from some normalizing constant.
This is well known in the case of O.P.W.s, but Morgan (1966) and Slater
(1966) have proved it also true for the KKRZ and A.P.W. secular
equations. Indeed the A.P.W. form gives the most rapidly convergent
plane-wave expansion of the wave-function in the interatomic region
(Johnson, 1966).

Quite a different approach to setting up the pseudo-potential is
illustrated by the model potential of Abarenkov & Heine (1965),

† A few words about nomenclature: originally the word 'pseudo-potential' was
applied only to those of a certain mathematical class involving projection onto the
core states as in (1.4) and (1.7). Such a distinction seems no longer appropriate, and
we apply the term to any potential which has the same phase shifts as the real
potential but with the multiples of π removed, i.e. it has been made as weak as
possible. In the case of d-bands, this is not weak enough for the band structure to
be N.F.E.-like. A 'model potential' is a pseudo-potential of a simple functional
form with adjustable parameters to reproduce the correct scattering.

namely, a simple square well of depth A inside some model radius R_M and the appropriate Coulomb potential outside (Fig. 1.7)

$$V_{ps}^b = -\sum_l A_l(\mathscr{E})\,\mathscr{P}_l \quad (r < R_M)$$

$$= -z/r \qquad (r > R_M). \qquad (1.12)$$

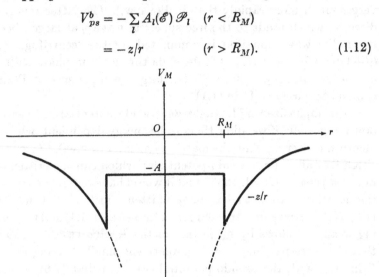

Fig. 1.7. A model pseudo-potential for an ion of charge z.

This is the pseudo-potential of the bare ion core of charge z to which has to be added the potential from the conduction electrons as detailed below. The well depth A can be adjusted so that (1.12) reproduces exactly the spectroscopically observed energy levels of one electron added to the ion. A depends on the angular momentum l, and $\mathscr{P}_l$ in (1.12) is a projection operator to pick out that component of the total wave-function. A_l also has to depend slightly on $\mathscr{E}$ to fit the whole series of levels of given l. The matrix elements, between plane waves, of the non-local operator (1.12) can be calculated analytically without difficulty and inserted in the secular equation. The model radius R_M may be so chosen that there is little discontinuity in V_{ps} at R_M, thus reducing the high Fourier components of V_{ps} as much as possible (Animalu & Heine, 1965) in order to get the best convergence of (1.3).

Before any of these methods can be applied to calculate $\mathscr{E}(\mathbf{k})$, the complete potential V or pseudo-potential V_{ps} in the solid has of course to be set up. The potential of the ion core is often taken from a Hartree–Fock calculation. On the other hand, the use of a model potential such as (1.12) fitted to the spectroscopic energy levels has the advantage of

c

automatically including exchange with the core orbitals and all internal correlation effects. In some cases, such as the d-states in noble metals, a different functional form inside R_M has been found better than the square well. To obtain A_l for $l \geqslant 1$ at the energy required in the solid, often entails some extrapolation from the atomic energy levels where A_l has been determined, and this can introduce a little uncertainty in the method.

As regards the *conduction electrons*, often the potentials of neutral free atoms are taken and simply superposed. These are usually from Hartree–Fock–Slater calculations (HFS, with the Slater (1951a) approximation for exchange), which have been published for all elements. After this potential of superposed atoms has been constructed, it has to be averaged to spherical symmetry inside the radius R for use in the A.P.W. or KKR methods. It is also approximated to by a constant (some average value) between the spheres for the KKR method, while variations about this constant can be incorporated with A.P.W.s if desired. Exchange with the core and conduction electrons is then usually treated by the HFS approximation. Although the procedure of superposing neutral atoms is somewhat arbitrary, it has yielded many useful answers (see, for example, Loucks, 1965b). Herman (1964) has used O.P.W.s to solve for the wave-functions and made the whole potential self-consistent, again within the HFS scheme.

The use of pseudo-potentials allows us to approach the whole problem of setting up a self-consistent potential for the conduction electrons in another way (Cohen & Phillips, 1961). We first need the pseudo-potential V_{ps}^b of a bare ion, usually calculated from (1.7) or the model potential (1.12), though there is no reason why other forms should not be used. The next step is to treat the conduction electrons as a uniform negative jelly into which the ions are planted. The pseudo-potential of the whole system is

$$(\text{const.}) + \sum_{\mathbf{g}}' v^b(g) \exp(i\mathbf{g}.\mathbf{r}), \qquad (1.13)$$

$$v^b(g) = \Omega^{-1} \int V_{ps}^b \exp(-i\mathbf{g}.\mathbf{r}) \, \mathbf{dr}. \qquad (1.14)$$

(We write $v^b(g)$, leaving the other variables of (1.5) understood.) The electron gas does not contribute to the Fourier components $v^b(g)$ in (1.14) because it is uniform. It only cancels the infinite $\mathbf{g} = 0$ component which the ions alone would give. This point can be proved in detail as follows. Let us think of V_{ps}^b as a real electrostatic potential, instead of as a pseudo-potential, and associate a charge density $\rho(\mathbf{r})$ with it: this step is immaterial to the mathematics of the proof but helps to visualize it. Clearly the system is neutral and the $\mathbf{g} = 0$ component of the total charge density is zero. The $\mathbf{g} \neq 0$ components $\rho_{\mathbf{g}}$ of the total charge density come purely from the ionic $\rho(\mathbf{r})$, and hence so do the Fourier components $V_{\mathbf{g}}$ of the total potential, since Poisson's equation in Fourier transform reads

$$g^2 V_{\mathbf{g}} = 4\pi\rho_{\mathbf{g}}. \tag{1.15}$$

We have therefore identified the $\mathbf{g} \neq 0$ components $V_{\mathbf{g}}$ of the total neutral potential with those $v^b(g)$ of a bare ion, q.e.d. Incidentally it follows from (1.15) that the constant in (1.13), which determines the absolute position of the bottom of the band, is left indeterminate as $0/0$ by the present approach. We will show how to calculate it from (1.12) in §1.6 in connection with the total cohesive energy.

In order to set up the *self-consistent potential*, we now allow the conduction electrons to react with the 'bare' pseudo-potential (1.14) and screen it. A Fourier component $\rho_{\mathbf{g}}$ of the conduction electron charge density is set up, proportional to the 'applied' potential $v^b(g)$ in lowest order of perturbation theory if we treat $v^b(g)$ as weak. (The method therefore only applies in N.F.E. cases.) The result of a self-consistent calculation is to reduce $v^b(g)$ to

$$v(g) = v^b(g)/\epsilon(g), \tag{1.16}$$

where $\epsilon(g)$ is the appropriate screening factor or '*dielectric constant*' (Ziman, 1964a, pp. 126–9). The pseudo-potential for the whole system is then

$$V = (\text{const.}) + \sum_{\mathbf{g}}' v(g) \exp(i\mathbf{g}.\mathbf{r}). \tag{1.17}$$

The point is that since the pseudo-potential acts on pseudo-wave-functions which are plane waves before being perturbed, we can take $\epsilon(g)$ from the theory of the free-electron gas.

Such is the principle of the method, but there are several points of detail to be inserted. Since the V_{ps}^b is non-local, the screening is not simply given by a multiplicative factor as in (1.16) and has to be calculated non-locally (Harrison, 1963; Animalu, 1965). The charge density of the conduction electrons is not uniform, even in the lowest approximation, because of the oscillations of $\psi_{\mathbf{k}}$ in the core (Fig. 1.3), resulting in a reduced density there (the 'orthogonality hole') and a heaping up to a density $z(1+\alpha)$ electrons per atom outside the core. Here α is the order of $0\cdot1$ and the effect may be incorporated with

V_{ps}^b in (1.15). Exchange and correlation with the core electrons is included in V_{ps}^b of the ion. Exchange and correlation with the conduction electrons produces a hole which moves with the electron and contributes an exchange and correlation energy $\mu_{xc}(\mathbf{k})$ for the state $\mathbf{k}$. To a first approximation it is uniform because the electron gas is uniform, but it fluctuates somewhat due to the charge density fluctuation ρ_g. That contribution can be included in (1.16), the best method at present probably being to calculated $\epsilon(g)$ with a short range screened exchange interaction treated in the Hubbard approximation (Sham, 1963, 1965). We obtain

$$\epsilon(g) = 1 - (1+\alpha)\left(\frac{8\pi e^2}{\Omega g^2}\right)\left[1 - \frac{\tfrac{1}{2}g^2}{g^2 + k_F^2 + k_s^2}\right]\chi(g), \qquad (1.18)$$

where
$$\chi(g) = -\frac{z}{2}(\tfrac{2}{3}\mathscr{E}_{FO})^{-1}\left(\frac{1}{2} + \frac{4k_F^2 - g^2}{8gk_F}\ln\left|\frac{2k_F - g}{2k_F + g}\right|\right). \qquad (1.19)$$

Here $\mathscr{E}_{FO}$ is the free electron Fermi energy $\hbar^2 k_F^2/2m$. The factor in square brackets in (1.18) comes from the exchange, and k_s is the screening parameter taken as $(2k_F/\pi)^{\frac{1}{2}}$ in atomic units. The effect of the orthogonality hole in modifying the screening is taken into account crudely by the factor $1+\alpha$ in (1.18).

These are small points. The usefulness of the dielectric screening method depends on the fact that $\epsilon(g) \to 1$ as $g \to \infty$ and is already of order 1·2 at the first reciprocal lattice vectors. Thus a 10 % error in the screening, i.e. in $\epsilon(g) - 1$, results only in a 2 % error in $v(g)$ in (1.16). The $\mu_{xc}(\mathbf{k})$ can be calculated at the Fermi level from formulae for the total exchange and correlation of a free electron gas. As nearly as theory or experiment can tell (i.e. to about 10–20 %, see, for example Pines, 1955), $\mu_{xc}(\mathbf{k})$ may be taken as constant throughout the band. Finally, there is a rather small and rather uncertain correlation correction which may be added, coming from the fact that correlation with core electrons and with conduction electrons is not additive as assumed implicitly so far. We merely mention these points to indicate how far it is possible now to treat all the interactions between nonlocality, self-consistency including the orthogonality hole, exchange and correlation (see, e.g. Animalu & Heine, 1965; Harrison, 1966). In one calculation on Si self-consistency, exchange and correlation were even computed with the calculated Bloch functions (Phillips & Kleinman, 1962).

A complete review of band-structure calculations up to 1962 may be found in Callaway (1964), with its useful list of references including the titles of all the papers to serve as one-line abstracts. Some further calculations are listed by Heine (1965a). A good general text is Slater (1965).

1.3 ANALYSIS OF N.F.E. BAND STRUCTURES

In §1.1 it was emphasized that the band structures of many metals and semiconductors can be described in N.F.E. terms with a pseudo-potential $v(g)$ for scattering by a reciprocal lattice vector $\mathbf{g}$. In §1.2 we saw how to calculate $v(g)$, but for really useful results one needs to achieve an accuracy of 0·01 Ry or better, which is difficult when one has cancellations between quantities inherently of magnitude 1 Ry in a complicated self-consistent many-body system. We shall therefore discuss the analysis of experimentally measured band structures to yield the '*observed*' pseudo-potential $v(g)$. These pseudo-potentials can then be correlated with atomic properties to describe trends in the band structures across the periodic table, or used to calculate other properties of the metals as in §1.5.

Let us start by treading the historical path, following the analysis of Harrison (1959) and Ashcroft (1963) on the shape of the *Fermi surface* of aluminium as determined by the de Haas–van Alphen effect (§2.2.1). From the general picture presented in §1.1, we would expect a N.F.E. model to apply. Only the {111} and {200} Brillouin zone planes cut the Fermi sphere, and around the corner $W_1 = (\pi/a)\,(2, 0, 1)$ of the Brillouin zone, only the plane waves $|\mathbf{k} - \mathbf{G}_n\rangle$ with

$$\left. \begin{array}{ll} \mathbf{G}_1 = 0, & \mathbf{G}_2 = (\pi/a)\,(\overline{2}, 2, \overline{2}), \\ \mathbf{G}_3 = (\pi/a)\,(\overline{4}, 0, 0), & \mathbf{G}_4 = (\pi/a)\,(\overline{2}, \overline{2}, \overline{2}), \end{array} \right\} \tag{1.20}$$

mix strongly into the pseudo-wave-function $\phi_{\mathbf{k}}$ because they lie near the corners W_1 to W_4 (Fig. 1.8) and are nearly degenerate. We need only investigate $\mathscr{E}(\mathbf{k})$ in $\frac{1}{48}$th part of the zone around W_1 because of the cubic symmetry. All other plane waves $|\mathbf{k} - \mathbf{g}\rangle$ have only a minor effect on $\mathscr{E}(\mathbf{k})$ in this region, and we therefore try to fit the band structure simply with the 4×4 secular equation

$$\det \|\{(\mathbf{k} - \mathbf{G})^2 - \mathscr{E}\}\,\delta_{\mathbf{GG'}} + \bar{v}_{\mathbf{GG'}}\| = 0, \tag{1.21}$$

$$\mathbf{G}, \mathbf{G'} = 1 \quad \text{to} \quad 4.$$

The matrix elements $\bar{v}_{\mathbf{GG'}}$ are presumably not exactly the pseudo-potential v, and we have added the bar to denote this, because (1.21) is only a small 4×4 piece out of the full secular equation (1.6). The $\bar{v}_{\mathbf{GG'}}$ are of two types according to whether $\mathbf{G} - \mathbf{G'}$ is a (111) or (200) type reciprocal lattice vector, and the whole of the complicated Fermi surface was well represented by the two parameters (Ashcroft, 1963)

$$\bar{v}(111) = 0 \cdot 0179 \, \mathrm{Ry}, \quad \bar{v}(200) = 0 \cdot 0562 \, \mathrm{Ry}. \tag{1.22}$$

The number of decimal places witnesses to the accuracy of the fit, much higher than one could expect from fundamental calculations of

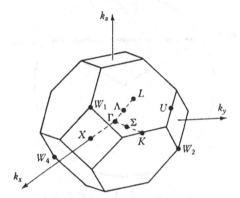

Fig. 1.8. Brillouin zone for face-centred cubic-lattice, showing points of high symmetry.

pseudo-potential except by chance. But it is at first sight very surprising that such a good fit was obtained because (a) non-locality† of the $\bar{v}$ has apparently been ignored, and (b) the secular equation truncated to 4×4. We shall now probe these two points in greater depth, and at the same time relate $\bar{v}$ more precisely to the pseudo-potential; this is essential if we are to apply the knowledge gained in (1.22) elsewhere.

The first point is easily disposed of. In a N.F.E. calculation, a particular matrix element $\langle \mathbf{k} - \mathbf{G} | \bar{v} | \mathbf{k} - \mathbf{G'} \rangle$ only affects the $\mathscr{E}(\mathbf{k})$ strongly if the unperturbed plane waves $| \mathbf{k} - \mathbf{G'} \rangle$ and $| \mathbf{k} - \mathbf{G} \rangle$ are nearly degenerate. The assumption therefore is that $\bar{v}_{\mathbf{GG'}}$ may be approximated over the region of $\mathbf{k}$ where it has an important influence by its value at the limiting point of degeneracy. Thus $\bar{v}(111)$ and $\bar{v}(200)$ in (1.22)

---

† A local operator F is a function of $\mathbf{r}$ only, and $F\phi$ denotes $F(\mathbf{r}) \, \phi(\mathbf{r})$. A non-local operator depends on two variables, $F(\mathbf{r}, \mathbf{r'})$, and $F\phi$ is $\int F(\mathbf{r}, \mathbf{r'}) \, \phi(\mathbf{r'}) \, d\mathbf{r'}$. A local operator has matrix elements $\langle \mathbf{k} + \mathbf{q} | F | \mathbf{k} \rangle$ which depend on $\mathbf{q}$ only, whereas those of a non-local F depend on $\mathbf{k}$ and $\mathbf{k} + \mathbf{q}$, or on $|\mathbf{k}|$, $|\mathbf{k} + \mathbf{q}|$ and $|\mathbf{q}|$ in the spherically symmetric case. Pseudo-potentials are in general non-local, as emphasized in connection with (1.5).

represent $\bar{v}$ for scattering through these reciprocal lattice vectors *on the Fermi sphere*. The geometry is shown in Fig. 1.9. This *on-Fermi-sphere approximation* can apparently remain very good, even though the pseudo-potential is really quite non-local (Fig. 1.10), and it is frequently used in the applications mentioned in §1.5.

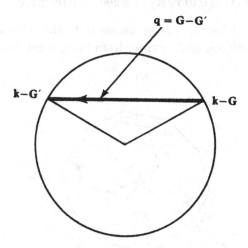

Fig. 1.9. Scattering on the Fermi sphere.

We turn now to the relation of $\bar{v}$ to v, and the truncation of the secular equation. The latter is in fact not an approximation but the result of a rigorous mathematical transformation. Consider first the infinite order secular equation (1.6) in terms of $v_{GG'}$ matrix elements, and partition it thus

$$\left\| \begin{array}{c|c} \mathbf{hh'} & \mathbf{hG'} \\ \hline \mathbf{Gh'} & \mathbf{GG'} \end{array} \right\| = 0 \qquad (1.23)$$

where $\mathbf{h}$ denotes a higher $\mathbf{g}$ not of the set (1.20). Apply now a transformation of type

$$\left| \begin{array}{cc} A & B \\ \tilde{B}* & C \end{array} \right| \times \left| \begin{array}{cc} I & -A^{-1}B \\ -C^{-1}\tilde{B}* & I \end{array} \right| = \left| \begin{array}{cc} A - BC^{-1}\tilde{B}* & 0 \\ 0 & C - \tilde{B}*A^{-1}B \end{array} \right|. \quad (1.24)$$

The original $A B \tilde{B}*C$ secular equation (1.23) is split into two separate ones in the upper left and lower right corners. In this way we can fold the infinite secular equation (1.6), (1.23) into a 4×4 secular equation like (1.21) which gives the lowest four eigenvalues *exactly*. The terms may be evaluated by calculating the A^{-1} in (1.24). In our case we expand it by perturbation theory which identifies $\bar{v}$ in (1.21) as

$$\bar{v}_{GG'} = v_{GG'} + \sum_{h} \frac{v_{Gh} v_{hG'}}{\mathscr{E} - (k-h)^2} + \cdots, \tag{1.25}$$

$$v_{gg'} = \langle k - g | v(\mathscr{E}) | k - g' \rangle. \tag{1.26}$$

The price that has to be paid for folding into a finite secular equation
is that the second- and higher-order terms in (1·25) introduce a k
and $\mathscr{E}$ dependence into the matrix elements of the secular equation.
In the case of pseudo-potentials such k and $\mathscr{E}$ dependence is there
already, so nothing is lost if the correction terms are small. For Al,
Animalu (1965) calculated them to be about 0·005 Ry. For calculating
the shape of the Fermi surface the $\mathscr{E}$ dependence is immaterial since
$\mathscr{E}$ is set equal to a constant $\mathscr{E}_F$. Incidentally, the $\mathscr{E}$ dependence of
$v(g)$ does contribute to the band gap which, at the centre of a zone face,
is equal to
$$\mathscr{E}_s - \mathscr{E}_p = \langle -\tfrac{1}{2}g | v(\mathscr{E}_s) | \tfrac{1}{2}g \rangle + \langle -\tfrac{1}{2}g | v(\mathscr{E}_p) | \tfrac{1}{2}g \rangle, \tag{1.27}$$

where $\mathscr{E}_s$ and $\mathscr{E}_p$ are the s-like and p-like states at the band gap.

Analyses of band structure need not, of course, be limited to the
Fermi surface, though much of the most accurate experimental information relates to that. In a study of Si and Ge (Brust, 1964; see also
similar work on compound semiconductors by Cohen & Bergstresser,
1966) prominent features of the band structures over a range of several
eV were inferred from *optical* and other data. These were then fitted
successfully using the extended secular equation (1.6) with $v(111)$,
$v(200)$, $v(311)$ as three (constant) parameters and all higher $v_{gg'}$
set equal to zero. The fact that constant matrix elements suffice, in
spite of the real non-locality of the pseudo-potential, presumably again
reflects the on-Fermi-sphere approximations similar to Fig. 1.9.
That no energy dependence was necessary in the matrix elements is
perhaps more remarkable, particularly in the diagonal ones. Suffice it
to note that the range of $\mathscr{E}(k)$ considered is not really very wide compared with $\mathscr{E}_F$, that there are indeed some small systematic discrepancies between the best fit and experiment which could be interpreted
as an energy dependence (M. L. Cohen, private communication), and
that the $\mathscr{E}(k)$ of diamond calculated with o.p.w.s deviates systematically from the pseudo-potential fit as one moves further from the Fermi
level (F. Herman, private communication).

When fitting band structures, it may be possible to treat the different
$v(g)$ or $\bar{v}(g)$ all as disposable parameters as in the cases discussed above.
However there may be situations when an even simpler model is
required, for example, in a rough first trial fitting, or when a more

complicated crystal structure introduces too many non-equivalent reciprocal lattice vectors, or when a pseudo-potential is being fitted to some other measured property (§1.5). Here the *one-parameter* '*empty core*' *model of Ashcroft* (1966a) may prove useful. The cancellation theorem and Fig. 1.6 suggest the bare pseudo-potential of an ion

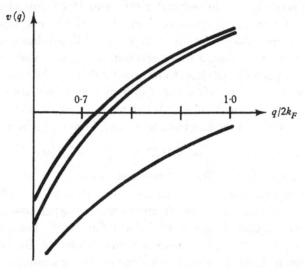

Fig. 1.10. Pseudo-potential $v(q)$ for Al, showing its non-local nature. The top and bottom curves are for backward and forward scattering respectively ($\mathbf{k}$ and $\mathbf{k}+\mathbf{q}$ parallel and antiparallel in (1.5)), whereas the middle curve is for scattering on the Fermi sphere, as in Fig. 1.9 (A. Animalu, private communication).

be taken as zero inside some adjustable radius R_A which is nearly that of the ion core, and $-z/r$ outside. This gives

$$v_{\text{loc, GG'}} = -\frac{4\pi e^2 z}{\Omega g^2 \epsilon(g)} \cos(g R_A), \qquad (1.28)$$

where $\epsilon(g)$ is the screening factor (1.18) and $g = |\mathbf{G} - \mathbf{G}'|$. A second parameter can of course be introduced by giving the well a depth as in Fig. 1.7.

As band structures become more precisely established, it is unlikely that the on-Fermi-sphere approximation of constant matrix elements will suffice. This has already been found in Mg where there is a large amount of very precise de Haas–van Alphen data (R. Stark, private communication), as well as in Bi (Golin, 1968) and K (L. Falicov & M. Lee, private communication). The need then arises to represent the non-locality simply but realistically. We can always split the pseudo-potential (in **r** space) into some mean local potential which acts equally on all l components of ϕ, plus some non-local parts. The local part gives constant matrix elements and may be treated as before. The empty core model and (1.12) suggest that we again choose a radius R_A approximately equal to the ion core radius, the non-locality arising from the differences between the well depths A_l, and if we regard only $l = 0$ and 1 as important at such small r, we may put

$$V_{\text{non-loc}} = A(\mathscr{P}_0 - \mathscr{P}_1) \quad (r < R_A),$$

$$= 0 \quad (r > R_A). \tag{1.29}$$

We write

$$\left.\begin{array}{ll} K = |\mathbf{k} - \mathbf{G}|, & K' = |\mathbf{k} - \mathbf{G}'|, \\ x = KR_A, & x' = K'R_A. \end{array}\right\} \tag{1.30}$$

Then the matrix elements of (1.29) for $K = K'$ are

$$v_{\text{non-loc } \mathbf{GG}'} = \frac{2\pi R_A^3 A}{\Omega} \Big\{ [j_0(x)]^2 - x^{-1} j_1(x) \cos x$$

$$- 3 \left(1 - \frac{g^2}{2K^2} \right) \{ [j_1(x)]^2 - j_0(x) j_2(x) \} \Big\}, \tag{1.31a}$$

and for $K \neq K'$

$$v_{\text{non-loc } \mathbf{GG}'} = \frac{4\pi R_A^2 A}{\Omega[K^2 - K'^2]} \Big\{ K j_1(x) j_0(x') - K' j_1(x)' j_0(x)$$

$$- 3 \left(\frac{K^2 + K'^2 - g^2}{2KK'} \right) \{ K j_2(x) j_1(x') - K' j_2(x') j_1(x) \} \Big\}. \tag{1.31b}$$

If the $l = 2$ potential is thought to be unusually strong, such as in Ca and perhaps K just before the $3d$ transition series (Vasvari, Animalu & Heine, 1967), then we can take the $l = 2$ potential as the local one and write instead of (1.29)

$$V_{\text{non-loc}} = A\mathscr{P}_0 - A'\mathscr{P}_1 \quad (r < R_A). \tag{1.32}$$

In (1.31) the first two terms in the curly bracket must then be multiplied by A and the remainder by A'. Alternatively, one can work directly with the $\mathscr{P}_2$ projection operator. In some problems x and x' may always be small because R_A is small, in which case the Bessel functions in (1.31 a, b) may be expanded in power series, and we obtain both for $K = K'$ and $K \neq K'$

$$v_{\text{non-loc } \mathbf{GG}'} \approx \frac{4\pi R_A^3 A}{3\Omega} \left\{ 1 - \left(\frac{2(K^2 + K'^2) - g^2}{2KK'} \right) \frac{KK' R_A^2}{5} \right\}. \tag{1.31c}$$

Again for (1.32) the two terms in (1.31c) should be multiplied by A and A', respectively.

For heavy atoms, *spin-orbit coupling* V_{SO} may be included by writing

$$\langle \mathbf{k} - \mathbf{G}, \nu | V_{SO} | \mathbf{k} - \mathbf{G}', \nu' \rangle = i \Lambda \mathbf{s}_{\nu\nu'} \cdot (\mathbf{k} - \mathbf{G}) \wedge (\mathbf{k} - \mathbf{G}'), \quad (1.33)$$

where ν, ν' denote the components of the Pauli spin-matrices s_x, s_y, s_z. Somewhat more complicated expressions may be derived from o.p.w.s (Weisz, 1966) or the square-well model potential (Animalu, 1966).

As already described, the Fermi surface of Al was fitted by a *folded* 4×4 secular equation with $\bar{v}$ matrix elements, whereas the band structure of Si with a large (essentially *infinite* order) secular equation and matrix elements of v. The former has the advantage of smaller size and incorporation of some higher-order correction in an effective matrix element. The arbitrariness of the pseudo-potential (subject always of course to its being a valid pseudo-potential) results in quite widely varying behaviour for $v(g)$ at large g. Comparison with (1.3) shows that if the summation in (1.25) were extended over all $\mathbf{g}$, it would give simply the t-matrix. In (1.25) one has therefore carried the transformation to the t-matrix one step further and included higher-order Born approximations to give a quantity $\bar{v}$ which should be more invariant, and more relevant for most purposes, e.g. in calculating the scattering for electrical resistivity, since all pseudo-potentials must give the same final scattering and band structure. However, the use of a small finite secular equation also has two disadvantages. First, the correction terms in (1.24) and $\bar{v}$ depend to some extent on the structure considered, whereas v is the pseudo-potential for one screened atom, so that there is some error in transferring values of $\bar{v}$ found by fitting the Fermi surface in one structure to another. Secondly, the 4×4 secular equation is only good near W. At K in the zone (Fig. 1.8), there are three plane waves in the lowest degenerate set, and two $(\pi/a)(-5/2, 0, 3/2)$, $(\pi/a)(3/2, 0, -5/2)$ in the highest set of which only the former is included in the four $\mathbf{k}$-$\mathbf{G}$s. In principle the $\mathbf{k}$ dependence of the higher-order corrections in the $\bar{v}$ matrix elements compensates for this, but in practice such an asymmetry is a serious drawback, and Melz (1966) could not get an unambiguous fit to his measurements on the change of Fermi surface around K with pressure.

We conclude it may be better to use a large secular equation with v-matrix elements, than a reduced one with $\bar{v}$-elements. In that case there is a problem because (1.28), (1.31) do not drop off very rapidly at high $\mathbf{g}$s due to the discontinuities in the potentials V_{ps}^b assumed in real space. We may remedy this by assuming instead a smoothed pseudo-potential

$$\int V_{ps}^{b}(\mathbf{r}') F(\mathbf{r} - \mathbf{r}') d\mathbf{r}' \qquad (1.34)$$

where F is a Gaussian smoothing function. The effect on the matrix elements (1.28), (1.31) is to multiply them by $D(g)$, the Fourier tranform of F, which is also a Gaussian

$$D(g) = \exp(-Bg^2). \qquad (1.35)$$

The limited experience at present suggests $v(g)$ *can be taken to cut off at about* $3k_F$, and B may be chosen accordingly. This is the kind of behaviour one expects, because the high Fourier components of $v(g)$ describes V_{ps} well inside the core at small r, and hence can be t-matrixed away by a transformation of the type (1.25) even when using the extended form of the secular equation.

Finally, it remains to mention what has been learnt by such fitting procedures about the pseudo-potentials of various elements. Table 1.1 gives the approximate hypothetical band gaps

$$\mathscr{E}_s - \mathscr{E}_p \approx 2v(g) \qquad (1.36)$$

Table 1.1. *Approximate band gaps* $2v(g)$ *for hypothetical bcc structures (in* Ry), *from calculations by Animalu & Heine* (1965)

Be	Mg	Zn	Cd	Hg	Al	Ga	In	Tl
0·19	0·06	0·05	0·04	0·02	0·05	−0·01	−0·03	−0·10

for the $z = 2$ and 3 metals if they occurred in the bcc structure with their normal atomic volume. This structure was chosen because the Brillouin zone is bounded by only one type of zone face, the {110}. Actually this table was compiled from $v(q)$ calculated consistently for all the elements from the model potential (1.12) by Animalu & Heine (1965). It seemed preferable to do that because two of the elements have not yet been fitted to experiment and the others by various methods with varying accuracy to different types of data. Such fits indicate an uncertainty of the order of 0·02 Ry in the absolute values of Table 1.1, but fully substantiate the trends. These are, first, that the $z = 3$ band gaps are systematically lower than the $z = 2$ ones. This reflects the fact that the ion cores of the $z = 3$ atoms are smaller relative to the atomic radii, making gR_A in (1.28) small and mostly less than $\frac{1}{2}\pi$ so that $v(g)$ is negative. Secondly, within elements of the same z, $\mathscr{E}_s - \mathscr{E}_p$ decreases as the atomic number z increases in accordance with the lowering of the s-state relative to the p-state in the free atoms (Cohen & Heine, 1958; Austin & Heine, 1966; Heine, 1966).

1.4 TRANSITION AND NOBLE METALS†

It is well known that the band structure $\mathscr{E}(\mathbf{k})$ of transition and noble metals consists of five narrow d-bands, crossing and hybridizing with a nearly free electron (N.F.E.) band of pseudo-plane-waves (ΨPW) in the sense of §§ 1.1–1.3 formed from the atomic s- and p-states. Clearly such a band structure cannot be described purely in terms of the N.F.E. model because of the d-bands. However, the general formulation propounded in §§ 1.1 and 1.2 still applies: an electron moves in more or less a constant potential from atomic centre to centre, where it is then scattered in a way characteristic of the atom. The secular equation of § 1.2,

$$\det \left\| \{ (\mathbf{k} - \mathbf{g})^2 - \mathscr{E} \} \delta_{\mathbf{gg'}} + \Gamma_{\mathbf{gg'}} \right\| = 0, \tag{1.6}$$

remains valid, but Γ is no longer a weak pseudo-potential. As already emphasized in § 1.2, the general expressions $\Gamma(\text{A.P.W.})$ and $\Gamma(\text{KKRZ})$ [(1.8), (1.9)] can encompass all situations including d-bands. The point is that the scattering becomes strong and highly energy dependent, passing through a *resonance* around the energy of the d-band. It is useful to focus attention on $\Gamma(\text{KKRZ})$, (1.9), expressed as it is in terms of phase shifts.

How then does the phase shift behave in the region of the d-band? Let us consider a single transition metal atom placed in a free-electron gas, with the d-level E_d of the atom a little above the bottom of the plane wave band, as it is in the solid. The d-state has a finite *life time* $\hbar/W$ because an electron in it may escape into a plane wave of the same energy, exactly as in the well-known Gamow theory of an alpha particle tunnelling out of the nucleus. In the present case the barrier the electron has to pass through is the centrifugal term $l(l+1)/r^2$ in the radial wave equation which is about 1 Ry positive at the edge of the atom where the atomic potential has become zero. It is the classic situation of a '*virtual bound state*', or a 'resonance' of an incoming

† This subject is also treated in detail in § 8.1.

plane wave with the atomic d-state, which leads to a resonance in the
$l = 2$ phase shift (Landau & Lifshitz, 1958, p. 443)

$$\tan \eta_2 = \frac{\frac{1}{2}W}{\mathscr{E}_d - \mathscr{E}}, \tag{1.37}$$

whose width W is related to the life time $\hbar/W$ of the virtual bound state
by the uncertainty principle.

Pseudo-potential theory indicates that the $l = 0$ and 1 phase shifts
should be relatively small, as for non-transition metals, and so should
other terms in the $l = 2$ phase shift which remain at energies well above
or below $\mathscr{E}_d$ where (1.37) becomes small (see e.g. Cohen & Heine, 1961).
We may write, therefore, $\Gamma = \Gamma_{\text{N.F.E.}} + \Gamma_{d\,\text{res}}$ (1.38)

where the second term arises from (1.37) and gives all the features
associated with the d-band. In particular we note that the width and
shape of the d-band, as well as the hybridization, must all be related
to the single constant W in (1.37), since $\mathscr{E}_d$ just specifies the position
of the d-band relative to the N.F.E. band.

What precisely we mean by *hybridization* is illustrated in Fig.
1.11. This shows schematically $\mathscr{E}(\mathbf{k})$ for a *fcc* metal along ΓL in the
Brillouin zone (Fig. 1.8). The d-bands have a tight-binding form, and
the two labelled Λ_3 (denoted by broken lines) are each doubly degenerate. In the absence of hybridization, the fifth d-band of Λ_1 symmetry
would follow the dotted curve from Γ'_{25} to L_1, and the N.F.E. band also
of Λ_1 symmetry from Γ_1 to L'_2 and then from L_1 higher. The effect of
hybridization is to split them apart where they would otherwise cross.
In general directions in the Brillouin zone the situation is similar,
except that the five d-bands are all non-degenerate and all affected
somewhat by hybridization. If we choose the direction of $\mathbf{k}$ as the axis
of quantization, the bands still consist predominantly of $m = 0$,
± 1, ± 2 atomic orbitals respectively in ascending order of energy,
and only the first hybridizes strongly with the plane waves (Phillips
& Mueller, 1967). Such a picture of $\mathscr{E}(\mathbf{k})$, already proposed qualitatively by Mott & Jones (1936), emerges from many band-structure
calculations (e.g. Mattheiss, 1964; Loucks, 1965 b). The only note-
worthy addition to Mott & Jones' description is to point out that the
states of the Λ_3-band are purely of d character, so that there are no
ΨPW-states in the hybridization gap between the two Λ_1-bands.
This is only rigorously correct along the ΓL and ΓX directions because of symmetry, but is still nearly true over the whole Brillouin

zone as already mentioned. The N.F.E. *sp* states do not therefore have even approximately a uniform density across the *d*-band, but are almost entirely excluded from a 'hybridization gap' in the lower part of the *d*-band.

One aside must be made here. To talk in terms of a band structure $\mathscr{E}(\mathbf{k})$ makes the implicit assumption that the electrons (including the *d* ones) are to be described by *itinerant*, or Bloch, states $\psi_{\mathbf{k}}$ and not

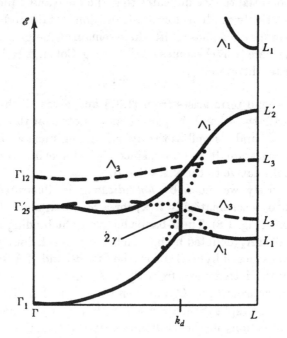

Fig. 1.11. Band structure of Cu in the (111) direction (schematic), 2γ is the hybridization gap where the N.F.E. band and the *d*-band would cross at $k = k_d$.

localized atomic orbitals. Actually in a many-body system the wave-function of one electron is not a meaningful quantity. It is better to accept that there is some complicated ground state and discuss the properties of the system in terms of its characteristic excitations (Pines, 1963). Our assumption therefore is that the excited states of the system may be described in terms of itinerant quasi-particles as for non-transition metals according to the Landau theory to be described further in §1.7. For low-energy excitations near the Fermi level, there appears adequate evidence for this in W, Re and Pd from Fermi surface studies, and also in ferromagnetic Ni (see, for example,

Nottingham Conference, 1964; Vuillemin & Priestley, 1965; Loucks, 1966; Thorsen *et al.* 1966).

However it is also true that the width of the d-band is relatively small compared with the quasi-particle interaction. There is no reason why among higher excitations such as contribute to optical properties, there should not be some states of the nature of excitons or exhibiting other many-body effects (see e.g. Phillips, 1965). Such excitons would overlap in energy with the continuum of quasi-particle excitations and appear as resonances rather than sharp levels. Very little is known of them with certainty, because they would look at least superficially similar to ordinary interband peaks in the optical spectra. One has to have a convincing interpretation of the latter in order to isolate the remaining many-body effects (Mueller & Phillips, 1967).

With these qualifications, therefore, we confine the following discussion to the one-quasi-particle $\mathscr{E}(\mathbf{k})$, and return to (1.37) and (1.38). To recapitulate, $\Gamma_{\mathbf{g}\mathbf{g}'}$ (KKRZ) in the secular equation (1.6) may be divided with the aid of (1.37) into a N.F.E. part $\Gamma_{\text{N.F.E.}}$ and a d-state resonant part $\Gamma_{d\,\text{res.}}$ These occur summed in each matrix element of (1.6). The secular equation may then be transformed (Heine, 1967) to a new form which separates the d-band from the N.F.E. ones

$$\det \|H_{M,\,ij} - \mathscr{E}\delta_{ij}\| = 0, \qquad (1.39)$$

where H_M is partitioned in the form

$$H_M \equiv \left[\begin{array}{c|c} \Psi PW - \Psi PW & \Psi PW - d \\ \hline d - \Psi PW & d - d \end{array} \right]. \qquad (1.40)$$

The lower right 5×5 part of H_M describes the d-bands, the upper left part the N.F.E. bands, and the off-diagonal part labelled $d - \Psi PW$ the hydridization between them.

H_M is termed the *model Hamiltonian*. Just as in the case of pseudo-potentials in § 1.3, the elements of H_M describe the net effective couplings between the bands after all processes have been 'renormalized' and summed up as far as possible. We shall not repeat here the steps necessary to derive (1.40) from the scattering behaviour (1.37), (1.38). The proof of a pudding is in the eating, and the usefulness of a pseudo-potential or model Hamiltonian lies in its describing real band structures reasonably accurately in physically perspicuous terms. Suffice it to note, therefore, that a model Hamiltonian of the form (1.40) has been successfully fitted to the band structures of Cu, Ag, Au, Ni and

Fe calculated previously by the A.P.W. or KKR method (Mueller, 1967; Hodges, Ehrenreich & Lang, 1966a; Jacobs, 1968).

With this fore-knowledge, we may confidently write down 'intuitively' the various parts of H_M and relate them to the scattering theory (1.37), (1.38). The top left part of (1.40) describes the N.F.E. part of the band structure in terms of pseudo-wave-functions expanded in plane waves $\mathbf{k} - \mathbf{g}$, exactly as in §§ 1.2 and 1.3. The expansion may extend to many plane waves, or be folded down to a 4×4 matrix as described for Al in § 1.3. For the present we shall assume the latter has been done. The pseudo-potential matrix elements may simply be treated as constants, or their energy and $\mathbf{k}$-dependence expressed in any of the forms such as the O.P.W. discussed in §§ 1.2 and 1.3. Practice varies, depending on the purpose of the calculation and the amount of theoretical or experimental information available. It is important to note that the band gaps in the N.F.E. part, e.g. $L_1 L_2$ in Fig. 1.11, contain some contribution from hybridization in addition to the N.F.E. $l = 0$ and 1 pseudo-potential. The latter is dominant in Cu and Ag, but the hybridization is the major contributor to the large band gap at L and the distortion of the Fermi surface in Au (Jacobs, 1968).

As regards the dd part of the model Hamiltonian, it is sufficient for the present to note that it has a conventional 'tight-binding' form in terms of five d-orbitals ϕ_m. We shall return later to the question of why this is so, for in a certain sense the d-levels are not bound at all, since they lie at positive energy in the N.F.E. band. The tight-binding matrix elements may be written down for the bcc and fcc structures from the tables given by Slater & Koster (1954). Most prominent are the two-centre nearest-neighbour terms, which may all be expressed in terms of the three overlap integrals $(dd\sigma)$, $(dd\pi)$, $(dd\delta)$. Three-centre nearest-neighbour terms and two-centre next-nearest-neighbour ones appear to be considerably smaller (Mueller, 1967).

Finally, H_M contains the d-ΨPW *hybridization matrix elements* $\gamma_{\mathbf{k}-\mathbf{g}, m}$. As already mentioned, the interaction arises from the fact that an atomic d-state ϕ_m can decay into the $l = 2$ component

$$4\pi\Omega^{-\frac{1}{2}}i^2 \sum_m j_2(|\mathbf{k} - \mathbf{g}|r) Y_{2m}(\mathbf{r}) Y_{2m}^*(\mathbf{k} - \mathbf{g}) \tag{1.41}$$

of the plane wave $|\mathbf{k} - \mathbf{g}\rangle$. The probability per unit time is

$$\frac{W}{\hbar} = \frac{2\pi}{\hbar} \sum_{\mathbf{k}} |\langle \phi_m | V | PW\mathbf{k} \rangle|^2 \delta(\mathscr{E}_d - \hbar^2 k^2 / 2m). \tag{1.42}$$

D

We therefore take for our interaction matrix element γ

$$\gamma_{k-g,m} = \langle \phi_m | V | k - g \rangle \qquad (1.43a)$$

$$= 4\pi \Omega^{-\frac{1}{2}} Y_{2m}(k-g) \int j_2(|k-g|r) \, V\chi_d r^2 dr \qquad (1.43b)$$

$$\approx (4\pi/15) \, \Omega^{-\frac{1}{2}} | k - g |^2 \, MY_{2m}(k-g), \qquad (1.43c)$$

where
$$M = \int V\chi_d r^4 dr \qquad (1.44)$$

and χ_d in the radial part of the orbital ϕ_m. In the last step we have used the expansion
$$j_2(x) = x^2/15 + O(x^4) \qquad (1.45)$$

for the spherical Bessel function: it is small for small r inside the atom, which partly explains the long life and smallness of the coupling. Substituting (1.43c) in (1.42) gives for W

$$W = \frac{2}{225} k_d^5 M^2 (2m/\hbar^2), \qquad (1.46)$$

where
$$\hbar^2 k_d^2/2m = \mathscr{E}_d.$$

We can verify that (1.43c) gives good agreement with band structures calculated by A.P.W. and KKR methods. The value of γ can be read off the calculated $\mathscr{E}(k)$ from the hybridization splitting at the point k_d where the N.F.E. and d-bands would otherwise cross (Fig. 1.11). Strictly speaking, this gives $\gamma(k_d, m = 0)$ with the direction of k_d defining the axis of quantization for m. From the $\mathscr{E}(k)$ of Segall (1962) for Cu we have along the ΓL direction $k_d = 0.47$ a.u. and $\gamma(k_d) = 0.087$ Ry. Using a reasonable potential V and atomic $3d$ function ϕ_m, we estimate from (1.44)

$$M = 6.1 \, \text{Ry} \quad (\text{Bohr radius})^{\frac{7}{2}}, \qquad (1.47)$$

and hence from (1.43c) $\gamma(k_d) = 0.080$ Ry.

The form (1.43a) for γ is only correct for $k - g$ well inside the fundamental Brillouin zone. For larger $k - g$, γ drops below (1.43a), has its maximum for $|k - g|$ about equal to the radius of the Brillouin zone, and then falls rapidly to zero (Mueller, 1967). The physical reason for this is not entirely clear. Mathematically it turns out that the interaction between the atomic d-states and the higher plane waves is already taken into account in the width and structure of the d-band. The higher plane waves are used to build up the correct Bloch functions in the overlap region between the atoms. By giving the d-band width and form in the model Hamiltonian (1.40), we have already 'decoupled' the N.F.E. and d-bands to the maximum extent possible, so that the hybridization matrix elements γ represent a residual interaction.

Away from the point k_d of cross-over (Fig. 1.11), the effect of the hybridization in general is to push the N.F.E. and d-bands somewhat apart. The upper (N.F.E.) level L_1 is raised by about 0.07 Ry and the L_1 level in the d-band lowered a similar amount. The level L_2' (Fig. 1.11) has p-like symmetry and remains unaffected. In second-order perturbation theory, the effect on the N.F.E. levels above the d-band is to raise the $\mathbf{k} - \mathbf{g}$, $\mathbf{k} - \mathbf{g}'$ pseudo-potential matrix elements by

$$\sum_m \frac{\gamma^*_{\mathbf{k}-\mathbf{g},\,m}\gamma_{\mathbf{k}-\mathbf{g}'\,m}}{\mathscr{E}_\mathbf{k} - \mathscr{E}_{d\mathbf{k}m}}, \tag{1.48}$$

where $\mathscr{E}_{d\mathbf{k}m}$ are the five d-levels obtained by diagonalizing first the $5 \times 5\,dd$ part of the model Hamiltonian at the point $\mathbf{k}$ in question. This is a major contribution to the difference in the radius of the Fermi surface of gold between the (100) and (110) directions, and probably to the anisotropy of relaxation time between the 'neck' and the 'belly' of the Fermi surface in AgAu alloys.

We turn now to the *width* and *structure* of the d-band. The basic problem is to define precisely the localized d-orbitals which are overlapping, and it is convenient to consider again a single atom in a free-electron gas. If we integrate the Schrödinger equation outwards from the origin at an energy $\mathscr{E}_d$, we will obtain a function Φ_m essentially identical with an atomic orbital ϕ_m inside the atom

$$\Phi_m = \chi_d Y_{2m}(\theta, \phi) \quad (r < R). \tag{1.49a}$$

However outside the atom Φ_m does not die away exponentially like an atomic orbital because $\mathscr{E}_d$ lies above the bottom of the plane wave band. Instead Φ becomes a spherical Bessel function which behaves asymptotically like

$$\Phi_m \sim (W/2k_d)^{\frac{1}{2}} r^{-1} \cos(k_d r) Y_{2m}(\theta, \phi). \tag{1.49b}$$

The amplitude of this tail is proportional to $W^{\frac{1}{2}}$ since W gives the probability of escape.

Since the Φ_m are unbound and stretch to infinity, it is really a misnomer to talk about the d-band in terms of 'tight binding'. However, if W is small, then so is the amplitude of the tails (1.49b), leading to a narrow band of interfering resonances. To this the mathematical machinery of the conventional tight-binding method may still be applied (Heine, 1967). The two-centre overlap integrals are

$$\langle \Phi_m(\mathbf{s} = 0)|\,V\,(\text{site } \mathbf{s})\,|\Phi_{m'}(\mathbf{s})\rangle. \tag{1.50}$$

Since $V(\mathbf{r}-\mathbf{s})$ is localized around the atomic site $\mathbf{s}$, $\Phi_{m'}(\mathbf{s})$ in (1.50) may be replaced by $\phi_{m'}(\mathbf{r}-\mathbf{s})$, and Φ_m $(\mathbf{s}=0)$ by the tail (1.49b). This tail is a solution of the wave equation for a constant potential, and can therefore be expanded in terms of spherical Bessel functions centred at $\mathbf{s}$

$$W^{\frac{1}{2}}\Sigma_{l'm'}\,b_{l'm'}j_2(k_d|\mathbf{r}-\mathbf{s}|)\,Y_{l'm'}(\mathbf{r}-\mathbf{s}). \tag{1.51}$$

The coefficients $b_{l'm'}$ depend only on the energy through k_d, and on the position vector $\mathbf{s}$. The overlap integral now becomes

$$W^{\frac{1}{2}}b_{2m'}\int j_2(k_d r)\,V(r)\,\phi_{m'}(r)\,r^2\,dr, \tag{1.52}$$

where the integral may be expressed from (1.42), (1.43) in terms of another factor of $W^{\frac{1}{2}}$. Thus all two-centre integrals, not only the nearest neighbour ones, are of the same order W, though distant neighbours contribute only a small coefficient due to destructive interference. In the limit of narrow bands, it may be shown that all three-centre integrals tend to zero more rapidly than W. In this limit, therefore, the d-band may be represented by a 'tight-binding' or narrow-band secular equation

$$\det\|WB_{mm'}-(\mathscr{E}-\mathscr{E}_d)\,\delta_{mm'}\| = 0. \tag{1.53}$$

The constants $B_{mm'}$ depend on $\mathbf{k}$, and through the $b_{2m'}$ on $\mathscr{E}_d$ and the structure of the crystal, but are independent of the potential or the magnitude of W. We conclude that the *structure* of the d-band depends only on the geometry of the crystal structure and on the position of $\mathscr{E}_d$ relative to the bottom of the N.F.E. band, whereas the *width* of the band is proportional to W and governed by the *intra*-atomic reduced matrix element M (1.44) through (1.46). It may appear strange that the bandwidth, normally thought of in terms of *inter*atomic overlap, should depend only on an *intra*-atomic matrix element: the reason is that we are not dealing here with the usual tight-binding situation of exponentially decreasing atomic orbitals, but with the interference between a whole set of resonances of width W in a free-electron band.

Equation (1.53) can be used for numerical calculation. Comparison of (1.53) with the theory of the KKR method (Kohn & Rostoker, 1954) shows that the $B_{mm'}$ in (1.53) are equal to the $l=l'=2$ KKR structure constants at energy $\mathscr{E}_d$, apart from a simple factor. Indeed (1.53) is just the $l=2$ part of the KKR secular equation with (1.37) substituted in for the phase shift. Since the structure constants have been tabulated [for *fcc* and *bcc* by Ham and Segall, see Ham (1962): for *hcp*,

Amar, private communication], the energy at least at symmetry points can be written down very simply. In the case of Cu, good agreement with Segall (1961) was obtained using the same value (1.47) of M as before (Heine, 1967). More detailed comparison with the KKR method shows that the limitation of (1.53) to a 5×5 secular equation is equivalent to including hybridization with a completely free-electron band with bottom of the band at the KKR 'muffin constant'.

We may derive a simple approximate formula for the overall d-band width. If we integrate the radial wave equation outwards from the origin for $l = 2$ at an energy near $\mathscr{E}_d$, we obtain the tail,

$$(\text{const.}) \, [j_2(kr) - \tan \eta_2 \, n_2(kr)] \, Y_{2m}(\theta, \phi), \qquad (1.54)$$

where $\tan \eta_2$ is given by (1.37). This expression should be valid as far in as the sphere of radius R inscribed in the atomic polyhedron. In the spirit of the Wigner–Seitz method, we expect the bottom of the d-band to correspond approximately to the bonding-boundary condition $\partial \psi / \partial r = 0$ and the top of the band to the antibonding condition $\psi = 0$ at the radius R half-way between nearest neighbours. Applying these conditions to (1.54) gives for the width of the d-band

$$\Delta = \tfrac{1}{2} M^2 R^{-5} (2m/\hbar^2). \qquad (1.55)$$

With the same value of M (1.47), as before, we obtain $\Delta = 0.227 \, \text{Ry}$, in fortuitously good agreement with the bandwidth $0.226 \, \text{Ry}$ calculated by Segall (1962) if we ignore the depression of the level X_1 by hybridization (Heine, 1967). We note Δ varies with interatomic spacing as R^{-5} as has been verified by Jacobs (1968) from actual band-structure calculations: it is not an exact law, but derived using the small x expansions of $j_2(x)$ and $n_2(x)$. We also note that (1.55) is independent of the level of the d-band relative to the N.F.E. one, and in this respect behaves like the familiar overlap of atomic orbitals.

In conclusion, therefore, we have described the energy bands in terms of a model Hamiltonian (1.40), where the shape and width of the d-band and its hybridization with the N.F.E. bands all originate from the resonance (1.37) in the $l = 2$ scattering. The bands may be fitted, up to and including the bottom of the second N.F.E. band, to a few millirydbergs with about a dozen parameters, some of which are related in a complicated way through W, the width of the resonance. The pseudo-potential and hybridization parameters represent single atom scattering processes, and as such may probably be transplanted to a good approximation to an alloy or another structure. The form of the d-band for another structure may be inferred either from (1.53) by calculating the change in the KKR structure constants, or from the standard tight-binding formalism with (1.55) indicating the variation of the nearest neighbour overlap integrals with distance. In principle, therefore, the change in band structure with strain or presence of a phonon can be set up, based on the same principles we shall use in § 1.5 for N.F.E. metals.

1.5 OTHER ELECTRONIC PROPERTIES OF N.F.E. METALS

The total energy and behaviour of metallic systems is almost entirely determined by the conduction electrons. In N.F.E. metals their interaction with the ions is completely describable by the pseudopotential. We might therefore hope to formulate in terms of $v(q)$ everything from the energy of a vacancy to the scattering of electrons in a liquid metal, from the electron–phonon interaction to the structure of alloys. Indeed some progress along these lines is slowly being made (Harrison, 1966; Ziman, 1964b).

We may set up the total pseudo-potential in the system for an arbitrary set of atomic sites s_n in the manner of (1.17):

$$V(\mathbf{r}) = (\text{const}) + \Sigma'_q S(\mathbf{q}) v(q) \exp(i\mathbf{q}.\mathbf{r}), \tag{1.56}$$

$$S(\mathbf{q}) = (1/N) \Sigma_n \exp(-i\mathbf{q}.\mathbf{s}_n), \tag{1.57}$$

where N is the total number of atoms in the system and $S(\mathbf{q})$ the same *structure factor* as in X-ray or neutron diffraction. There is no difficulty generalizing (1.56) to more than one atomic species.

In order to make further progress we clearly need to know $v(q)$ as a function of q. We have shown in § 1.2 how it may be calculated, but have emphasized repeatedly in §§ 1.2 and 1.3 that it is difficult to achieve the last 0·01 Ry or so in accuracy though this is important for the applications. The wise thing is therefore to let nature tell us the answer and take $v(q)$ from an analysis of, say, the observed Fermi surface determined from de Haas–van Alphen measurements in the manner of § 1.3. This is illustrated for aluminium in Fig. 1.12. The two points on the right are the values (1.22) at the (111) and (200) reciprocal lattice vectors, and the point at $q = 0$ is the limit

$$v(q \to 0) = -z/\mathcal{N}(\mathscr{E}_F) \tag{1.58a}$$

$$\approx -\tfrac{2}{3}\mathscr{E}_{FO} \tag{1.58b}$$

fixed by basic theory (Ziman, 1964a, pp. 130, 177). Here $\mathcal{N}(\mathscr{E}_F)$ is the density of states at the Fermi level per atom, and $\mathscr{E}_{FO}$ the free-electron Fermi energy. A whole curve for $v(q)$ was calculated from fundamentals. It missed passing through $v(111)$ and $v(200)$ by 0·01 Ry, and was then adjusted slightly to fit them. We may therefore take this as a reliable interpolation of $v(q)$ from the observed points (Fig. 1.12) (Ashcroft & Guild, 1965).

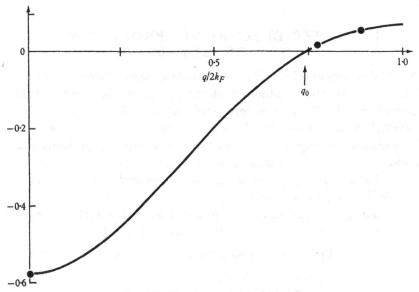

Fig. 1.12. Pseudo-potential of Al in Ry.

The resistivity of molten aluminium can now be calculated. It is proportional to (Chapter 6)

$$\int_0^{2k_F} |S(q)|^2 \, |v(q)|^2 \, q^3 \, dq, \qquad (1.59)$$

where $S(q)\, v(q)$ is the matrix element of the potential (1.56) for scattering by wave-vector q. Here $|S(q)|^2$ for the liquid was taken directly from X-ray measurements, and excellent agreement with the observed resistivity was obtained (Ashcroft & Guild, 1965). It is important to observe that in (1.59) one needs the matrix element of the non-local pseudo-potential relevant to scattering on the Fermi sphere, and this is precisely the geometry to which the experimentally determined points $v(111)$ and $v(200)$ on Fig. 1.12 relate, as shown in Fig. 1.9. In fact many properties of metals are concerned with processes around

$\mathscr{E}_F$, for example, the resistivity of the solid at high temperature due to phonon scattering, and the enhancement of the effective mass at low temperature by the electron–phonon interaction (§ 1.7). Both these were calculated for Al and good agreement with experiment obtained (Ashcroft & Wilkins, 1965). The self-consistently determined potential (1.14), (1.16), (1.56) solves then the old question of 'rigid ion' versus 'deformable ion' in the electron–phonon interaction (see Chapter 5, also Sham & Ziman, 1963). The pseudo-potential, moreover, becomes the intermediate vehicle for interpreting one set of properties in terms of another. A further example is the deformation potential in Si calculated by Kleinman (1963) with the o.p.w. form.

The success of such calculations depends of course on the *transferability* of $v(q)$ between one material and another. A rather more drastic test of this was made for Sb. The pseudo-potential of Sb, as determined from the optical properties of InSb, was used to calculate the band structure of Sb semimetal. In antimony several bands are nearly degenerate around $\mathscr{E}_F$ and a good starting approximation for $v(q)$ is necessary to obtain even qualitatively a unique picture. Only small adjustment of $v(q)$ was required to fit the observed Fermi surface (Falicov & Lin, 1966). In transferring pseudo-potentials a correction should be made for change in atomic volume, the Ω^{-1} in (1.14). Also $\epsilon(q)$ in (1.16) depends on the electron density. For indium in InSb, should one take the mean density of four electrons per atom, or just three? These are non-linear effects not included in the simple dielectric screening method: it is probably best to take a density of three electrons. Fig. 1.13 shows $v(q)$ for indium determined by Cohen & Bergstresser (1966) from the optical properties of InP, InAs, InSb, after scaling to the atomic volume of indium metal but without adjusting the screening. Three points derived from the Fermi surface of indium metal are also shown. There is clearly some scatter about a smooth curve indicating variations in the screening. More remarkable is the smallness of the variations. Part of the explanation is that $\epsilon(q)$ is about 1·1 to 1·2 for these qs, so that the valence electron density already contributes little, and variations in environment have only very small effect.

We shall now consider the box of volume $N\Omega$ with N atoms placed at arbitrary sites $\mathbf{s}_n$ as in (1.56), and calculate the *total energy* U of the whole system up to second order in the pseudo-potential. Its most important use will come in the next section in discussing the cohesive energy of a metal and the occurrence of various structures. But there are other applications, such as the energies of defects and phonon spectra, which may appropriately be considered here, and in any case the general theory is important in showing how various aspects of electronic structure are connected via the pseudo-potential.

The total energy U is conveniently divided into two terms:

$$U = U_0 + U_s, \qquad (1.60)$$

where the first term comes from treating an atomic polyhedron as an *atomic sphere* of radius R_a and the electrons as a free-electron gas, or rather as single O.P.W.s with an orthogonality hole. U_0 is the major

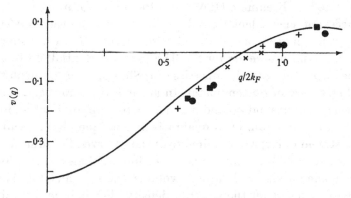

Fig. 1.13. Pseudo-potential $v(q)$ for In in Ry, as fitted from ● InP, ■ InAs, and + InSb, all scaled to the atomic volume of metallic In (Cohen & Bergstresser, 1966), and × from In metal (J. R. Anderson, private communication). The curve is calculated from the model potential (1.12) and (1.16) (Animalu & Heine, 1965).

contribution to the cohesive energy, and is discussed under this heading in §1.6 (see (1.73)). It is independent of structure and for many purposes irrelevant when we are concerned with the relative energies of different atomic arrangements.

We are here interested in the *structure dependent energy* of atomic rearrangements at constant volume, which is (see, e.g. Cohen, 1962, 1963; Harrison, 1963, 1966; Sham, 1963; Blandin, 1963; Pick & Sarma, 1964; Heine & Weaire, 1966), per atom,

$$U_s = U_E + U_{bs}$$
$$= U_E + \sum_{\mathbf{q}}{}' \, |S(\mathbf{q})|^2 [v(q)]^2 \beta(q) \chi(q). \qquad (1.61)$$

Here U_E is the *Ewald* or Fuchs (1935) energy of point ions of charge $z^* = z(1+\alpha)$ (including the orthogonality hole) in a uniform negative background, less $-0.9\, z^{*2}e^2/R_a$ already included in U_0. The remainder of (1.61) is termed the *band-structure energy* and comes from second-order perturbation theory applied to (1.56); we shall define β and χ as we derive it. The largest contribution to U_{bs} is the sum of one-electron energies $\mathscr{E}(\mathbf{k})$. The second-order contribution to $\mathscr{E}(\mathbf{k})$ from a single $\mathbf{q}$ is

$$\frac{|S(\mathbf{q})\,v(\mathbf{q})|^2}{\mathbf{k}^2 - (\mathbf{k}+\mathbf{q})^2}. \tag{1.62}$$

This has to be summed over all occupied states inside the Fermi surface, which to second order may be taken as the unperturbed Fermi sphere. We obtain

$$|S(\mathbf{q})\,v(q)|^2\,\chi(q) \tag{1.63}$$

where χ is the *perturbation characteristic*

$$\chi(q) = \sum_{k\,<\,k_F} \frac{1}{\mathbf{k}^2 - (\mathbf{k}+\mathbf{q})^2} \tag{1.64}$$

whose value has already been given in (1.19). As is well known in Hartree–Fock theory, certain electrostatic and exchange terms have to be subtracted from the sum of the eigenvalues $\mathscr{E}(\mathbf{k})$ as otherwise they would be counted twice in the total energy U. That is the origin of β in (1.61). When exchange and correlation in the total energy are treated by the Hubbard–Sham approximation (Sham, 1963, 1965), it turns out β is the same as ϵ (1.18), but this would not be so in a more complete many-body treatment. Summation over $\mathbf{q}$ then gives (1.61).

In the step from (1.62) to (1.63), we have treated $|Sv|$ as a constant factor and applied the summation only to the denominator. Since $v(q)$ is in fact non-local, this is an approximation which need not be made but is not as serious as might at first appear (see Harrison, 1966, p. 43). In (1.64), all terms with $\mathbf{k}$ and $\mathbf{k}+\mathbf{q}$ both inside the Fermi sphere cancel exactly, since they correspond to mixing wave-functions inside the single Slater determinant. The largest contributions to (1.61), (1.63) comes when the energy denominator goes to zero at the limiting point where $\mathbf{k}$ is just inside and $\mathbf{k}+\mathbf{q}$ just outside the Fermi sphere. The error is therefore not large if we use the limiting point approximation of §1.3 and take $v(q)$ for *scattering on the Fermi sphere* as before. The reader is referred to Harrison (1966) and Kleinman (1966) for various details concerning the contribution of the orthogonality hole, exchange, etc., to U.

One of the most fruitful applications of (1.61) has been to the calculation of *phonon spectra* (Sham, 1963). The phonon frequency is simply a measure of the energy of a lattice wave distortion. Fig. 1.14

shows one recent such calculation for Al (Animalu, Bonsignori & Bortolani, 1966), computed from a pseudo-potential based on the model (1.12) and quite close to Fig. 1.12. With the same pseudo-potential Hodges (1967) has calculated the stacking fault energy of Al as 195 ergs cm^{-2} compared with the experimental value 280 ± 50 ergs cm^{-2} (Edington & Smallman, 1965).

All workers have found that the results of such calculations depend rather sensitively on $v(q)$. The mean structural weight $|S(q)|^2$ in q-space is given simply by the mean atom density, and any structural

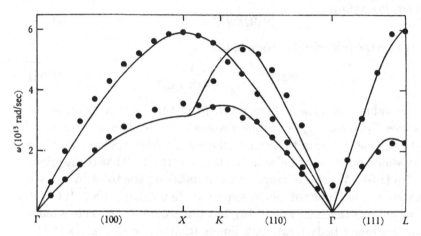

Fig. 1.14. Calculated (continuous curve) and measured (circles) phonon spectrum of Al. (After Animalu *et al.* 1966.)

change merely shifts it around. One is effectively differentiating $[v(q)]^2$. Moreover, there is always a cancellation between the two terms in (1.61), leaving a net small quantity. The Ewald term always opposes distortions from a regular symmetrical structure, whereas the band-structure term reduces this since it describes the screening by the electron gas. A pseudo-potential fitted to two experimental points as in Fig. 1.12 leaves open considerable uncertainties particularly about the behaviour at large q and the non-local corrections. There is also doubt about the exchange and correlation energy in $\beta(q)$, and the orthogonality hole. The most fruitful procedure may therefore be to consolidate all the uncertainties into the *energy-wave-number charac-teristic*

$$\Phi_{bs}(q) = [v(q)]^2 \beta(q) \chi(q). \tag{1.65}$$

With a calculated or experimentally determined pseudo-potential as starting approximation, $\Phi_{bs}(q)$ can be fitted to the measured phonon spectra, and then applied to stacking fault calculations, phase changes (§ 1.6), and the structure of liquids (Johnson *et al.* 1964; Meyer *et al.* 1968).

The energy (1.61) may be Fourier transformed to an effective *pair potential* $\Phi(r)$ between ions (Cohen, 1962; Harrison, 1966):

$$\Phi(r) = \frac{z^{*2}}{r} + \frac{\Omega}{\pi^2} \int_0^\infty \Phi_{bs}(q) \, r^{-1} \sin (qr) \, q \, dq. \qquad (1.66)$$

At large r it turns into the Friedel wiggles

$$\Phi \sim (\text{const.}) \, [v(2k_F)]^2 \frac{\cos 2k_F r}{(2k_F r)^3}. \qquad (1.67)$$

In the present formulation, the *asymptotic form* (1.67) arises from the region $q \approx 2k_F$ in the integration (1.66), where χ has its most rapid variation (see § 1.6). It corresponds therefore to neglecting the variation of $v(q)$. Blandin *et al.* (1967) have applied this approximation to stacking fault calculations, among other things, where the arrangement of nearest neighbours does not change. While Hodges (1967) has improved on it for detailed numerical calculation, it does give an illuminating picture of the broad trends with valency.

Everything so far has been based on perturbation theory and the assumption of a weak pseudo-potential. In a regular crystalline structure, even a complicated one, this condition is satisfied, as it is too in a crystal distorted by a lattice wave or a stacking fault, because the atom density is very uniform. However, in a vacancy, near an interstitial, or at a surface, the potential becomes strong due to the large $v(q)$ at small q (Fig. 1.12). ['Strong' and 'large' compared with $v(q)$ at the reciprocal lattice vectors, not with bare atomic potentials.] Perturbation theory no longer suffices. In some cases the *phase shifts* at the Fermi level only are needed, e.g. in the resistivity of liquid metals and liquid alloys (Dickey *et al.* 1967) or the oscillating long-range Friedel interaction between atoms. In other situations it is necessary to solve for the *t-matrix* (Messiah, 1961) and carry through the self-consistency calculation in terms of it. Bennemann has developed a new iteration procedure for calculating t especially for potentials peaked in q-space around $q = 0$ as $v(q)$ is (Fig. 1.12), and applied it to problems of bond formation, work function, and formation and migration energies of vacancies and interstitials in group 4 semiconductors (Bennemann, 1964, 1965).

1.6. COHESION AND STRUCTURE

Of all the wealth of phenomena known about the structure of metals and alloys, only a tiny fraction is understood in fundamental terms. The present section is therefore a selection of topics which can be treated on the basis of the theory set up in preceding sections. Most of it will relate to *pure metals*, but there are of course ample empirical and theoretical grounds for extending it at least qualitatively to *alloys*. For example, the properties of N.F.E. alloys, including those of the noble metals with elements to their right in the periodic table, correlate with *electron per atom ratio* (e/a) as is well known from the work of Hume-Rothery and others. Among transition metals a *rigid band model* works remarkably well over quite a range of alloys as shown by Beck and others (see e.g. Shrinivasan *et al.* 1966, and references given there). By appealing to such empirical generalizations as well as to theoretical ideas, we can fit together into a coherent picture a certain range of facts about alloys, without claiming to have a complete, exact, or fundamental theory.

In most of the following we shall therefore concentrate on the pure N.F.E. metals, for there we are on safest ground. In §1.5 the total energy was already written

$$U = U_0 + U_s \tag{1.60}$$

and we first show how to calculate the *one-atom* U_0 which is the major contribution to the total energy. We then consider the structure-dependent U_s, comparing the relative energies of different structures to account for those observed in fact. We shall illustrate this with two types of phenomena; first, those depending on *variation of the band gap* $2v(g)$ in (1.61), i.e. on the shape of the pseudo-potential curve $v(q)$; and, secondly, those predominantly due to the shape of the perturbation characteristic $\chi(q)$ in (1.61), i.e. to the *position of the Brillouin zone planes* relative to the Fermi sphere.

As mentioned in §1.5, the *structure-independent term* U_0 is obtained by replacing the atomic polyhedron with an atomic sphere (radius R_a) of equal volume. It is then sufficient to consider the energy of one sphere since, being spherical and neutral, there is no interaction between them. We treat the electrons as a free-electron gas of density z per atom. The total energy *per electron* is then

$$U_0 = \tfrac{3}{5}\mathscr{E}_{OF} + U_x + U_c + \mathscr{E}_H(k = 0)$$

$$- \frac{e^2}{2z} \int \frac{\rho(\mathbf{r})\rho(\mathbf{r}')}{|\mathbf{r} - \mathbf{r}'|}\, d\mathbf{r}\, d\mathbf{r}'. \tag{1.68}$$

The first term is the kinetic energy, where $\mathscr{E}_{OF}$ is the Fermi level $\hbar^2 k_F^2/2m$ of a non-interacting Fermi gas. U_x and U_c are the exchange and correlation energies of the free-electron gas. $\mathscr{E}_H(0)$ is the bottom of the band in the Hartree sense, since exchange and correlation are included completely in U_x and U_c: it is the mean pseudo-potential in the sphere. The last term subtracts the electrostatic self-energy of the electron gas, as always in Hartree-type expressions for the total energy, since it is otherwise counted twice. The first three terms are conventionally expressed in terms of the radius

$$r_s = z^{-\frac{1}{3}} R_a \qquad (1.69)$$

of the sphere containing on the average one electron. We have

$$\tfrac{3}{5}\mathscr{E}_{OF} + U_x = \frac{2\cdot21}{r_s^2} - \frac{0\cdot916}{r_s}\,\text{Ry.} \qquad (1.70)$$

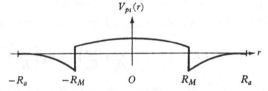

Fig. 1.15. Total pseudo-potential of ion and uniform electron gas in one atomic sphere.

A convenient table of U_c, based on what are probably the best available calculations, may be found in Animalu & Heine (1965). The pseudo-potential in the sphere is that of a bare ion, for which we take the form (1.12), plus the potential

$$\frac{ze^2}{R_a}\left[\frac{3}{2} - \frac{1}{2}\left(\frac{r}{R_a}\right)^2\right] \qquad (1.71)$$

due to the electron gas. It should be noted how closely the bare ion pseudo-potential and (1.71) cancel (Fig. 1.15), which incidentally explains why the Fourier components $v(g)$ of the pseudo-potential in the solid are small for $g \neq 0$. From (1.12) and (1.71) the mean value is

$$\mathscr{E}_H(0) = \frac{-0\cdot6z}{R_a} + \frac{3z}{R_a}\left(\frac{R_M}{R_a}\right)^2 - A_0\left(\frac{R_M}{R_a}\right)^3\text{Ry.} \qquad (1.72)$$

If in the spirit of the Ashcroft model (§1.3) we put $R_M \approx R_{\text{core}} \approx \frac{1}{2}R_a$ and $A_0 = 0$, we obtain $\mathscr{E}_H(0) \approx 0\cdot1z/R_a\,\text{Ry}$, which is typically of order $+0\cdot1\,\text{Ry or less}$.

It should be emphasized that it is quite wrong in the present context to take for $\mathscr{E}_H(0)$ the $q \to 0$ limit (1.58), for two reasons. First, (1.58) is the screened pseudo-potential *including* exchange and correlation with the conduction electrons, and therefore includes large parts of U_x and U_c. Secondly, (1.58) is really the answer to a different problem. It is the deformation potential for a long wavelength density variation in a uniform crystal. Alternatively if we take a large box with a suitable number of atoms in it, and form the atoms into a solid occupying only a fraction of the box, then (1.58) is the difference in potential, including exchange and correlation but calculated only in lowest-order perturbation theory, between the inside and the outside of the metal. It is therefore related to the work function, and includes the surface dipole moment which does not contribute to the cohesive energy. Not that perturbation theory is a good way of calculating the surface dipole anyway, but that is another matter.

The last term of (1.68) integrates to $-1 \cdot 2 z^2 / R_a$, so that the total energy *per electron* (relative to separated ions and electrons) is

$$U_0 = \frac{2 \cdot 21}{r_s^2} - \frac{0 \cdot 916}{r_s} + U_c - 1 \cdot 8 \frac{z}{R_a} + \frac{3z}{R_a} \left(\frac{R_M}{R_a} \right)^2 - A_0 \left(\frac{R_M}{R_a} \right)^3 \text{Ry.} \quad (1.73)$$

Of course analogous formulae may be derived for other forms of the pseudo-potential, or $\mathscr{E}_H(k = 0)$ evaluated by integrating the radial wave equation in the manner of Wigner and Seitz.

There are two errors in the result (1.73), even within the approximation of atomic spheres. The first is the orthogonality hole. As discussed in §1.2, the conduction electron wave-functions should be represented by single O.P.W.s rather than single P.W.s, with a consequent reduction of the charge density over the core and heaping up outside. It affects all electrostatic terms including the exchange and correlation. Secondly, it is not strictly correct to write $\mathscr{E}_H(k = 0)$ in (1.68) and use only the $l = 0$ pseudo-potential. We really require the value of

$$\langle \mathbf{k} | V_{psH} | \mathbf{k} \rangle \quad (1.74)$$

averaged over the band. Its effect can sometimes be incorporated approximately as an m^* in the kinetic energy. Both these corrections are difficult to calculate reliably, but a few attempts have been made (see, for example, Kleinman, 1966; Weaire, 1967).

Some results calculated with the simple formula (1.73) are shown in Table 1.2. The atomic radius was obtained by minimizing the energy, and the compressibility evaluated at this radius. The table is interesting both for its successes and its failures. As a zero-order approximation that completely ignores deviations from a free-electron gas, (1.73) must be regarded as quite successful.

Table 1.2. *Observed values of atomic radius R_a, cohesive energy U, and compressibility C for 23 elements, compared with values calculated from the simple formula* (1.73) *and the model potential parameters of Animalu & Heine* (1965). (*Calculations by D. Weaire, private communication*)

Element	Ra (atomic units)		$-U$ (Rydbergs)		C $(10^{-13} \text{ cm}^2 \text{ dyne}^{-1})$	
	Obs.	Calc.	Obs.	Calc.	Obs.	Calc.
Li	3·26	3·76	0·512	0·513	87	109
Na	3·93	4·24	0·460	0·457	156	180
K	4·86	5·36	0·388	0·369	350	456
Be	2·35	3·07	1·13	0·995	79	155
Mg	3·34	3·70	0·892	0·852	30	31
Zn	2·90	3·09	1·05	0·991	17	16
Cd	3·26	3·30	0·993	0·937	23	20
Hg	3·35	2·88	1·10	1·05	37	12
Ca	4·12	4·48	0·733	0·722	57	64
Ba	4·66	5·41	0·617	0·613	102	132
Al	2·98	3·26	1·38	1·32	14	9
Ga	3·15	3·09	1·47	1·38	20	7
In	3·47	3·38	1·36	1·28	25	10
Tl	3·58	3·09	1·43	1·38	36	7
Si	3·18	2·99	1·96	1·82	3	4
Ge	3·31	2·96	1·97	1·83	14	4
Sn	3·51	3·26	1·77	1·69	19	5
Pb	3·65	3·18	1·81	1·73	23	5
As	3·26	2·87	2·55	2·29	32	2
Sb	3·65	3·14	2·24	2·12	24	3
Bi	3.85	3·18	2·21	2·10	30	3
Se	3·51	2·71	3·23	2·84	12	1
Te	3·79	3·03	2·73	2·58	51	2

The most serious shortcoming of using (1.73) is the neglect of what we have termed in § 1.5 the *structure-dependent energy*

$$U_s = U_E + U_{bs}, \qquad (1.75)$$

and in particular the *band structure* part (per *atom*)

$$U_{bs} = \Sigma'_{\mathbf{g}} |S(\mathbf{g})|^2 [v(g)]^2 \beta(g) \chi(g). \qquad (1.76)$$

Essentially what this says is that the contribution to the cohesive energy is proportional to the *square* of the band gap $2v(g)$. Now it is clear from Figs. 1.12 and 1.13 and Table 1.1 that the main reciprocal lattice vectors lie rather close to q_0 where $v(q)$ passes through zero. It also follows from (1.14), (1.16) that q_0 is determined in our method of setting up the pseudo-potential purely by V_{ps} of the bare ion, and thus does not change with volume of the metal. If the band gap in the sense of (1.36) is positive, i.e. g lies to the right of q_0, a decrease in volume increases g, hence increases the band gap and lowers the total energy. If on the other hand g is less than q_0 and $v(g)$ negative, an increase in volume lowers the energy. Stated simply, g wants to get away from q_0

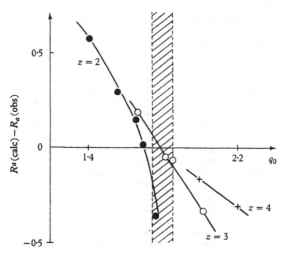

Fig. 1.16. The discrepancy between observed atomic radius and that calculated with the zeroth-order formula (1.73), in atomic units. The shaded area is the region of the first few reciprocal lattice vectors of simple structures. (The scale for q_0 is explained in Fig. 1.19 below.)

so as to increase the magnitude of the band gap $|2v(g)|$. Thus when the band gap is positive (negative), we expect the calculated value of R_a in Table 1.2 to be too large (small).

Comparison with the band gaps shown in Table 1.1 shows this to be so: the calculated R_a are too large for Be, Mg, Zn, Cd, Al as expected from the positive $v(g)$, and too small for Ga, In, Tl in agreement with their negative band gaps. Only Hg is odd, and its $v(g)$ is nearly zero. (The important gs of all simple structures are approximately equal, and we may take the values for the bcc structure in Table 1.1 as typical for the present argument.) Fig. 1.16 shows how the discrepancy in R_a correlates with the position of q_0 relative to the main gs. The slopes of the lines in the figure decrease with increasing valency z, because the compressibility decreases and because the magnitude of χ in

E

(1.76) decreases, since it has a factor of $\mathscr{E}_{FO}$ in the denominator (see (1.19), (1.64)).

The discrepancies in Table 1.2 become larger for the covalently bonded elements with $z = 4$ and 5 where the pseudo-potential is quite strong. Also the agreement for the compressibility is relatively poor, because differentiating any quantity twice highlights its deficiencies, and because the important gs lie close to the position of maximum slope of $v(q)$ (see Figs. 1.12, 1.13) which makes the band-structure effect quite large. In aluminium, for example, compression increases $v(g)$, making the metal more compressible than expected from U_0 alone. Ashcroft & Langreth (1967) have found the effect to double the compressibility. Indeed, by including U_{bs} they obtain very good values for the compressibilities as well as the atomic radii of a large range of elements.

Evaluation of (1.76) shows U_{bs} contributes a few times 10^{-2} Ry to the energy, and is always negative; significant but small compared with U_0. One consequence of this, incidentally, is that in alloys of N.F.E. metals the volume available to an atom is a more significant quantity than the nearest neighbour distance which is affected by structure. This includes alloys of the noble metals in cases where the large ion cores are not in contact (Nevitt, 1966: see also the α, α' and ζ phases of Au In (W. Hume-Rothery, private communication)).

We turn now to *structural* phenomena arising from the *shape* of $v(q)$, which is almost always similar to Fig. 1.12. The significant feature is its passing through zero at $q = q_0$, and the basic principle is again that *the energy is lowered by the reciprocal lattice vectors avoiding* q_0 *and hence making the band gaps as large as possible.* It is convenient to rewrite the band-structure energy (1.76) in terms of the energy-wave-number characteristic Φ_{bs} of § 1.5:

$$U_{bs} = \Sigma'_g W(g) \, \Phi_{bs}(g), \tag{1.77}$$

$$\Phi_{bs}(q) = [v(q)]^2 \beta(q) \chi(q), \tag{1.78}$$

where, as in (1.61) and (1.76), $g = 0$ is excluded in the summation. The structural weight $W(g)$ is $|S(g)|^2$ multiplied by the number of equivalent reciprocal lattice vectors with the same $|g|$. The $\chi(q)$ has already been defined in (1.19), (1.64) and is depicted in Fig. 1.17. As also mentioned in § 1.5, $\beta(q)$ is sufficiently near unity in the range of q of interest that it may be ignored in qualitative discussions (though not in calculations!). $\chi(q)$ therefore combines with $v(q)$ to give a $\Phi_{bs}(q)$ of the form shown in Fig. 1.18. It has a maximum at q_0, a minimum beyond that, and rises again to zero at large q because of the rapid slope of $\chi(q)$ around $q = 2k_F$ (Fig. 1.17) and the decrease of $v(q)$ at large q (§1.3). That structure is favoured whose structural weight clusters closest to the minimum, or otherwise avoids the maximum of Φ_{bs} by lying at $q < q_0$. Since U_{bs} is small compared with U_0, the volume change induced

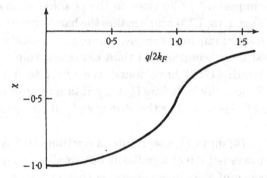

Fig. 1.17. The perturbation characteristic $\chi(q)$ in units of $\frac{2}{3}\mathscr{E}_{F0}/z$.

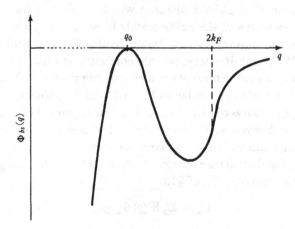

Fig. 1.18. The energy-wave-number characteristic (schematic).

by U_{bs} contributes a second-order small quantity to the total energy. We can therefore confine ourselves to structural changes at constant volume and consider the distribution of $W(g)$ on a fixed curve of $\Phi_{bs}(q)$.

As the first application we consider the *c/a ratios of the hexagonal divalent metals*. Figure 1.19 shows the structural weights of the simple structures, and q_0 and $2k_F$ for several metals. For the *hcp* structure the lowest three *g*s are relevant, and they all move as the *c/a* ratio varies from the ideal. The change in $W(g)$ can easily be evaluated from the geometry, and the combined contribution to U_{bs} is approximately proportional to the curvature of $\Phi_{bs}(q)$ at $q \approx 1 \cdot 75$ units (Heine & Weaire, 1966). For Zn and Cd, this falls in the strongly convex region around q_0 (Fig. 1.18), resulting in large c/a ratios. The q_0 for Be and Mg on the other hand is much smaller (Fig. 1.19), so that $q = 1 \cdot 75$ falls near the point of inflection of $\Phi_{bs}(q)$ beyond q_0, and c/a is near the ideal ratio.

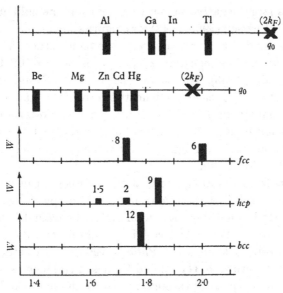

Fig. 1.19. Positions of q_0 and $2k_F$ of divalent and trivalent metals, together with structural weights of *fcc*, *hcp* (ideal c/a) and *bcc* structures, all in units of $2\pi/A$ where $A = (16\pi/3) \frac{1}{3} R_a$ (A would be the lattice constant if all the metals crystallized in the *fcc* structure with the same atomic volume as observed.)

The structural weight can often avoid q_0 through a symmetrical structure such as *fcc* becoming *unstable*. A distortion is energetically favourable if by lowering the symmetry it splits a set of gs falling around q_0 in such a way that some of the gs move to the left and some to the right. The lowering of the energy contains a term proportional to the curvature at the point in question, and the splitting may therefore take place for any set g falling in the convex region around q_0 (Fig. 1.18). We note that q_0 for Hg, Ga and In falls close to the main gs for all three structures in Fig. 1.19, and this qualitatively accounts for the occurrence of distorted *fcc* structures in α-Hg, β-Hg, In and Ga II. Presumably the complex structures Ga I and Ga III also have distributions of structural weight which better avoid q_0 than those of the simple structures in Fig. 1.19. The effect may be seen already in the elastic constants of Al, where q_0 lies close to $g(111)$. The Ewald term in U_s always opposes distortions from a symmetrical structure, and contributes $+17\cdot2 \times 10^{11}$ dyne cm^{-2} to C_{44}. This elastic constant relates to the rhombohedral distortion which splits the (111) set of gs. The observed C_{44} is $1\cdot5 \times 10^{11}$ dyne cm^{-2}, from which we conclude that the band-structure contribution to it must be $-15\cdot7 \times 10^{11}$ dyne cm^{-2}. Most of this large negative amount comes from the splitting of the $g(111)$'s : see Harrison (1966, p. 43) for a figure showing how sharp the maximum around q_0 is. For comparison the band-structure contribution to the other shear constant $\frac{1}{2}(C_{11} - C_{12})$ is $+1\cdot5 \times 10^{11}$ dyne cm^{-2} (Heine & Weaire, 1966). The balance between the Ewald and band-structure parts of C_{44} is clearly a delicate one and Al is just stable with respect to rhombohedral distortion. In Hg the Ewald term is smaller in the ratio 4:9 because z is 2 instead of 3, and calculations indicate the *fcc* structure as being unstable for reasonable assumed pseudo-potentials (Weaire, 1968).

It appears from the foregoing that the numbers of nearest neighbours or their relative directions in these complex structures of Ga, In, and Hg has nothing to do with directed covalent bonds in the conventional sense. The latter arise, in pseudo-potential terms, when $v(g)$ is very strong (Kleinmann & Phillips, 1962), whereas the origin of the complex structures in Ga, In and Hg lies in $v(g)$ being small, more precisely from $v(q)$ passing through zero near the main gs. However, the complex structures can probably be related to *bond lengths* in real space by considering the pair interaction $\Phi(r)$ (1.66). Since the scale of $\Phi_{bs}(q)$ is largely determined by q_0, when q_0 is relatively large then the first main minimum of $\Phi(r)$ occurs at a rather low value $r = d_m$ (Fig. 1.20). In a simple molecule, we may suppose the interatomic distance is near d_m: although our whole dielectric screening approach is unsatisfactory for molecules, the Φ_{bs} at large q which we are concerned with is mostly determined by V_{ps} of a bare ion, and we might expect the minimum for the pair potential in a molecule to fall near d_m. In a densely packed medium, however, we have seen that the atomic volume and hence the mean interatomic spacing $\bar{d}$ is largely fixed by the term U_0 in (1.60) which does not enter $\Phi(r)$. There is no need for $\bar{d}$ to fall near d_m. Indeed we may conjecture that when d_m is abnormally low, e.g. in Hg among the divalent metals, $\bar{d}$ may fall near the maximum of $\Phi(r)$. In that case a rearrangement at constant volume would be energetically favourable, in which some near neighbours move closer to a distance near $d_m < \bar{d}$ and others move outwards to conserve the volume per atom (Fig. 1.20). Detailed results are given for mercury by Weaire (1968) and for gallium by Heine (1968).

While we have concentrated so far on features related to $v(q)$, there are other phenomena arising from the *perturbation characteristic* $\chi(q)$ in $\Phi_{bs}(q)$. These occur when q_0 is small, as in the noble metals, Li, Na, Be, Mg, and Al, so that the main gs fall in the relatively flat region of $v(q)$ around its maximum. It may then be a reasonable first approximation to put $v(q) = $ constant, so that $\Phi_{bs}(q)$ is simply proportional to $\chi(q)$

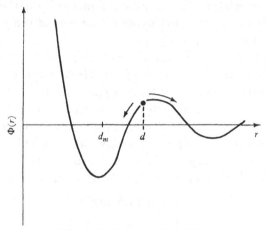

Fig. 1.20. Pair potential $\Phi(r)$.

depicted in Fig. 1.17. The shape is a function of $q/2k_F$ only, so the effects arising from it depend purely on the value of $2k_F$ relative to the gs, i.e. on the electron per atom ratio z. Note the rapid rise of $\chi(q)$ around $q \approx 2k_F$. It clearly pays to have Brillouin zone planes cutting the Fermi sphere rather than outside it, as recognized by Mott & Jones (1936). However, there is no particular value in having it *just* touching the Fermi sphere: the energy is still lower if g is considerably less than $2k_F$. Let us take the exceedingly crude further step of approximating to χ by a step function, -1 for $q < 2k_F$ and zero beyond: U_{bs} is then proportional to the total structural weight W below $2k_F$. For monovalent metals ($z = 1$), the $2k_F$ is less than any of the gs of the *bcc*, *hcp* and *fcc* structures, which therefore all have equal energy in this approximation. For $z = 2$ the *fcc* is quite unfavourable and the *hcp* structure with a weight of 12·5 below $2k_F$ wins marginally over the *bcc* (Fig. 1.19), in agreement with the observed *hcp* structure for Be, Mg,

Zn, and Cd. For $z = 3$ in this approximation the fcc structure has lowest energy with $W = 14$ below $2k_F$ (Fig. 1.19), as found in Al. The bcc structure would be favoured for $2k_F \approx 1\cdot8$ (in the units of Fig. 1.19), corresponding to $z \approx 1\cdot5$ and the structure of β-brass type alloys. Returning now to (1.78) with the complete $\chi(q)$, we note that with the assumption of constant $v(q)$ the integration (1.66) for Φ_{bs} can be performed analytically, and all calculations conveniently carried out in real space. In this way Blandin (1966) found the hcp structure to be stable or metastable for $z = 1\cdot68$ to $2\cdot30$, and the fcc structure from $z = 2\cdot12$ to $3\cdot5$. Although $v(q)$ is not really constant, there is considerable correlation in the above remarks with the observed structures, which witnesses to the important role of the χ factor and the electron per atom ratio in U_{bs}.

It is instructive to rephrase in terms of our present formalism what is believed to be the explanation of the Hume-Rothery rule about the *α-phase boundary of noble metal alloys* (Mott & Jones, 1936, p. 171; Jones, 1937). Let us treat $v(q)$, $\beta(q)$, U_E, and the contribution of higher gs, all as constant. Of course they are not, but they can plausibly be regarded as slowly varying compared with the rapid variation of $\chi(q)$ (Fig. 1.17) around $2k_F$ and hence as not affecting the qualitative argument. The energies of the fcc and bcc structures then become (in arbitrary units)

$$\left.\begin{aligned} U_{fcc} &= 8\chi(g_{111}) + (\text{const.}), \\ U_{bcc} &= 12\chi(g_{110}), \end{aligned}\right\} \tag{1.79}$$

where the numbers 8 and 12 are the structural weights from Fig. 1.19 and we have added a constant to U_{fcc} to represent the {200} reciprocal lattice vectors which are also close to $2k_F$. Using (1.19) we have calculated U_{fcc} and U_{bcc} as shown in Fig. 1.21 with an arbitrary suitable choice of constant in (1.79). U_{fcc} has a kink around an electron per atom ratio $z = 1\cdot33$ where $g_{111} = 2k_F$, due to the kink in χ. Strictly if one differentiates the formula (1.19), one finds χ has an infinite gradient at $q = 2k_F$ but the logarithmic divergence is sufficiently weak that it never shows on a figure (Harrison, 1966, p. 50). Similarly U_{bcc} is seen to have a kink at $z \approx 1\cdot48$ where $g_{110} = 2k_F$, and the α-phase boundary is defined in the usual way by drawing the common tangent between the two curves. It is clear that whatever other smoothly varying terms one may add to (1.79), the α-phase boundary will tend to come at z a little greater than $1\cdot33$ on the right edge of the kink.

It is important to note that the kinks in U_{fcc} and U_{bcc} occur when the *free* electron sphere touches the zone planes, not when the actual

Fermi surface does. Since this is at first surprising, it might be thought an artifact of the perturbation theory used, particularly if one imagines that most of U_{bs} comes from the lowering of $\mathscr{E}(\mathbf{k})$ around the zone planes. However, that is not so. It is easy to verify that the change in the shape of the Fermi surface and in $\mathscr{E}(\mathbf{k})$ immediately around a zone plane contribute terms of order $[v(g)]^3$ or $v^3 \ln v$ which are smaller than the second-order parts already taken into account. Figure 1.22 shows the results of a more complete calculation of U_{bs} for

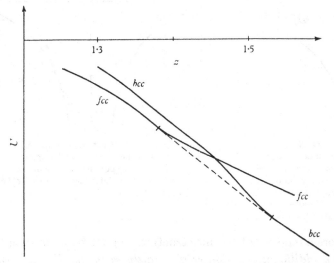

Fig. 1.21. The energy of $\alpha(fcc)$ and $\beta(bcc)$ brass according to (1.79), with the common tangent defining the phase boundaries.

the *fcc* structure. Ignoring β in (1.78) is equivalent to calculating the sum of eigenvalues $\mathscr{E}(\mathbf{k})$ without subtracting the electrostatic interaction (see §1.5), and this can be done exactly if we adopt the 8-cone model of Ziman (1961 b) for $\mathscr{E}(\mathbf{k})$. We have plotted the results as a function of $g_{111}/2k_F$ to compare with Fig. 1.17, where k_F is still the free electron radius. With $v(q)$ assumed constant, Fig. 1.22 is the same as χ but including *all* orders of perturbation. We see that the point of inflexion still occurs close to the 'ideal' value $g_{111} = 2k_F$, i.e. $z = 1.33$. The curve has a discontinuity in the third derivative where the real Fermi surface first touches the zone face at $g_{111}/2k_F \approx 1.15$, but this is too weak to see on the figure.

We have seen, therefore, that it is the free-electron sphere rather than the actual Fermi surface which is most relevant to the total energy U_{bs}, as long as the Fermi surface is basically spherical apart from the distortion caused by the zone plane in question. However, the situation is quite different if the Fermi surface has appreciable areas which are nearly *flat*. The change in $\mathscr{E}(\mathbf{k})$ around the zone plane now contributes of order V_g^2 or $V_g^2 \ln |V_g|$ to U_{bs}, as it does in one dimension, where $2V_g$ is the band gap. The shape of $\chi(q)$ in one dimension is quite different, having a deep minimum at $q = 2k_F$ (Fig. 1.23). There

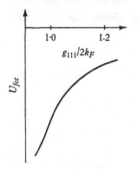

Fig. 1.22. Band structure energy of α-brass for varying electron per atom ratio, assuming constant band gap, as calculated from 8-cone model (D. Weaire, private communication).

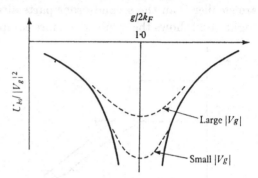

Fig. 1.23. $U_{bs}/|V_g|^2$ for the one-dimensional case, as a function of $g/2k_F$ (schematic only). Full line—perturbation calculation: broken line, complete calculation for two different values of $|V_g|$.

is therefore a tendency for some structural change to set in such that *one of the additional superlattice g's exactly spans the Fermi surface.* Actually it is not necessary that the Fermi surface be flat, but that there be two parts of the Fermi surface which are sensibly *parallel* so that a single $\mathbf{g}$ connects corresponding points on them. Indeed U_{bs} varies as $V_g^2 \ln |V_g|$ in the limit as the band gap goes to zero. This variation, which is stronger than quadratic, may dominate over other forces in the system and result in some structural, magnetic, or other change.

Of course pieces of Fermi surface are never parallel with mathematical exactitude, but a number of examples of the phenomenon have been investigated. Best known is the antiferromagnetic spin density modulation in chromium (see Chapter (8) and Slater, 1951 b; Lomer, 1964; Falicov & Penn, 1967). In CuAu alloys the flat regions of the Fermi surface are responsible for a long period structure modulation (Tachiki & Teramoto, 1966; Sato & Toth, 1961). In both cases the wavelength of the modulation follows the expansion or contraction of the Fermi surface on changing the electron per atom ratio. The spin density modulations in some rare earth metals are thought to have the same origin, and there is some experimental evidence for flat areas of Fermi surface (Williams *et al.* 1966; Keeton & Loucks, 1967).

Ordered β-brass affords another example. The atomic positions form a *bcc* array, but since the Cu and Zu atoms alternate, the primitive unit cell is simple cubic of side a. Figure 1.24 shows a section of the Brillouin zone and Fermi surface in periodic k space (Pearson, 1966). The vector connecting parallel regions of Fermi surface happens to be very nearly $\frac{1}{3}g(100)$, corresponding to a superlattice dimension of

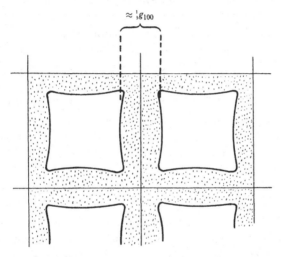

Fig. 1.24. The Fermi surface of β-brass in the second band, shown in the periodic zone scheme. With three electrons per simple cubic unit cell, the second band is about half full, and the Fermi surface is formed from portions of the free electron sphere overlapping each zone face.

length $3a$. The same applies in the y- and z-directions, so that it is not surprising to find among the β-brass type alloys at low temperatures some complex structures with a unit cell based on $3 \times 3 \times 3$ simple cubic cells (Lomer, 1966).

Finally, what can be said about the cohesion of the *transition and noble metals*?† We can distinguish four relevant causes of cohesion, without venturing to speculate about their relative importance. (i) As discussed in §1.4, the structure of the d-band is related to bonding and antibonding overlap between neighbours. This obviously contributes to cohesion when the band is partially full, the maximum

† A detailed discussion of this topic is given in §8.2.

occurring when it is just half full. However, in the noble metals with full d-shells the bonding and antibonding terms cancel, but (ii) overlap can still contribute to the mean position of the d-band. The absence of strong cohesion in the rare gases suggests the latter may not be an important contribution. (iii) The off-diagonal hybridization terms in the model Hamiltonian raise the levels above the cross-over (Fig. 1.11) and lower those below by equal and opposite amounts. Since the effect extends appreciably up to 10 eV or so in the N.F.E. band and these states are never all occupied, there is a net lowering of the total energy. The maximum effect is achieved when $\mathscr{E}_F$ lies in the hybridization gap at the cross-over. (iv) The correlation energy of a closed shell in an atom is commonly about $\frac{1}{2}$ eV per electron, whereas the correlation energy of an electron gas of comparable density is about 1 eV per electron. In an atom only short wavelength charge density fluctuations occur, limited by the diameter of the atom, whereas in a free electron gas about half the correlation energy comes from the long wavelength fluctuations. As atoms are grouped into larger units, a fraction of this potential correlation energy starts to be realized. For instance, Van der Waals forces may be regarded in this light. From the point of view of perturbation theory, the magnitude of the effect depends on the excitation energy in the denominator needed for appropriate excitations of the system. With $\mathscr{E}_F$ not far above the d-band in the noble metals, or in the d-band for transition metals, the energy differences are quite small. One might expect therefore a sizeable contribution to the cohesive energy, particularly for metals with a goodly number of d-electrons.

1.7 LANDAU QUASI-PARTICLES

In metals the Coulomb interaction between conduction electrons is typically of order 1 Ry, quite comparable with the bandwidth. It is therefore at first sight surprising that a model of independent particles in one-electron Bloch orbitals $\psi_{\mathbf{k}}$, such as we have been using, is at all relevant. How can that be? Let us picture an extra electron added to the metal. Its charge will rapidly tend to repel the gas of other conduction electrons from its immediate vicinity, creating a denuded region, the *correlation and exchange hole*. We may intuitively suppose that the band structure $\mathscr{E}(\mathbf{k})$ and the Bloch states $\psi_{\mathbf{k}}$ refer to the motion of the whole quasi-particle consisting of an electron together with its surrounding correlation and exchange hole. Such a picture can indeed be justified: see Nozières & Pines (1966), particularly chapter 1, to which the reader is referred for further details about all aspects of the present section. It may be shown that the energy $\mathscr{E}(\mathbf{k})$ of a quasi-particle can be defined by the zero of the inverse Green function $G^{-1}(\mathbf{p}, \mathbf{r}, \mathscr{E})$ (Nozières, 1964), and we can write a Schrödinger equation for the motion of the whole *quasi-particle*

$$(-\nabla^2 + V(r) + M - \mathscr{E}) \psi = 0, \qquad (1.80)$$

where M is the 'proper self-energy' operator describing the exchange and correlation with the other conduction electrons. For a free-electron gas it reduces to $\mu_{xc}(k)$ defined in §1.2. Since the quasi-particle moves through a periodic medium, all the concepts of $\mathbf{k}$ space, the Bloch theorem and Brillouin zones, remain valid.

There is, however, one qualification: the quasi-particle has a *finite lifetime* and its energy an imaginary component. There is a residual screened Coulomb interaction between quasi-particles, so that a quasi-particle $\mathbf{k}$ above the Fermi surface may collide with one $\mathbf{k}_1$ in the Fermi sea which is scattered to $\mathbf{k}_2$ while the original quasi-particle recoils to $\mathbf{k}_3$. Both energy and momentum have to be conserved in the process, the former implying that $\mathbf{k}_1$, $\mathbf{k}_2$ and $\mathbf{k}_3$ must all lie within an energy $|\mathscr{E}(\mathbf{k}) - \mathscr{E}_F|$ of the Fermi level where $\mathscr{E}(\mathbf{k})$ is the original quasi-particle energy. This requirement puts a severe limitation on the range of collisions which are possible when $\mathscr{E} - \mathscr{E}_F$ is small, and the lifetime of the quasi-particle *tends to infinity* as $(\mathscr{E} - \mathscr{E}_F)^{-2}$. The Fermi surface itself is therefore *perfectly sharp* (Mott, 1956), as is indeed found in the de Haas–van Alphen effects where features of the Fermi surface on the scale of 10^{-3} to $10^{-4} \mathrm{Ry}$ may be studied. The effect is well known in the soft X-ray emission spectra where the cut-off at the Fermi level is sharp, but the transitions from states near the bottom of the band are broadened of the order of $0 \cdot 1 \mathrm{Ry}$. The broadening of states well above $\mathscr{E}_F$ is presumably comparable.

But in many phenomena one is only concerned with low energy excitations very close to $\mathscr{E}_F$ produced by electric and magnetic fields and thermal excitation. Such situations may be described by the *Landau theory* in which the states of the whole electron systems are specified by a distribution $(f_{\mathbf{k}} - f_{\mathbf{k}}^0)$ of quasi-particles just above $\mathscr{E}_F$ and quasi-holes just below. Here $f_{\mathbf{k}}^0$ is the Fermi–Dirac distribution at $T = 0$, and we write the quasi-particle distributions as $(f_{\mathbf{k}} - f_{\mathbf{k}}^0)$ to preserve similarity to the ordinary discussion of independent particles.

The total energy of the system (to second order in $f_k - f_k^0$) may be written (Landau, 1956)

$$U = \sum_k \mathscr{E}(\mathbf{k})\,(f_k - f_k^0) + \tfrac{1}{2}\sum_{kk'} \eta(\mathbf{k},\mathbf{k}')\,(f_k - f_k^0)\,(f_{k'} - f_{k'}^0), \qquad (1.81)$$

where $\eta\,(\mathbf{k},\mathbf{k}')$ represents the interaction energy of quasi-particles, and the variables $\mathbf{k}, \mathbf{k}'$ are taken to include a specification of spin. Before we discuss the contribution of η to various properties, we must say precisely what we mean by '*many-body corrections*'. As mentioned in §1.2 and in connection with (1.80), the one-quasi-particle energy $\mathscr{E}(\mathbf{k})$ contains an exchange and correlation term $\mu_{xc}(\mathbf{k})$ which was stated to be almost independent of $\mathbf{k}$. This is true to within 10 % for a N.F.E. metal and in any case becomes a bit meaningless well below $\mathscr{E}_F$ because of the lifetime broadening. However, near $\mathscr{E}_F$ where $\mathscr{E}(\mathbf{k})$ is well defined, detailed calculations (Rice, 1965 and references there) suggest that the k dependence of μ_{xc} for a free electron gas contributes up to 10 % to the electron velocity $\partial\mathscr{E}/\partial\hbar k$. These are certainly many-body corrections but they are incorporated in the $\mathscr{E}(\mathbf{k})$ and hence *already included* in the independent quasi-particle model. We are concerned with *further* corrections arising from the interaction $\eta(\mathbf{k},\mathbf{k}')$ in (1.81).

The best-known example is the enhancement of the spin paramagnetism. The simple Pauli result $\mu_B^2 \mathscr{N}(\mathscr{E}_F)$ for the susceptibility counts only the single-quasi-particle energy $\mathscr{E}(\mathbf{k})$ (Ziman, 1964a, p. 286), whereas turning some spins over alters the total exchange energy of the system in a way that enhances the susceptibility χ_{mag}. It becomes

$$\chi_{mag} = \frac{\mu_B^2 \mathscr{N}(\mathscr{E}_F)}{1 - \nu} \qquad (1.82)$$

where μ_B is the Bohr magneton, and

$$\nu = -2/[\mathscr{N}(\mathscr{E}_F)\,(2\pi)^6]$$
$$\times \iint \frac{dS_{\mathbf{k}}}{\partial\mathscr{E}/\partial k_n} \frac{dS_{\mathbf{k}'}}{\partial\mathscr{E}/\partial k_n'} [\eta(\mathbf{k}\uparrow,\mathbf{k}'\uparrow) - \eta(\mathbf{k}\uparrow,\mathbf{k}'\downarrow)].$$

When $\partial\mathscr{E}/\partial k_n$ is small as in transition metals, ν may exceed unity, resulting in a formally infinite susceptibility and hence a spontaneous ferromagnetic polarization.

Another important application of (1.81) is to the *electronic specific heat*. In thermal equilibrium there are equal numbers of excited quasi-particle just above any element $dS_{\mathbf{k}}$ of Fermi surface, and quasi-holes below. Their contributions exactly cancel in the second term of

(1.81) and the electronic specific heat is given by the usual formula $\frac{1}{3}\pi^2 k^2 \mathcal{N}(\mathcal{E}_F)$ in terms of the one-quasi-particle density of states $\mathcal{N}(\mathcal{E}_F)$.

The energy of a quasi-particle may be considered modified in the molecular field sense by the interaction from the other quasi-particles present

$$\tilde{\mathcal{E}}(\mathbf{k}, \mathbf{r}) = \mathcal{E}(\mathbf{k}) + (2\pi)^{-3} \int [f_{\mathbf{k}'}(\mathbf{r}) - f_{\mathbf{k}}^0] \, \eta(\mathbf{k}, \mathbf{k}') \, d\mathbf{k}'. \qquad (1.83)$$

The *local energy* $\tilde{\mathcal{E}}$ of a quasi-particle may depend on $\mathbf{r}$ since $f_{\mathbf{k}}$ in general does, and we can define a 'local Fermi surface' or surface of equal potential by $\tilde{\mathcal{E}}(\mathbf{k}, \mathbf{r}) = \mathcal{E}_F$. As usual in transport theory we write

$$f_{\mathbf{k}}(\mathbf{r}) - f_{\mathbf{k}}^0 = - \phi_{\mathbf{k}}(\mathbf{r}) \, \partial f_{\mathbf{k}}^0 / \partial k_n \qquad (1.84)$$

where k_n is the component of $\mathbf{k}$ normal to the Fermi surface. Then ϕ measures the distance in k space that the Fermi surface has been distorted from equilibrium. Alternatively the quasi-particle distribution may be defined in terms of the distortion $\psi_{\mathbf{k}}(\mathbf{r})$ (not to be confused with the wave-function ψ) from the 'local Fermi surface' (Heine, 1962)

$$\psi_{\mathbf{k}} = \phi_{\mathbf{k}} + (8\pi^3 \hbar v_{\mathbf{k}})^{-1} \int \eta(\mathbf{k}, \mathbf{k}') \, \phi_{\mathbf{k}'} \, dS_{\mathbf{k}'}. \qquad (1.85)$$

The *Boltzmann equation* is

$$-\frac{\partial \phi_{\mathbf{k}}}{\partial t} \frac{\partial f^0}{\partial k_n} - \mathbf{v}_{\mathbf{k}} \cdot \frac{\partial \psi_{\mathbf{k}}}{\partial \mathbf{r}} \frac{\partial f^0}{\partial k_n} + e\mathbf{n}_{\mathbf{k}} \cdot \mathbf{E} \frac{\partial f^0}{\partial k_n} = f_{\mathbf{k}}^0 \Big]_{\text{scattering}} \qquad (1.86)$$

(n_k is the unit normal on the Fermi surface). In order to solve it, one must express it completely in terms of ϕ or ψ, but we have written a hybrid form to exhibit each term at its simplest. The time-dependent term must depend on the complete quasi-particle distribution ϕ and not ψ, since $\tilde{\mathcal{E}}$ in terms of which ψ is defined also changes with time. The drift term, however, involves ψ because the other quasi-particles exert accelerations through $\eta(\mathbf{k}, \mathbf{k}')$ which keep $\tilde{\mathcal{E}}$ constant. The scattering term involves $\phi_k / \tau(\mathbf{k})$ in a relaxation time approximation since the relaxation is of the whole system to equilibrium. However, if we write a collision integral for elastic scattering, that depends on ψ since $\tilde{\mathcal{E}}$ is conserved (Silin, 1958).

The total *current* may be written

$$\mathbf{J} = e(2\pi)^{-3} \int \mathbf{v}_{\mathbf{k}} \psi_{\mathbf{k}} \, dS_{\mathbf{k}}. \qquad (1.87)$$

In the case of time-independent processes, (1.86) and (1.87) in terms of

ψ are formally identical with the equations in the independent quasi-particle model if we assume elastic scattering. There are then no Landau corrections to the transport properties. This conclusion still holds for thermal currents and in the presence of static magnetic fields, and applies, for example, to the Wiedemann–Franz law. Time-dependent transport properties in general do have corrections through the first term in (1.86). In the case of the anomalous skin effect, the corrections tend to zero in the extreme anomalous limit (Silin, 1957) because the displacement is such a sharp spike around the Fermi equator as pictured by the ineffectiveness concept (Ziman, 1964*a*, p. 244). The same applies to the cyclotron frequency under Azbel'–Kaner conditions, which remains as normally defined in terms of the Fermi velocity $\partial \mathscr{E}(\mathbf{k})/\partial \hbar \mathbf{k}$ (Ziman, 1964*a*, pp. 250, 255). In the opposite limit when the electric field is uniform over the cyclotron orbit as realized in semiconductors, there are Landau corrections. It would be interesting if with ultrasonic waves one could cover the crucial range between the two extremes.

A more fundamental formula for the total current is

$$\mathbf{J} = (2\pi)^{-3} \int \mathbf{j}_{\mathbf{k}} (f_{\mathbf{k}} - f_{\mathbf{k}}^0) \, d\mathbf{k} \qquad (1.88)$$

in terms of the current $\mathbf{j}_{\mathbf{k}}$ carried by a quasi-particle. This is not $e \, \mathbf{v}_{\mathbf{k}}$ but

$$\mathbf{j}_{\mathbf{k}} = e \left[\mathbf{v}_{\mathbf{k}} + (8\pi^3 \hbar)^{-1} \int \mathbf{n}_{\mathbf{k}'} \eta(\mathbf{k}, \mathbf{k}') \, dS_{\mathbf{k}'} \right]. \qquad (1.89)$$

In the case of a free-electron gas, i.e. with interaction but zero periodic potential the momentum of a quasi-particle must be $\hbar \mathbf{k}$ and the current $(e/m) \hbar \mathbf{k}$. Then (1.89) reduces to the Landau identity

$$\frac{1}{m} = \frac{1}{m^*} + \frac{k_F}{2\pi^2 \hbar^2} \int_0^{\pi} \eta_s(\theta_{\mathbf{k}\mathbf{k}'}) \cos\theta \sin\theta \, d\theta \qquad (1.90)$$

where $m^* = \hbar k / v$ and η_s is the mean interaction for parallel and anti-parallel spins. This result means that for a free-electron gas certain exact cancellations of many-body corrections can take place. For example, the 'semiconductor' cyclotron mass becomes the bare electron mass m, which is what the usual formula for the cyclotron mass would give if applied to the *Hartree* band structure $\hbar^2 k^2/2m$ without the exchange and correlation terms $\mu_{xc}(k)$. That may partly explain why $\mathbf{k} \cdot \mathbf{p}$ perturbation theory is so successful in fitting the observed cyclotron masses in semiconductors because the $\mathbf{k} \cdot \mathbf{p}$

perturbation theory also breaks down for the exchange and correlation potential.

The $\eta(\mathbf{k}, \mathbf{k}')$ considered so far is the screened Coulomb interaction. There is another one, the *interaction via virtual phonons*. Since it is the origin of superconductivity, it is well known that an electron moving through the lattice attracts the positive ions in its immediate neighbourhood, and their small displacement to the centre results in an attractive potential for other electrons (Ziman, 1964a, p. 324). This

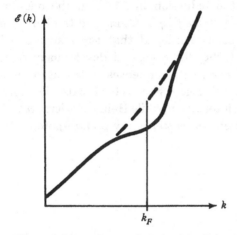

Fig. 1.25. Effect of electron interaction via virtual phonons on the band structure near the Fermi level at absolute zero of temperature.

contributes to the self-energy $\mathscr{E}(\mathbf{k})$, analogously to $\mu_{xc}(\mathbf{k})$. At absolute zero of temperature we may write

$$\mathscr{E}(\mathbf{k}) = \mathscr{E}_e(\mathbf{k}) + \frac{1}{(2\pi)^3} \int \frac{|\mathrm{M}(\mathbf{k}, \mathbf{k}+\mathbf{q})|^2}{\mathscr{E}_e(\mathbf{k}) - \mathscr{E}_e(\mathbf{k}+\mathbf{q}) - \hbar\nu_{\mathbf{q}}} \, d\mathbf{q}, \qquad (1.91)$$

where $\mathscr{E}_e$ is the purely electronic energy and M the matrix element for emission of a phonon $\mathbf{q}$ of energy $\nu_{\mathbf{q}}$ with the electron being scattered to $\mathbf{k}+\mathbf{q}$. If we consider $\mathbf{k}$ just below the Fermi surface, the scattering can only contribute as in (1.64) if $\mathbf{k}+\mathbf{q}$ is an empty state above $\mathscr{E}_F$, the contribution being large only if $\mathbf{k}$ and $\mathbf{k}+\mathbf{q}$ lie within θ_D (the Debye temperature) of $\mathscr{E}_F$. The distortion of the band structure is shown in Fig. 1.25. The reduction in the Fermi velocity is about 35 % in Al and a factor of two in Pb, as observed in the cyclotron mass and electronic specific heat (Ashcroft & Wilkins, 1965). At high temperatures the effect disappears because with the thermal excitation of quasi-particles there are terms with positive energy denominator which cancel the negative ones. The general framework of the Landau theory still applies, even at finite temperature, and it turns out that time-independent transport processes are unaffected by the phonon contributions, i.e. the interaction $\eta(\mathbf{k}, \mathbf{k}')$ cancels the effect of the distortion (Fig. 1.25) of the quasi-particle spectrum $\mathscr{E}_e(\mathbf{k})$ (Prange & Kadanoff, 1964). The anomaly of Fig. 1.25, being intimately connected with the discontinuity at the Fermi surface, rides up and down with $\mathscr{E}_F$ if the electron density is altered, which allows one to derive an identity between the reduction in $\partial\mathscr{E}/\partial k$ in the one-quasi-particle energy and the mean value of $\eta(\mathbf{k}, \mathbf{k}')$ around the Fermi surface.

It has already been mentioned that the reader should refer to Nozières & Pines (1966) for a general development of the Landau theory and much more, with references to the literature. Prange & Kadanoff (1964) and Prange & Sachs (1967) extend its formulation to include the phonon interaction, while Heine, Nozières & Wilkins (1966) give some applications to scattering by pseudo-potentials.

CHAPTER 2

ELECTRONIC STRUCTURE: THE EXPERIMENTAL RESULTS

by D. SHOENBERG†

In the last chapter the term 'electronic structure' was used in a very comprehensive sense to mean not only the specification of $\mathscr{E}(\mathbf{k})$, but a more fundamental interpretation of this specification in terms of parameters such as the pseudo-potentials from which other properties of the metal can be inferred. Here we shall be mainly concerned with a more restricted aspect of electronic structure, namely, the form of $\mathscr{E}(\mathbf{k})$ only close to the Fermi surface, or in other words the form of the Fermi surface and the immediately neighbouring surfaces. In fact we shall see that this is not really as restricted as it seems, because it is sometimes possible to describe the experimentally determined surfaces most simply in terms of the already mentioned fundamental parameters from which the electronic structure in its more general sense can be inferred. Alternatively, even where a pseudo-potential scheme has not been worked out, there is sometimes a detailed band-structure calculation available which can be compared with the experimentally determined Fermi surface. If the agreement is good, or can be made good by adjusting some of the parameters in the band-structure calculation, there is some basis for supposing that the band structure is a reliable one. The question of the correctness of the assumption that the band structure is reliable or that parameters such as the pseudo-potentials really have the significance they are intended to have, can only be settled by direct comparison of predictions based on these assumptions, for instance, as regards the optical properties, with experiment. This has already been discussed to some extent in the last chapter and will not be pursued here.

In this chapter we shall review first the methods developed during the last 10 or 15 years for experimental determination of the Fermi surface and the immediately neighbouring surfaces of constant energy, and then review the results that have actually been achieved on the various metals. We shall also occasionally touch on another

† Dr Shoenberg is Reader in Physics at the Royal Society Mond Laboratory, Free School Lane, Cambridge, and Fellow of Gonville and Caius College, Cambridge.

aspect of electronic structure, namely, the measurement of the relaxation time of conduction electrons and its k-dependence, though little has as yet been done in this direction.

The basic physical principles of the various experimental methods are mostly discussed in Chapters 3 and 4 and here we shall be mainly concerned with their application to the interpretation of the experimental measurements in terms of electronic structure. In the presentation of results we shall not attempt a detailed discussion of every metal which has been studied but discuss only a few examples to illustrate typical situations. The alkalis and the noble metals provide the simplest illustrations of the use of various methods and will therefore be discussed in rather more detail than metals with more complicated Fermi surfaces. Only a few key references to the original literature will be given; for comprehensive bibliographies see Shoenberg (1965) and Mackintosh (1967).

2.1 EXPERIMENTAL METHODS

In principle the experimental value for any physical property of a metal provides information about the electronic structure but, except for a few rather special properties, the information is of a very limited kind. Properties which can be completely specified by a small number of parameters are evidently of little direct use, since from them only a small number of integral properties of the electronic structure are determined, and it is hardly possible for instance to deduce the form of a Fermi surface merely from a single value of some integral over it. To give some concrete examples, the value of the electronic specific heat gives the total density of states which is expressed by an integral of the reciprocal of the velocity over the Fermi surface; the electrical conductivity is specified by a small number of constants (only 1 for a cubic metal) which are expressed theoretically by integrals over the Fermi surface, but with the further complication that the k-dependence of the relaxation time is also involved; the optical Faraday effect is determined by an integral over the Fermi surface of an expression involving the velocity and the principal radii of curvature at each point (McGroddy, McAlister & Stern, 1965). At best, measurement of physical properties such as these may give a hint as to the character of the Fermi surface and permit some check on an electronic structure which has been determined in other ways, though usually it is as much a check of the assumptions implicit in the theoretical formula for the property, as of the electronic structure which is fed into the formula.

Somewhat more direct information about electronic structure can be obtained from the optical properties in the ultra-violet in as far as the frequency dependence of absorption depends on interband transitions, but though some qualitative features of the band structure (such as band gaps, etc.) may emerge from the results it is again usually only the reverse process which is possible, i.e. knowing the electronic structure to check to what extent it explains the optical behaviour. To be able to go directly from the experimentally measured quantities to some definite aspect of the electronic structure such as the Fermi surface requires properties with rather special features. First, the property must involve some directional dependence (for instance, dependence on the orientation of a magnetic field with respect to the crystal lattice) which cannot be expressed by a simple tensorial relation, secondly, the relation between the observed quantity and the Fermi surface must be sufficiently simple to enable a reconstruction of the surface to be unambiguously or nearly unambiguously made, and finally the theoretical basis of the relation must be really reliable. We shall now consider briefly a number of phenomena which satisfy these criteria to varying extents and discuss what information can be obtained from them.

2.2   MAGNETIC PHENOMENA

First, we consider a group of phenomena based on the behaviour of a metal in a magnetic field. All of these are essentially low temperature phenomena, since they require that the electron should not be appreciably scattered during one orbit of its motion in the field. This requires that at the highest field used, $\omega_c \tau \gg 1$, where ω_c is the cyclotron frequency and τ is some appropriate average of the scattering time (we shall refer to it loosely as a relaxation time, though this may not always be strictly permissible). The higher is this value of $\omega_c \tau$, the lower is the field down to which the effect can be observed, and the more accurately can the relevant experimental parameter be measured. It is only at low temperatures (usually below 10 °K) that phonon scattering become unimportant and τ becomes long enough to satisfy this requirement, and even then only for very pure samples. In the quantum oscillations discussed in § 2.2.1 there is the additional requirement that kT should be small compared with the energy separation $\hbar\omega_c$ of the quantum levels and this is usually an even more severe restriction, requiring temperatures as low as 1 or 2 °K.

The essential link with electronic structure in all these phenomena is that when a magnetic field is applied, the representative point in **k**-space moves round a curve which is a cross-section of a constant energy surface normal to the field. This leads to relatively simple relations between the property measured for a particular direction of the magnetic field and various geometrical features of the Fermi surface or of immediately neighbouring constant energy surfaces.

2.2.1 de Haas–van Alphen and allied effects (see §3.8)

These arise essentially from the quantization of the motion in a magnetic field and take the form of oscillations in various properties as the magnetic field **H** is varied. The most studied of these effects, and

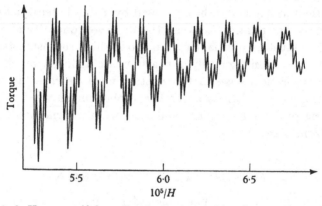

Fig. 2.1. de Haas–van Alphen effect in Zn recorded by the torque method (Joseph & Gordon, 1962). Two frequencies can be seen; they are approximately $F = 5 \times 10^6$ G and 6×10^5 G and are associated with different sections of the arms of the monster of Fig. 2.19.

often the most convenient, is the de Haas–van Alphen effect (Fig. 2.1) in which it is the magnetic properties which oscillate, but quantum oscillations of the same periodicity occur also in the electrical conductivity (Shubnikov–de Haas effect), the temperature (magneto-thermal effect), the ultrasonic attenuation (not to be confused with the geometrical resonances to be discussed in §2.2.4) and many other properties. The oscillations usually consist of a small number of periodic components, each of which is nearly simple harmonic if expressed as a function of $1/H$, and has a phase† which apart from an additive constant which need not concern us, is given by

$$\phi = c\hbar A/2\pi eH \equiv F/H. \tag{2.1}$$

† Units of 2π will be used for phase throughout the chapter; thus the phase specifies the serial number of the oscillation counting from high fields downwards.

Here, A is an extremal cross-sectional area of the Fermi surface normal to the direction of H, and for any given direction there will be as many periodic components in the oscillations as there are extremal areas, each with its own 'de Haas–van Alphen frequency' F defined by (2.1).

It can be seen that measurement of F gives A absolutely (i.e. without any adjustable parameters†), and a knowledge of A for every extremal cross-section as a function of orientation is sufficient, if not always to specify the Fermi surface uniquely and entirely on its own, at least to specify it on the basis of a few additional assumptions about its general form.

The great power of these quantum effects, compared with the other oscillatory effects considered below, as tools for determining the Fermi surface, is the very high phase of the quantum oscillations at the lowest fields for which they still have appreciable amplitude. The amplitude of the oscillations is controlled by a factor

$$\exp\left(-\frac{2\pi^2 kT}{\hbar\omega_c} - \frac{\pi}{\omega_c \tau}\right) \qquad (2.2)$$

and usually the first term in the bracket is the dominant one, even for T as low as $1\,°K$ and for relatively impure specimens ($\tau \sim 10^{-11}$ s). If we substitute

$$\omega_c = 2\pi e H / c\hbar^2 (dA/dE) \qquad (2.3)$$

(the derivative to be taken at the Fermi surface) and suppose that the oscillations fade out when the argument of the exponential is $-\alpha$ (α is in practice of order 5 or so), we can express the highest phase of the oscillations as

$$\phi = \frac{\alpha A/(dA/dE)}{2\pi^2 kT} \qquad (2.4)$$

and bearing in mind that $A/(dA/dE)$ is of the order of E_F, the Fermi energy, this gives a highest phase of order 2×10^4 at $T = 1\,°K$ for a major piece of Fermi surface. The importance of this high phase is first, that of the sections of the Fermi surface normal to the field, it is only the ones very close to the extremal ones which contribute appreciably to the oscillations, so it is a very well-defined extremal area that is measured, and, secondly, that, in principle at least, the frequency and hence the area can be very accurately measured. If, for instance, the sample is rotated slowly in a steady magnetic field, the magnetization will oscillate as a function of angle, and each such

† At one time it was thought that many body effects might appreciably change the value of the constant of proportionality in (2.1), but it has been shown by Luttinger (1960) that this is not so and that (2.1) should be exact.

oscillation corresponds to a unit change of phase, i.e. to a change of area of only 1 part in 10^4 if the phase is 10^4, so that small area changes can be measured extremely accurately. The absolute accuracy of area measurement is limited rather by the field calibration than by the phase and it is difficult to achieve much better than 1 part in 10^3.

One disadvantage of the de Haas–van Alphen effect is that rather high magnetic fields are needed if the amplitude of the oscillations is to be observable. Thus if the highest phase F/H is of order 2×10^4, as estimated above, the lowest field at which the oscillations can be observed is of order $F/(2 \times 10^4)$, which for $F = 6 \times 10^8$ (the value for Cu), is 30 kGs and, of course, appreciably higher fields are necessary if the oscillations are to be observed over a significant range. For the smaller orbits of polyvalent metals, ω_c can be a good deal higher and F a good deal lower, so fields as low as 5 or 10 kGs can be used. However, the general scale of fields required is always rather higher than for most of the other magnetic methods to be discussed below.

Although the determination of A has been the main use of the de Haas–van Alphen effect so far, it can also provide other valuable information. As can be seen from (2.2) and (2.3), the decay of amplitude of the oscillations with increasing temperature at fixed H depends on dA/dE, and because of the high phase it is rather precisely the derivative of the *extremal* area which is implied. The measurement of dA/dE in this way is, however, tedious and requires great care if an accuracy of order 1 % is to be achieved. In practice it is simpler to study dA/dE by cyclotron resonance, though it may not be quite exactly the extremal area whose derivative is measured.

Again it can be seen from (2.2) and (2.3) that if ω_c is already known, measurement of the decay of the oscillations with falling field should give the relaxation time τ, and in this way permit a study of how the electron scattering depends on the orbit concerned and on the particular type of scattering process (see for instance King-Smith, 1965). In general, however, the situation is complicated by field dependence due to other causes, such as crystal imperfections and magnetic breakdown, and so far this aspect of the de Haas–van Alphen effect has not received as much attention as it deserves.

2.2.2 Azbel'–Kaner cyclotron resonance (see § 4.6)

The microwave impedance of a metal at frequency ω might be expected to show some sort of resonance if a magnetic field is varied through the region for which the cyclotron frequency ω_c is just ω, but what happens depends critically on the detailed geometry. If the field is applied perpendicular to the metal surface, the effect is usually slight and at best gives only indirect information about the electronic structure. But if applied parallel to the surface and perpendicular to the electric vector of the microwave field, we have a striking effect known as Azbel'–Kaner cyclotron resonance (Fig. 2.2), consisting of oscillations of the impedance periodic in $1/H$, whose amplitude decays as the field falls. The phase of the oscillations can be expressed as

$$\theta = \omega/\omega_c = \omega mc/eH = \omega c\hbar^2(dA/dE)/2\pi eH \qquad (2.5)$$

so that measurement of the oscillatory 'frequency' gives the cyclotron mass m, and hence dA/dE.

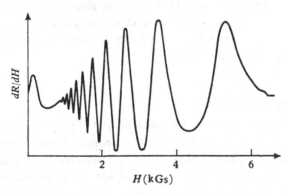

Fig. 2.2. Cyclotron resonance in Cu at 24 kmc/s (Kip, Langenberg & Moore, 1961).

The decay of these cyclotron resonance oscillations as the field decreases is governed by a factor which is $\exp(-2\pi/\omega_c\tau)$ approximately, so the oscillations fade out at a field such that $\omega_c\tau \sim 1$ and it follows that the highest phase that can be achieved is of order $\omega\tau$. Since it is difficult to use microwaves with ω greater than $5 \times 10^{11}\,\mathrm{s^{-1}}$ (4 mm waves), it can be seen that even for very pure samples, with $\tau \sim 10^{-10}\,\mathrm{s}$ (mean free path ~ 0.1 mm), the highest phase of the oscillations can only be 50 or so; in practice, in most of the studies so far carried out, it has rarely exceeded 20. Because of this low phase the dA/dE that is measured is no longer necessarily exactly the derivative of the extremal area, since sections of the Fermi surface appreciably away from the extremal one may contribute appreciably, thus producing some kind of average value. In practice the difference is probably not very important, since dA/dE is not likely to vary drastically with k_H, and indeed for band structures in which the energy surfaces are quadratic and geometrically similar, dA/dE is exactly independent of k_H, so that cyclotron resonance should give exactly the same dA/dE as does the temperature dependence of the de Haas–van Alphen effect. In the few situations where a critical experimental comparison is possible it turns out that the values of dA/dE measured by the two methods do in fact agree within the experimental uncertainties of a few per cent, but the evidence is still rather meagre.

Just as for the de Haas–van Alphen effect, the decay of oscillation amplitude with field can be used for a study of electron scattering, and additional information can also be obtained from the line shape of the individual oscillations. In principle cyclotron resonance should be the more powerful method since the decay is governed entirely by τ, while in the de Haas–van Alphen effect the major part of the decay is due to the finite temperature, and it is more difficult to separate out the smaller contribution due to τ.

2.2.3 Size effects (see § 4.8)

The path of an electron in a magnetic field **H** if projected on a plane normal to **H** has exactly the shape of an appropriate cross-section of an energy surface in k-space, but is scaled by the factor $c\hbar/eH$ and turned through $90°$ (see §3.3). Thus measurement of a dimension of an orbit in real space gives information about a corresponding dimension of a constant energy surface; for phenomena involving energy absorption, such as we shall consider, the constant energy surface is in fact the Fermi surface. There are various phenomena in which characteristic effects appear when the field is such that the orbit just fits into the thickness of the specimen, and two of these phenomena have been successfully used for Fermi surface determinations.

2.3.3 a Extinction of cyclotron resonance (Khaikin, 1962 a). If a cyclotron resonance experiment is carried out in a thin parallel-sided plate, the resonance becomes extinguished when H is just low enough for the electron orbit to fit exactly into the thickness d. For lower fields an electron starting from the upper surface of the plate is scattered at the lower surface and no longer returns to contribute to the resonance. The value H_0 of the field at which extinction occurs, immediately gives a 'caliper' dimension p of the Fermi surface; it can be shown that it is essentially the *extremal* caliper dimension taken perpendicular to the plate thickness and to **H** which is relevant. Thus we have

$$d = pc\hbar/eH_0. \tag{2.6}$$

2.2.3 b *The* R.F. *size effect* (Gantmakher 1963). Here the impedance of the thin plate is measured at a frequency much lower than the cyclotron frequency, but high enough for the skin depth to be small compared with the thickness; alternatively, the transparency of the plate to the R.F. can be measured. It is then found (Fig. 2.3) that when H is such that the orbit fits into the thickness, i.e. when $H = H_0$ as given by

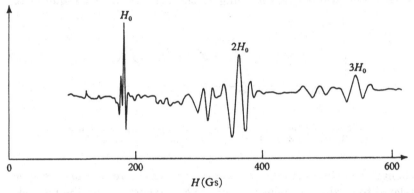

Fig. 2.3. Gantmakher effect in Sn at 3 mc/s (after Gantmakher, 1962). The ordinate is effectively the field derivative of the surface reactance of the specimen; H_0 is given by (2.6).

(2.6), there is a characteristic 'splash' in the curve of impedance of transparency against **H** and, moreover, the splash is repeated whenever $H = nH_0$, i.e. whenever an integral number of orbits fit into the thickness.

Although these methods have the advantage that they require only modest magnetic fields (typically only a few hundred gauss), they are difficult to apply in practice because they demand extremely high sample purity. In fact it is evident that the electron mean free path must be larger than the thickness d if the electron is not to be appreciably scattered in its orbit. So far it is only in a few metals that it has been possible to achieve a purity such that the mean free path is even as long as a millimetre or so. This means that d has had to be rather smaller than a millimetre to get a marked effect and this involves considerable technical difficulty if d is to be sufficiently accurately constant over the sample. The accuracy of measurement is also limited by the finite breadth of the splashes in the R.F. size effect and by vagueness of the extinction point in the other method.

2.2.4 Magneto-acoustic effect ('geometric resonance')

The principle here is similar to that of the size effects just discussed except that the orbit size is fitted into an odd number of half wavelengths of an ultrasonic sound wave instead of into a specimen thickness. Thus the attenuation of an ultrasonic wave shows oscillations as a function of field (Fig. 2.4), with the interval between successive oscillations corresponding to the change of field required to

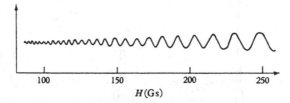

100   150   200   250
$H(\text{Gs})$

Fig. 2.4. Magneto-acoustic effect in ultra-pure Mg at 260 mc/s (Ketterson & Stark, 1967). The ordinate is attenuation.

change the orbit size by one wavelength. As before, the orbit size is related to a caliper dimension p of the Fermi surface and the phase ψ of these 'magnetoacoustic' oscillations is given approximately by

$$\psi = \hbar c p / e H \lambda \qquad (2.7)$$

where λ is the ultrasonic wavelength. Thus measurement of the period in $1/H$ of the oscillations immediately gives the caliper dimension p in terms of λ.

Just as in the size effect methods, the mean free path must be long enough for the electron not to be much scattered in one orbit, and it is easy to see that the highest phase for which the oscillation amplitude is appreciable is of order l/λ or $\omega \tau v / 2\pi v_s$, (where l is the mean free path, τ the relaxation time l/v, and v_s the sound velocity). The highest angular frequency that can conveniently be used is of order $2 \times 10^9 \, \text{s}^{-1}$, and $v/v_s \sim 300$, so we see that even for rather pure specimens with $\tau \sim 10^{-10} \, \text{s}$ (mean free path $\sim 0 \cdot 1 \, \text{mm}$), the highest phase is only 10 or so, just as in cyclotron resonance. The field corresponding to this phase is only of order $1000 \, \text{Gs}$, so once again, only modest fields are required. As before, the low phase means that the measured caliper dimension may be appreciably different from the extremal one, especially if some non-extremal orbit is emphasized in the averaging because of a strong interaction between the electrons and the lattice vibrations. The low phase also limits the accuracy with which the periodicity of the oscillations can be measured to about 1 or 2 %.

2.2.5 Galvanomagnetic effects (see §3.10)

In a magnetic field the transverse magneto-resistance has a different field dependence according as the Fermi surface does or does not support any open orbits normal to the field direction, and where open orbits do exist, their direction can be determined from the variation of resistance with the current direction. From detailed studies of this kind, the topology of the Fermi surface can be inferred and in favourable conditions it is possible to estimate the linear dimensions of particular features, such as necks linking the units of a multiply connected surface. Measurement of the Hall effect gives the volume of the Fermi surface, if it is a closed surface, or the difference of the volumes of certain slabs in an open surface from which again dimensions of necks can be inferred. Magneto-resistance measurements also give information about the relaxation time, most directly from longitudinal rather than transverse studies (Pippard, 1964b), but as yet this possibility has been little exploited.

2.2.6 Miscellaneous magnetic effects

There are several variants of the experiments already mentioned and some others, in which effects occur depending on an extremum of dA/dk_H, where k_H is measured along the field direction. Such an extremum† occurs in particular at an elliptic limiting point, where it is easily seen that

$$|dA/dk_H| = 2\pi(\rho_1\rho_2)^{\frac{1}{2}} = 2\pi K^{-\frac{1}{2}}. \qquad (2.8)$$

Here ρ_1 and ρ_2 are the two principal radii of curvature, the reciprocal of whose product is K, the Gaussian curvature. In principle a different type of extremum can also occur in a surface for which the A–k_H curve has a point of inflection (e.g. a pear-shaped surface) and in some conditions oscillatory effects occur when the field is normal to a periodic open orbit direction (the period of these oscillations then depends on the repeat periodicity of the orbit), but we shall not consider these more difficult cases. Some experiments may be mentioned:

† Strictly speaking this a 'cut-off' rather than an extremum.

2.2.6a *Cyclotron resonance* (see § 4.7). If the microwave field is polarized parallel instead of perpendicular to the steady magnetic field, the cyclotron resonance should in principle be dominated by the sections of the Fermi surface close to the elliptic limiting point in the field direction. Thus the cyclotron mass measured in this case should be proportional to dA/dE at the limiting point and this in turn is proportional to $(dA/dk_H)/v$, where v is the electron velocity at the limiting point (which is of course proportional to dE/dk_H and directed along H). Since, as we have seen, dA/dk_H at the limiting point is just $2\pi K^{-\frac{1}{2}}$, we have in principle a method of determining v at every point of the Fermi surface if the Fermi surface has already been so well determined that K is everywhere known. In practice very stringent conditions are required for observing this kind of cyclotron resonance and it has been observed only for one or two special directions of H.

2.2.6b *Doppler shifted cyclotron resonance* (see § 4.9). In helicon wave propagation, if the magnetic field H is small enough, there will always be some electrons whose component of velocity along H is such that as they move through the much slower helicon wave they feel an alternating field at the appropriate cyclotron frequency, and consequently absorb energy out of the helicon wave sufficiently to attenuate it seriously. There is, however, an absorption edge for this Doppler shifted resonance, since if H is too large, no electrons move fast enough along the field to see the cyclotron frequency. The field at the absorption edge is such that

$$qv = \omega_c, \qquad (2.9)$$

where ω_c and v refer to the limiting point and q is the wave-number of the helicons. It follows at once that the critical field is proportional to q and to dA/dk_H, so that measurement of q and the critical field determines the Gaussian curvature at the limiting point of the Fermi surface in the field direction.

2.2.6c *The Kjeldaas edge and related effects* (see Kjeldaas, 1959). This is very similar to the effect just considered except that it is ultrasonic acoustic waves that are propagated along the magnetic field instead of helicons. The absorption edge is again determined by (2.9), where q is now the acoustic wave-number and once again the Gaussian curvature can be measured.

Strictly speaking, equation (2.9) applies only if the wave velocity

ω/q (i.e. v_s for sound) is negligible compared with the electron velocity v. More generally we should have

$$qv \pm \omega = n\omega_c \qquad (2.10)$$

where the sign of ω depends on the sense of rotation of a circularly polarized wave, corresponding to electrons moving parallel or anti-parallel to the field; moreover, the edge is repeated for integral values of $n > 1$ if the Fermi surface is not axially symmetrical along the field direction.† Thus the edge should be 'doubled' if linearly polarized waves are used, and repeated at integral submultiples of the field for which $n = 1$. Normally the doubling cannot be resolved because v_s/v is too small (typically 4×10^{-3}), but if the wave-vector is at considerable angle θ to H, so that qv is replaced by $qv\cos\theta$, and, moreover, v is small, as in the bismuth type metals, the doubling is observable (Eckstein, 1966). The *difference* of the fields at which the two edges occur gives dA/dE at the limiting point, i.e. $(dA/dk_H)/v$, while the *mean* field gives dA/dk_H. Thus in principle both the Gaussian curvature and v can in this way be obtained in the same experiment, though in practice the method is of limited application and it is difficult to achieve much accuracy. The 'tilt' effect (Spector, 1960), in which an edge occurs in the variation of attenuation with θ is somewhat similar in principle but again of limited application.

2.2.6d *The tilted* R.F. *size effect* (see § 4.8). If in the Gantmakher experiment the magnetic field is not exactly parallel to the surface of the plate but tilted at a small angle ϕ, electrons setting off nearly parallel to $\mathbf{H}$ from one side of the plate will refocus at the other side if they complete an exact number of turns on the way, i.e. if

$$\omega_c(d/v\sin\phi) = 2\pi n, \qquad (2.11)$$

where v is the component of velocity along $\mathbf{H}$ and d is the thickness of the plate. If the R.F. electric field has a component parallel to $\mathbf{H}$, such a refocusing produces a 'splash' in the R.F. impedance and (2.11) shows that the splashes will come at equal intervals ΔH of field which are predominantly determined by the extremum of $v(dA/dE)$, i.e. once again by dA/dk_H or the Gaussian curvature. Evidently the splashes will be appreciable only if the mean free path is not too short compared with the oblique path length $d/\sin\phi$ and by studying

† The detailed theory depends also on whether the sound waves are longitudinal or transverse, but this need not concern us here.

how their magnitude varies with ϕ, it is possible to estimate the mean free path. Potentially this is rather a powerful method for studying mean free paths, since it picks out the mean free path for electrons travelling in a particular direction rather than some average, and encouraging results have been obtained for Sn and In; since, however, even more extreme purity is required than for the measurement of orbit calipers, it is at present of rather limited application.

Oscillations of the R.F. impedance (sometimes called 'Gantmakher waves') occur also for **H** perpendicular to the plate. They are much more sinusoidal than for **H** nearly parallel to the plate, and as explained on p. 249 they are somewhat akin to the Sondheimer oscillations discussed in § 2.2.6f. Although the theory of these oscillations is different from that of the more spiky oscillations for small ϕ, their periodicity is still given by (2.11) with $\phi = 90°$, and is therefore again determined by the Gaussian curvature.

2.2.6e Sharvin's focusing experiment (see § 4.4). If a magnetic field **H** is directed exactly along the line joining two point contacts applied to the opposite faces of a thin parallel-sided plate and current is fed through the contacts, the voltage between the contacts shows marked oscillations as H is varied. The situation is similar to that in the R.F. size effect when the field is perpendicular to the plate. Electrons setting out nearly parallel to the field from one contact will focus at the other contact if their helical paths contain an integral number of loops; the condition for this is exactly as before, i.e. (2.11) with $\sin \phi = 1$ and once again the Gaussian curvature is determined.

2.2.6f D.C. size effects (see § 4.4). The ordinary D.C. resistance of a thin plate shows oscillatory field dependence if **H** is perpendicular to the plate (Sondheimer oscillations), and the periodicity is again determined by (2·11), i.e. by the Gaussian curvature. As discussed in § 4.4, these oscillations are rather feeble, but stronger in compensated than in uncompensated metals.

2.3 NON-MAGNETIC PHENOMENA

2.3.1 Anomalous skin effect (see § 4.5)

This is the skin effect in conditions where the mean free path is much longer than the skin depth, i.e. in very pure samples at low temperatures and high frequency. The surface resistance is proportional to the integral round a zone of the Fermi surface of a radius of curvature in a plane containing the normal to the metal surface and the electric vector of the microwave field; the zone is defined by points on the Fermi surface whose normals lie parallel to the metal surface (i.e. points which represent electrons moving parallel to the metal surface). By cutting crystals with plane surfaces normal to various crystal directions, and for each sample varying the direction of the electric vector in the metal surface, a considerable amount of geometrical information about the Fermi surface can be obtained. If the surface consists of a single sheet this information can provide a valuable guide to the nature of the surface, and indeed the first strong indication that the Fermi surface of Cu is an open one was obtained in this way (Pippard, 1957). If, however, there are several sheets the measured quantity is a sum of contributions from all the sheets and it becomes impossible to disentangle the information at all unambiguously.

A rather different application of the anomalous skin effect is that under certain simplifying assumptions the surface resistance of a polycrystalline specimen gives the total surface area of the Fermi surface and this can sometimes be useful in distinguishing between alternative proposed models for the surface based on other considerations.

2.3.2 Positron annihilation

If a positron enters a metal it rapidly comes to rest (in a time of order 10^{-12} s) and after a time of order 10^{-10} s annihilates with one of the electrons in the metal, emitting two photons. The two photons would be emitted in exactly opposite directions if it were not for the component $\hbar k_z$ of momentum of the annihilated electron in the direction perpendicular to the mean direction of the photons and lying in the plane of their paths. Thus since each photon has momentum mc, we see that the small angle θ between the nearly oppositely emitted photons is given by

$$\hbar k_z = mc\theta. \qquad (2.12)\dagger$$

On the simplest interpretation, then, the number $N(\theta)\,d\theta$ of photon coincidences at angles between θ and $\theta + d\theta$ should be proportional to

---

† This result only applies to completely free electrons; in a more realistic treatment (see Berko, 1962) there are contributions to the photon momentum not only corresponding to k, but also to $k + G$, where G is one reciprocal lattice vector. These contributions, however, rapidly decrease in weight as G increases, so the simple treatment still gives a qualitative picture.

the number of electrons with k_z between k_z and $k_z + dk_z$. There should also be a cut-off in the curve of $N(\theta)$ corresponding to the maximum value of k_z. By varying the direction of the crystal axes relative to the plane in which θ is measured, it should be possible to measure all the radius vectors of the Fermi surface directly from the cut-off values, and moreover to have a further check on the shape of the surfaces from each $N(\theta)$ curve, since this should give the variation of area with k_z. Potentially this would seem a very powerful method since the geometrical information it could give is so direct and so copious, and moreover it does not require low temperatures or high purity as do all the other methods. Unfortunately, however, the interpretation of the $N(\theta)$ curves is not as simple as outlined above and because of the uncertainties of how to allow for the various theoretical complications such as the effect of core electrons, correlation effects, etc., it has so far proved of only limited value; we shall see, however, that used with care it can sometimes provide answers of a semiquantitative kind in situations where the other methods fail.

2.3.3 Compton effect

In the Compton scattering of X-rays in the high momentum transfer region the spectral distribution of the scattered line is given by the projection of the total electron momentum distribution onto the scattering vector. Thus the profile of the scattered line should in principle give information very similar to that given by positron annihilation experiments, but without the complications of interpretation caused by electron-positron correlation prior to annihilation (Platzman & Tzoar, 1965). Unfortunately technical difficulties associated with the intensity of the scattered radiation severely limit the usefulness of this method and although some preliminary experiments on Li confirm the general soundness of the idea (Cooper, Leake & Weiss, 1965) it is unlikely to be a serious contender for some time to come except perhaps for metals of low atomic number, when the X-ray absorption is low.

2.3.4 Kohn effect (see Ziman, 1964a, pp. 133, 173)

According to Kohn (1959a) an 'image' of the Fermi surface should be apparent in the ω–$\mathbf{q}$ relation for the lattice waves of the metal. Whenever the phonon vector $\mathbf{q}$ has the length of a maximum or minimum chord of the Fermi surface in the direction of $\mathbf{q}$, a slight kink should

G

appear in ω–$\mathbf{q}$ relation and in principle it should be possible to map out the Fermi surface by following the $\mathbf{q}$ surface over which such kinks appear. Such an effect has been found in neutron diffraction studies of Pb and Al (Brockhouse *et al.* 1962; Stedman & Nilsson, 1965) but the kinks are very slight, particularly in Al and have been observed for only a few directions of $\mathbf{q}$. A similar effect has been found in the thermal diffuse scattering of X-rays (Paskin & Weiss, 1962) and again the kinks are only just noticeable above the background scatter.

Because of the slightness of the kinks and the difficulty of the experimental techniques, the Kohn effect is unlikely to have much general application in the determination of Fermi surfaces, though it may be of value in special situations such as the study of alloys (Woods & Powell, 1965) where other methods fail.

2.4 SUMMARY AND COMPARISON OF METHODS

If we leave on one side methods such as positron annihilation and Kohn effect which at our present level of theoretical understanding and technical achievement are not as useful as they might at first sight appear to be, we are left essentially with methods which provide the following kinds of information:

(*a*) Either extremal areas of cross-section A, or extremal caliper dimensions p, of the Fermi surface.

(*b*) More complicated geometrical information about the Fermi surface: (i) averages of radii of curvature round zones of the surface from the anomalous skin effect, (ii) Gaussian curvature at points on the surface, and (iii) topological information from magneto-resistance.

(*c*) Values of cyclotron masses which give dA/dE for extremal, or nearly extremal, cross-sections of the Fermi surface.

(*d*) Information about electron scattering. Since very little systematic work has been done this topic will not be further considered.

It is really only from information of type (*a*) that it is possible to determine accurately and reliably the size and shape of a Fermi surface, and even then it is often necessary to appeal to theoretical ideas about the Fermi surface to interpet the data in a meaningful way.

Although at first sight it would seem that the most powerful methods of determining a surface should be those which measure linear dimensions rather than areas, since an area is already an integral over the surface and inevitably irons out the detailed form of the surface to some extent, in fact the measurement of areas by the de Haas–van Alphen effect has so far proved to be the more powerful method. This is mainly because the de Haas–van Alphen effect does not make such great demands on sample purity and sample geometry as do the caliper methods, so that it is more widely applicable, at least at the present level of technical possibilities. Moreover, because of the very high phase of the oscillations the accuracy of measurement is appreciably higher, and there is less uncertainty about the interpretation of the measured quantity as an extremal property of the surface. There is also the rather paradoxical reason that if the surface is at all complicated the number of extremal calipers for any given direction of **H** is often greater than the number of extremal areas and this excess of information tends to be an embarrassment rather than a help, because it leads to ambiguities of interpretation unless the general form of the surface is very reliably known already.

Information of type (b) can sometimes be used in conjunction with a theoretical model to give a qualitative and occasionally a semi-quantitative picture of the Fermi surface, but its main function so far, has been to resolve ambiguities which may arise in interpreting the area or caliper data, to provide checks on detailed features and of course to help in understanding the theory of the phenomena themselves.

Values of dA/dE do not in themselves lead to a Fermi surface determination, though from examination of the orientation ranges over which particular cyclotron masses occur, topological information about the surface is provided of the same kind as that from magneto-resistance; and of course some qualitative picture of the Fermi surface is given if the mass data are examined in conjunction with a theoretical band structure. If, however, the Fermi surface is already known from other data, the values of dA/dE should, at least in principle, lead to a value of the electron velocity v at each point of the Fermi surface. Thus if A is the extremal area of the Fermi surface normal to a particular direction, the extremal area of an immediately neighbouring surface of energy $(E_F + \epsilon)$ will be $(A + \epsilon \, dA/dE)$ and from a knowledge of all values of this area (for a given small ϵ) the form of the neighbouring surface can be determined. If now the normal distance between the Fermi surface and the neighbouring one is found to be Δk_n at some point of the Fermi surface, we have at once that the velocity at that point is given by

$$v = \epsilon/\hbar \Delta k_n, \qquad (2.13)$$

This procedure has been carried through only for a few relatively simple situations apart from the trivial one of a spherical Fermi surface (where A and dA/dE are isotropic, so that v is also isotropic).

2.5 THE DERIVATION OF A FERMI SURFACE FROM GEOMETRICAL DATA

We shall now consider how a Fermi surface can actually be determined from a knowledge of either its extremal areas of cross-section or its caliper dimensions. It turns out that in either case it is only in rather restricted conditions that it is possible to go unambiguously from the areas or caliper dimensions to a specification of the surface. It is only if the surface has a centre of symmetry and is everywhere convex that there is no ambiguity; it is then always the plane through the centre of symmetry normal to the given direction of **H** which has the extremal area and which contains the extremal caliper dimensions perpendicular to **H**. A knowledge of all the central areas for such a surface completely specifies the surface according to a theorem of Lifshitz & Pogorelov (1954), and likewise the caliper dimensions of each such central section specify its 'tangent-polar' equation and so the whole surface can be constructed if this information is available for every direction of **H**. Note, incidentally, that in principle while the area method requires only one measurement for each direction of **H**, the caliper method requires a set of measurements for all directions normal to each particular direction of **H**.

Unfortunately these general geometrical considerations have comparatively little relevance to real Fermi surfaces, which are often not everywhere convex and, moreover, sometimes consist of a set of sheets, which though related by crystal symmetry, so that the whole set is centro-symmetric, are not individually centro-symmetric. Even in the few situations where it is fairly certain that the surface is both convex and centro-symmetric the Fermi surface is in practice determined from the areas, not by using the Lifshitz–Pogorelov theorem, but by fitting to a suitable equation containing a small number of coefficients to be determined from the data. This of course requires far fewer data (both to determine the coefficients and to check that the equation is a reasonable fit) than would the application of the Lifshitz–Pogorelov theorem, which involves integrals over all directions of **H**. Similarly an empirical equation of this kind could be established from a quite limited set of caliper data (though in practice this has never been done).

If the surface is not everywhere convex but still has a centre of symmetry, there is no longer a unique surface specified either by area or caliper dimensions and it is essential to start from a plausible model, usually based on theoretical ideas and of course consistent with crystal symmetry. If the model involves a multiply connected sheet of Fermi surface it usually implies a variety of qualitative features such as extremal areas and caliper dimensions in particular

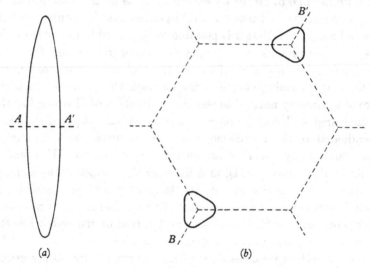

(a)   (b)

Fig. 2.5. (a) Side view of a cigar (looking along BB' of (b)); (b) section through the plane AA' showing the relation of a pair of complementary cigars; the cigars at the other hexagon corners have been omitted.

non-central sections and the first step both in area and caliper studies is to verify that these do indeed occur. Confidence is gained too if other experiments, such as galvano-magnetic measurements, confirm the topological features of the model and once the model can be regarded as reliable, its detailed parameters can be fixed by quantitative fitting of the geometrical data.

Where a sheet of the surface has no centre of symmetry, the determination of its shape from its areas or its caliper dimensions becomes even more difficult. This is illustrated by the example of Fig. 2.5 which shows schematically a pair of 'cigars' forming part of the Fermi surface of certain hexagonal metals. The pair of cigars has a centre of symmetry but the separate cigars have none, and it is impossible to deduce their shapes uniquely from area or caliper data alone unless some special assumption is made about the analytical form of the surface. In principle a check on such an assumption would be to verify that the same values of the parameters of the assumed analytical form will correctly give both extremal areas and extremal caliper dimensions, but in practice the data (or some of them) may not be accurate enough to make this a very stringent check. Often indeed, there are insufficient data for any real check and the results are interpreted in terms of a 'symmetrized' surface, even though this may not be the true surface.

We have seen that in practice the determination of a Fermi surface consists essentially in finding the values of a fairly small number of parameters which specify some suitably chosen analytical description of the surface fitting the data, and before coming to the results for individual metals we shall outline the types of analytical description which have been used (we consider only the problem of fitting areas). In some ways the most satisfactory situation is where the choice of form is guided only by considerations of crystal symmetry and the nature of the area data. Examples of this type of approach are expansions in cubic harmonics for nearly spherical surfaces or in Fourier series for the open surfaces of the noble metals, and occasionally the fitting of ellipsoids to particular sheets of Fermi surfaces. It should be emphasized that the fitting parameters have no direct physical significance in themselves, but this kind of specification has the advantage that it is not tied to any particular theory of the band structure and the surface is available as an objective summary of the experimental data for comparison with any theoretical Fermi surface which may be proposed.

An alternative approach which has had considerable success, especially with the rather complicated Fermi surfaces of polyvalent metals, is based on the 'nearly-free electron' model in which the band structure is specified by a small number of pseudo-potential coefficients, supplemented where appropriate by spin-orbit coupling parameters and by parameters specifying the d-electrons. Provided the total number of parameters is not too great, values of them may be found which produce a Fermi surface having the observed areas. This then specifies the surface, but in a rather indirect way, since fairly elaborate computing is needed to go from trial values of the parameters to the areas. This approach has the advantage that the parameters are now intended to have direct physical significance, so that they can be used to discuss other physical properties of the metal. Considerable caution is needed, however, before it can be reliably assumed that they really do mean what they are intended to mean, because it may turn out that several different schemes of parameters fit the area data equally well. Studies of how the Fermi surface is modified when the dimensions of the unit cell are changed by mechanical stress (e.g. hydrostatic pressure), may help to resolve such ambiguities, but because of technical difficulties, this possibility has not yet been much explored.

It should be noticed that the determination of a Fermi surface can be accomplished at a variety of levels. In the early studies it was considered quite an achievement merely to get a general idea of the nature of the surface with perhaps approximate values of some of the important dimensions, and indeed there are many metals for which even this primitive kind of determination has not yet been achieved. More recently, however, with improved techniques of measurement, much greater accuracy has been achieved for at least a few metals, thus making possible ever more critical comparisons with band-structure calculations and also more critical checks of the theory of the phenomena on which the determinations are based.

For some metals, even though good accuracy in area measurements has been achieved, only partial results have been obtained and the analysis has not been taken beyond a rather qualitative level, such as a rough comparison with a band-structure calculation. Also it should not be forgotten that most of such determinations as have been made are based not, as ideally they should be, on area measurements normal to directions over the whole solid angle, but only those lying in symmetry planes. This is because for directions not lying in symmetry planes small errors in the exact crystal orientation can produce changes of area larger than the intrinsic error in the area measurement. If the technique of orienting the crystal with respect to the field is appreciably improved and reliable transverses can be obtained in other planes, it may yet prove that the surfaces thought to be really accurately known still need appreciable modification.

2.6 THE ALKALIS

Of the monovalent metals the alkalis have much more nearly spherical Fermi surfaces than the noble metals, mainly because in the former the d-bands are much further removed from the Fermi level and so cause much less distortion from the spherical surface of a free-electron metal. The departures from sphericity have been rather accurately measured in Na, K and Rb using the method outlined in §2.2.1 in which the change of phase of the de Haas–van Alphen oscillations is followed as the sample is rotated in a constant magnetic field. The results for K are shown in Fig. 2.6 and it can be seen that the area differences are determined with an accuracy of rather better than 1 part in 10^4 of the diametral area of the sphere. Since the surface is nearly spherical and has cubic symmetry the relative differences of area from those of a sphere should be described by an expansion in cubic harmonics

$$\Delta A / A = a\alpha_4 + b\alpha_6 + c\alpha_8 + \dots \qquad (2.14)$$

and it was found possible to obtain a reasonable fit to the experimental points by using only the first 3 terms with suitable values of a b and c. From the theory of cubic harmonics it then immediately follows that the departures from the radius k_0 of a sphere with the same volume as the actual Fermi surface are given by

$$\Delta k / k_0 = \tfrac{4}{3}a\alpha_4 - \tfrac{8}{5}b\alpha_6 + \tfrac{64}{35}c\alpha_8 + \dots \qquad (2.15)$$

and the Fermi surface is determined. Contour diagrams of the orientation dependence of $\Delta k / k_0$ for the alkalis are shown in Fig. 2.7 and it can be seen that the departures from a sphere are indeed very small: only of order 1 in 1000 for Na and K, nearly 1 % for Rb, and several per cent for Cs (Okumura & Templeton, 1965). For all the alkalis, the mean radius is within experimental error the same as for a free-electron sphere.

The area data from the de Haas–van Alphen effect may also be analysed more significantly in terms of pseudo-potentials (Ashcroft, 1965; Lee, 1966; Lee & Falicov, 1968). If the pseudo-potentials V_G are treated as sufficiently small (so that powers of V_G/E_F beyond the second can be neglected) explicit expressions for the radius and cross-sectional area of the Fermi surface in any direction can be written down. For Na this proves to be an adequate approximation and a good fit to the de Haas–van Alphen data can be obtained with

$$|V_{110}| = 0{\cdot}23\,\text{eV}$$

ignoring the effect of V_{200}, etc. A more elaborate calculation based on a 43×43 secular determinant, i.e. taking account of all the {110}, {200} and {211} reflection planes, gives nearly the same answer, though the answer is slightly different according as V_{110} is positive ($0{\cdot}25\,\text{eV}$)

or negative (-0.21 eV). It should be noticed that although a good fit is obtained by putting V_{200}, V_{211}, etc., zero, the observed distortions are rather insensitive to the values of V_{200}, V_{211}, etc., and it is possible only

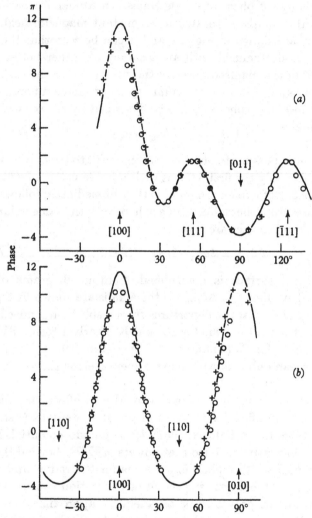

Fig. 2.6. Orientation dependence of cross-sectional area of Fermi surface of K (a) in (110) plane, (b) in (100) plane. A phase change of π corresponds to an area change of 1 part in 7400. The curves are calculated from (2.14) with suitably chosen values of a, b and c (Shoenberg & Stiles, 1964).

to set upper limits for them, e.g. the fit begins to be poor if V_{200} exceeds 0.3 eV. Thus for Na the Fermi surface can be quite reasonably speci-fied by only the single parameter V_{110}, and it is satisfactory that the

value found for it agrees well with theoretical estimates, e.g. Heine &
Abarenkov's (1964) 'model potential' calculation which gives
$+0.25 \pm 0.10$ eV.

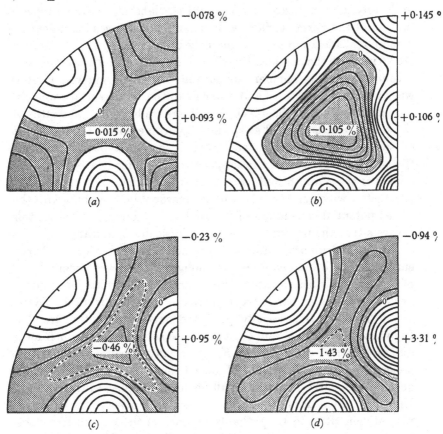

Fig. 2.7. Contours of the Fermi surface of the alkali metals; the numbers are the
values of k/k_0 at $\langle 100 \rangle$, $\langle 110 \rangle$ and $\langle 111 \rangle$ and the shaded regions indicate negative values.
(a) Na (Lee, 1966); contours k/k_0 at intervals of 0.02 %. (b) K (Lee & Falicov, 1968);
contours at intervals of 0.02 %. (c) Rb (Shoenberg & Stiles, 1964); contours at intervals of 0.2 % (with an extra one, dashed, at -0.3 %). (d) Cs (Okumura & Templeton,
1965); contours at intervals of 0.5 % (with an extra one, dashed, at -1.25 %).

The position for K is less simple, basically because the d-like bands
are rather closer to the Fermi energy and the use of a local pseudo-
potential is no longer a good approximation. Ashcroft obtained
some measure of agreement with the experimental results by ignoring
this difficulty and choosing a suitable value (0.23 eV) of the single
local pseudo-potential coefficient $|V_{110}|$, but attempts to improve

this fit by introducing $|V_{200}|$ and $|V_{211}|$, require values of these coefficients of several eV, which are physically implausible. Indeed it is possible to specify values of $|V_{110}|$, $|V_{200}|$ and $|V_{211}|$ which give a reasonable fit to the area data, but this is merely a rather elaborate analytical specification of the Fermi surface without any real physical significance. A more realistic approach has recently been made by Lee & Falicov (1968) in which the non-locality of the pseudo-potential was taken into account by adding angular-momentum dependent square-potential wells within the ion cores and a fair fit to the Fermi surface was obtained. The depth of the d-well, the difference between the depths of the s- and p-wells, as well as V_{110} and V_{200} (representing the local part of the pseudo-potential) were treated as adjustable parameters, though the fit was rather insensitive to the difference between the depths of the s- and p-wells. It is encouraging that the values of the parameters which gave the best fit are reasonably consistent with the model potential calculation of Abarenkov & Heine (1965) since this suggests that the procedure is indeed physically significant.

No detailed interpretation has yet been attempted of the Fermi surfaces of the other alkali metals, though the observed general trend of increasing distortion with atomic number is qualitatively to be expected in terms of band-structure theory (see, for instance, Ham, 1962). The sole theoretically predicted exception to this trend is Li, whose B.C.C. (high temperature) phase should have a much more distorted Fermi surface than K, essentially because there are no p-electrons in the ionic core of Li, so that the cancellation which keeps the effective potential small no longer applies to the p-states. Unfortunately it has not so far been possible to observe the de Haas–van Alphen effect in Li, probably because of the martensitic transformation to an H.C.P. phase at about 75° K. At the low temperatures necessary to observe the de Haas–van Alphen effect, a sample of Li which was a good crystal at room temperature, is probably highly strained and no longer monocrystalline, and may contain a mixture of the B.C.C. and H.C.P. phases. A similar transformation occurs in Na (at 30 °K) but although the amplitude of the de Haas–van Alphen oscillations was indeed much less than it should be for a perfect crystal, measurements were possible and the internal evidence suggested that it was the B.C.C. phase which gave the oscillations. Optimistically, an effect may yet be found in Li if higher fields and lower temperatures are used.

In the meantime there is evidence from positron annihilation experiments at temperatures above the transformation point that the Fermi surface of Li is indeed considerably distorted (Donaghy & Stewart, 1967) and this provides a good example of the usefulness of the method. The evidence is that for Li the $N(\theta)$ curves (see § 2.3.2) with the main photon direction normal to $\langle 100 \rangle$, $\langle 110 \rangle$ and $\langle 111 \rangle$ are appreciably different (Fig. 2.8), while in Na these curves coincide within experimental error. Although it is not possible to derive the shape of the Fermi surface from an individual $N(\theta)$ curve, because of

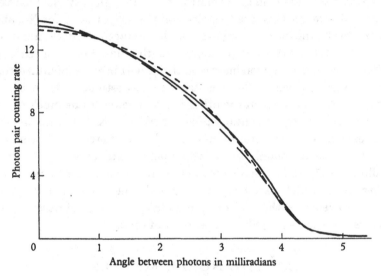

Fig. 2.8. Positron annihilation for lithium (Donaghy & Stewart, 1967).
————, 100, – – – – 100, — — — 111.

velocity dependence of the annihilation cross-section and other corrections (so that even for the almost spherical Fermi surfaces of Na and K the $N(\theta)$ curves are not parabolic), the fact that the $N(\theta)$ curves for Li are appreciably different for different crystal orientations, clearly indicates anisotropy of the Fermi surface. Detailed considera-tion of the results suggest that k_{110} is about 5 % greater than k_{100}.

Many of the other experimental methods of studying electronic structure have also been applied to some of the alkali metals, but as would be anticipated, their accuracy is inadequate to show up the very small distortions from a spherical surface and their main value has been in checking the theory of the phenomena. Thus the magneto-acoustic effect and the R.F. size effect have shown that within experi-mental accuracy the Fermi surface is a sphere of the expected radius, while studies of Doppler shifted cyclotron resonance, the Kjeldaas edge and the tilted field R.F. size effect, give a Gaussian curvature corresponding to the expected radius of the free-electron Fermi sphere. One interesting point is that these experiments do not seem to support Overhauser's (1964) spin-density wave theory, which predicts that the Fermi surface should be appreciably pear-shaped with an extremum of dA/dk_H not at an elliptic limiting point.

Cyclotron resonance experiments are also not precise enough to reveal any anisotropy in the cyclotron mass, though they yield masses appreciably larger than the free-electron mass m_0 ($1\cdot24m_0$ for Na, and $1\cdot21m_0$ for K; the latter is consistent with a less accurate determination from the temperature dependence of the de Haas–van Alphen effect). Since the cyclotron mass is isotropic within experimental error, so too is the velocity. The cyclotron masses estimated from band-structure calculations, on the other hand, turn out to be much closer to m_0 than the experimental values for Na and K. The discrepancy between the experimental values and the band-structure estimates appears to be due mainly to electron–phonon interaction (Ashcroft & Wilkins, 1965), and a discrepancy in the same sense is also found for all other metals which have been investigated. One general result is that usually, even in rather anisotropic metals, the ratio of observed to calculated mass is roughly independent of direction.

2.7 THE NOBLE METALS

The next simplest Fermi surfaces are those of the monovalent noble metals, Cu, Ag and Au. Pippard's (1957) anomalous skin effect study of Cu had suggested that the Fermi surface is so much distorted from a sphere that the Fermi surface makes contact with the (111) faces of the Brillouin zone (Fig. 2.9a) and this immediately implies a multiply-connected surface with qualitatively new features. Thus normal to each ⟨110⟩ direction there should be an extremal hole orbit (the 'dogs' bone' (Fig. 2.9b)), but no central extremal 'belly' orbit; normal to ⟨100⟩ there should be an extremal non-central hole orbit (the 'rosette' (Fig. 2.9c)) as well as the central belly orbit; and normal to ⟨111⟩ there should be a small minimum area 'neck' orbit (Fig. 2.9a), a central belly orbit, and a large central 'six-cornered rosette' hole orbit (Fig. 2.9d). Each of these orbits should exist for finite ranges of directions around the relevant symmetry direction. Soon after the de Haas–van Alphen effect had been observed in these metals it was possible to confirm that oscillations of the appropriate frequencies corresponding to nearly all these predicted features† did in fact occur in all three metals when the magnetic field was in the relevant range of directions. This is very strong evidence that the

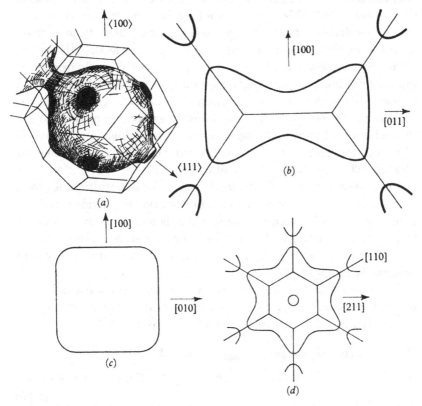

Fig. 2.9. (a) Fermi surface of noble metals shown in a Brillouin zone (schematic); (b) the 'dogs bone' orbit: (c) the 4-cornered rosette; the name was chosen somewhat inappropriately before the exact shape was known; (d) the 6-cornered rosette (approximately on 3 times reduced scale compared with (b) and (c)).

model of Fig. 2.9 is indeed correct for all three metals and further support for the model comes from its excellent agreement with the results of other experiments to be discussed below.

Once the model could be regarded as reliable, the next step was to determine the detailed shapes of the Fermi surfaces, and the most

† The frequency corresponding to the six-cornered rosette, about 1·9 times the belly frequency, has, however, not yet been observed, probably because of insufficient amplitude in the experimental conditions so far used.

precise determinations have been made by a detailed study of the orientation dependence of the de Haas–van Alphen frequencies. Since the central belly areas differ by only a few per cent from those of a free-electron sphere a high accuracy of frequency measurements is essential and in the original investigation by the pulsed field method (Shoenberg, 1962) this was achieved by beating the de Haas–van Alphen oscillations from the specimen against those from a fixed reference specimen of nearly equal frequency. Thus as the orientation of the specimen was varied, the small variations of the belly frequency showed up as much larger variations of the beat frequency, and so could be measured with a relative accuracy of order 0·2 %. With the development of steady high field methods it has become possible to use the technique of rotating the specimen in a steady field and a relative accuracy of frequency measurement of order 0·05 % has been achieved (Joseph, Thorsen, Gertner & Valby, 1966; Halse, 1967). Moreover, with this technique the orientation is varied continuously rather than in steps, so that much more detailed information is provided, and the accuracy of angular measurement, which is as important as that of frequency measurement, has also been improved. A typical rotation curve for Cu is shown in Fig. 2.10 and the variation of frequency with orientation in Fig. 2.11.

The simplest analytical description of the Fermi surface which takes account of the cubic symmetry and the multiple connectivity of the surface is a Fourier expansion of the form (Roaf, 1962)

$$W = 3 - \Sigma \cos \tfrac{1}{2}ak_y \cos \tfrac{1}{2}ak_z + C_{200}(3 - \Sigma \cos ak_x)$$

$$+ C_{211}(3 - \Sigma \cos ak_x \cos \tfrac{1}{2}ak_y \cos \tfrac{1}{2}ak_z) + C_{220}(3 - \Sigma \cos ak_y \cos ak_z) + \text{etc.}$$
$$(2.16)$$

where a is the lattice constant, and W and the Cs are adjustable parameters to be fitted to the data. It should be emphasized that this is a purely empirical formula, and although it has the same form as the energy surface of a tight-binding calculation there is no implication that the parameters have any direct physical significance. The extremal cross-sectional areas of (2.16) can be computed and by a process of trial and error, values of the parameters are chosen which give the experimentally observed features. Roaf found that the original de Haas–van Alphen data could be quite well fitted using only 4 parameters for Cu and Ag and 5 for Au, but the recent more accurate data require at least 5 for all the metals. Halse (1967) has made a detailed analysis of these and as can be seen from Fig. 2.11 his 7-term formula gives an excellent fit to the experimental points. An interesting feature of the Fermi surface is the existence of non-central extremal belly orbits for certain directions; these occur when the central section is a maximum rather than a minimum with respect to k_H, since then the area must have a minimum for a non-zero k_H before it starts to grow as the plane approaches the nearest necks. These non-central extremal areas vary more rapidly with orientation than the central ones and so produce the high frequency angular variation which can be seen in

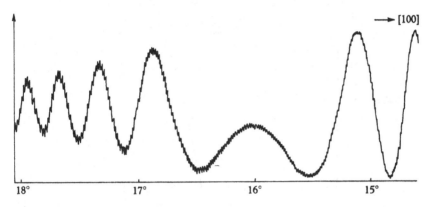

Fig. 2.10. de Haas–van Alphen oscillations in Cu for rotation in a (110) plane at 49·5 kGs (Halse, 1967); note the stationary point at 16° corresponding to the minimum of the B_1 curve in Fig. 2.11. The faster oscillations come from the non-central belly section.

Fig. 2.10 on top of the lower frequency variation due to the central orbits. The reliability of Halse's formula is rather impressively confirmed by the accurate prediction it makes of nearly every detail of these non-central areas (see, for instance, Fig. 2.11), even though no use was made of them in choosing the parameters of the formula.

The magneto-acoustic caliper data (Bohm & Easterling, 1962) are in good agreement with the surface deduced from the de Haas–van Alphen data, but this is not a very stringent test of the detailed shape, since the accuracy of the caliper data is only of order 1 or 2 % which is insufficient to fix the shape as precisely as does the de Haas–van Alphen effect. Magneto-resistance measurements confirm the topology of the surface and give about the right sizes for the necks, while Hall effect studies agree well with the expected effect of the open orbits in the Fermi surface.† Positron annihilation measurements (Fujiwara & Sueoka, 1966) in Cu also indicate necks of about the right size and this is of considerable interest since it opens up the possibility of studying the Fermi surfaces of alloys in which the short relaxation

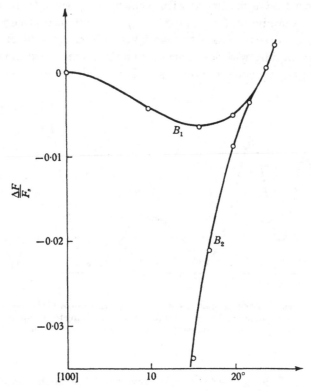

Fig. 2.11. Angular variation of F for Cu in the (110) plane. The curves are experimental, derived from records such as that of Fig. 2.10; the points are calculated from a 7-term formula of the type of equation (2.16) (Halse, 1967). B_1 refers to the central and B_2 to the non-central belly section; ΔF is $F - F_{100}$ and F_s is the value for the free-electron sphere.

time rules out the use of magnetic methods. Preliminary results of Fujiwara, Sueoka & Imura (1966) suggest that addition of 10 % (atomic) Al to Cu roughly doubles the neck diameter.

Cyclotron resonance measurements (e.g. Koch, Stradling & Kip, 1964) again confirm the general form of the surface and on the whole

† Some puzzling discrepancies between the thermal Hall effect measurements of Lipson (1966) and the predictions from the form of the surface probably indicate some inadequacy in the theory of the thermal Hall effect rather than any inaccuracy in the surface.

H

agree within experimental uncertainty with the rather limited data deduced from the temperature variation of the de Haas–van Alphen effect. In Cu a rather more crucial comparison is possible with a different feature of the de Haas–van Alphen data at a few special field directions. This comes about because of the spin factor $\cos(\pi g m / 2m_0)$ in the de Haas–van Alphen amplitude, where g may differ appreciably from 2 if spin-orbit coupling is important; thus whenever gm/m_0 is an odd integer the amplitude of the de Haas–van Alphen oscillations should vanish. Such a vanishing is indeed observed both in the neck oscillations ($gm/m_0 = 1$) and the belly and other oscillations ($gm/m_0 = 3$) for H is several different well-defined directions in the (100) and (110) planes. Thus from the cyclotron resonance values of m/m_0 for these special directions of H we can deduce values of g for the particular orbits concerned. These values mostly come within 1 or 2 % of 2·03, the average g value indicated by spin resonance experiments, but for the belly orbits near $\langle 111 \rangle$ and $\langle 100 \rangle$ g appears to be about 2·13, i.e. 5 % high. It is not clear if this really means that g has this high value or if it indicates that the cyclotron resonance mass is low because it refers to an average over k_H rather than to the extremal orbit picked out by the de Haas–van Alphen effect (see § 2.2.2).

Recently Halse (1967) has obtained the velocity distribution over the Fermi surface from the cyclotron masses (see § 2.4) for Cu and Ag. Figure 2.12 shows the results. His analysis also gives values of the density of states at the Fermi surface in excellent agreement with electronic specific heat data.

As mentioned earlier (§ 2.3.1) the anomalous skin effect data can give only a qualitative indication of the form of the Fermi surface, but once an accurate surface has been obtained from the de Haas–van Alphen effect it is possible to compute what the anomalous skin effect should be. Roaf (1962) found that the 4-term formulae, which fitted the original de Haas–van Alphen data for Cu and Ag, did not seem to reproduce the anomalous skin effect data very well, but that by modifying the formulae (putting in 2 extra terms and adjusting the parameters of the other terms) he could fit the anomalous skin effect data rather better, without appreciably spoiling the fit to the de Haas–van Alphen data. This conclusion has, however, been put in doubt by Halse's analysis of the more accurate de Haas–van Alphen data for Cu, which are definitely inconsistent with Roaf's modified formula. Moreover, Halse finds that his own 5- or 7-term formulae based on the de Haas–van Alphen effect alone, lead to anomalous skin effect results which do not differ from the experimental values by much more than the experimental uncertainties (see Fig. 2.13). Such discrepancies as there are could probably be removed by slight flattening of the surface at appropriate places without spoiling the fit to the de Haas–van Alphen data, though it is doubtful if this can be done by adjusting the coefficients of the Fourier expansion formula. It would evidently be desirable to improve the accuracy of the anomalous skin effect measurements so that a more crucial test could be made, but at the

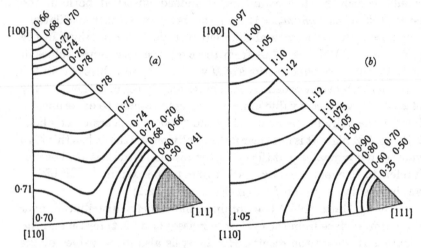

Fig. 2.12. Variation of Fermi velocity over the Fermi surface for (a) Cu, (b) Ag. The velocities are expressed as fractions of the velocity for a spherical Fermi surface with $m/m_0 = 1$, i.e. of $1·58 \times 10^8$ cm/s for Cu and of $1·40 \times 10^8$ cm/s for Ag ; note that the intervals between contours are not always the same.

present level of accuracy, there is no reason to doubt that the same Fermi surface adequately describes both the magnetic effects and the anomalous skin effect, without taking into acount any possible many-body effects.

The experimentally determined Fermi surfaces of the noble metals are in very good agreement with various 'first-principles' band-structure calculations, to well within the intrinsic uncertainties of these calculations, thus giving confidence in the reliability of the basic assumptions of the theoretical calculations. Recently there have been several calculations on the lines discussed in the last chapter which describe the band structure in terms of a fairly small number of physically significant parameters, specifying the s-, p- and the d-bands and their interactions. The values of these parameters have been

D. SHOENBERG

chosen to produce a band structure which agrees with that calculated from first principles, and since the 'first principles' calculation gives approximately the same Fermi surface as experiment, it can be said that the parameters also fit the experimental Fermi surface. The situation is, however, complicated because the Fermi surface is

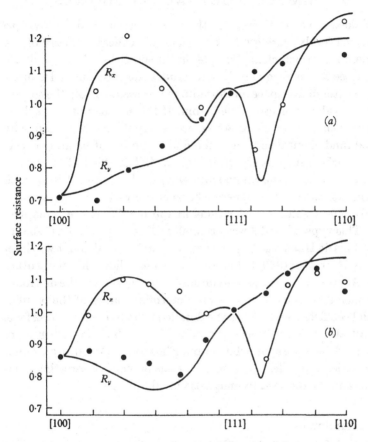

Fig. 2.13. Anomalous skin effect in: (a), Cu, (b) Ag. The full curves are the values computed from Halse's 7-term formulae for the Fermi surfaces, the points are experimental (Pippard, 1957; Morton, 1960). One unit of the ordinate represents the surface resistance for a spherical Fermi surface.

sensitive to only a few of the parameters involved and so in itself is insufficient to determine all of them. No doubt, when this 'parametric' method is further developed and refined, it will pay to use the accurate experimentally determined Fermi surface as part of the fitting information, rather than the dimensions of a theoretical Fermi surface, but so far the usefulness of the parameters in predicting other properties of the metal has been little more than qualitatively evaluated and the extra accuracy is hardly worth while.

2.8 'SIMPLE' POLYVALENT METALS

As explained in the last chapter, the free-electron model proves to give a remarkably close guide to the Fermi surfaces of many of the polyvalent metals which are 'simple' in the sense that the d-bands are not too close to the Fermi level. In its crudest form this model is just a sphere in extended k-space, of the volume necessary to hold the correct number of valence electrons per atom. If the sections of this sphere falling in each of the various zones, are remapped in the periodically repeated fundamental zone, we then find a number of separate sheets of Fermi surface such as shown in Fig. 2.14. The effect of the lattice potential, or rather of the equivalent weak pseudo-potential, is to break the surface of the original sphere wherever it crosses zone boundaries and this causes some modifications in the details of the remapped sheets. This type of model was originally suggested by Gold (1958) to interpret his de Haas–van Alphen experiments on lead and it was later shown by Harrison (1960) to have a fairly wide validity for many other metals. Al provides perhaps the simplest example of how the de Haas–van Alphen data can be analysed to determine values of the pseudo-potential coefficients as well as the detailed form of the Fermi surface. We shall also briefly discuss the case of Pb which is rather more complicated, because spin-orbit coupling has to be taken into account, and the series of divalent hexagonal metals in which interesting complications are introduced by magnetic breakdown.

2.8.1 Aluminium

The special feature of Al is that the 3-electron sphere passes very close to the corners W of the zone, so that quite a small perturbing potential can modify the detailed form of the sharp points of the 2nd zone free-electron hole surface and the detailed form of the rather thin arms of the 3rd zone 'monster' (Fig. 2.14). Using only two pseudo-potential coefficients V_{200} and V_{111} and a 4-o.p.w. calculation, Ashcroft (1963) showed that different topological situations arose for different ranges of these two parameters. Thus in region A of the $V_{200}|V_{111}|$ plane (Fig. 2.15) the monster has the same connectivity as for $V_{200} = V_{111} = 0$

7

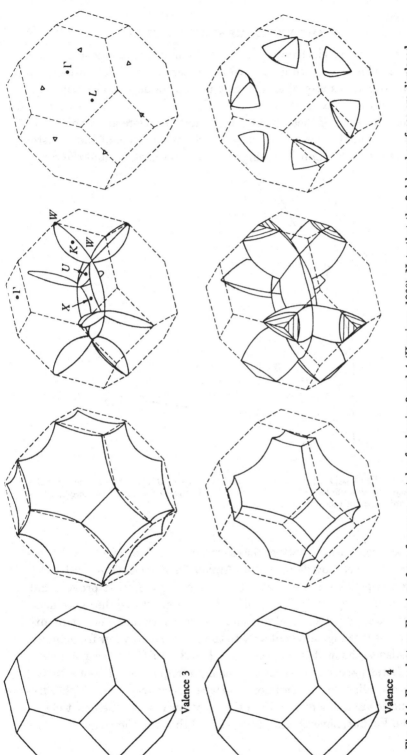

Valence 3

Valence 4

Fig. 2.14. Free-electron Fermi surfaces for f.c.c. metals of valencies 3 and 4 (Harrison, 1960). Note that the 3rd band surfaces are displaced by half a reciprocal lattice vector in the direction ΓX to exhibit the form of the 'monster' more clearly and that the 4th band surfaces are displaced by half a reciprocal lattice vector in the direction ΓL.

(the strictly free-electron case), in C_1 the monster breaks into 'rings of four' (Fig. 2.16), while in D the monster is completely broken up into separate 'sausages' and thin tunnels form between the 2nd zone hole surfaces, and so on.

Now for each of these topological situations there are several very small extremal areas, either close to where the arms of the monster join or along the tunnels joining the 2nd zone surfaces, and these give

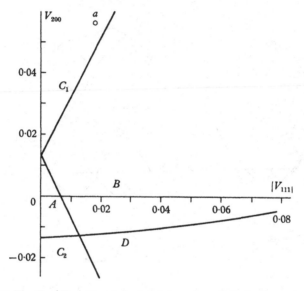

Fig. 2.15. The $V_{200}|V_{111}|$ plane is broken into regions A, B, C_1, C_2 and D, each of which corresponds to a characteristic topology of the Fermi surface of Al (Ashcroft, 1963). The units for both axes are Rydbergs.

rise to a complicated pattern for the orientation dependence of the corresponding very low de Haas–van Alphen frequencies ($F \sim 3 \times 10^5$ Gs), which depends on the particular type of topology. It proved that only values of V_{200} and $|V_{111}|$ in the region C_1 reproduced the experimentally observed pattern and the point 'a' gave an almost perfect quantitative fit (see Fig. 2.17) not only to the observed very low frequencies, but also to the medium frequencies ($F \sim 3 \times 10^6$ Gs) corresponding to the fatter parts of the arms and the high frequencies ($F \sim 4 \times 10^8$ Gs) corresponding to the 2nd zone surface. Of course, these higher frequencies are less sensitive than the very low ones to the exact choice of the Vs, but, nevertheless, it was found that the choice indicated by fitting the very low frequencies was appreciably better than, say the strictly free-electron model. Thus, paradoxically, it is the very low frequencies, associated with the very fine detail of the Fermi surface, which provide the key to an accurate determination of the whole surface—a case of the tail wagging the dog.

The extent to which the values of V_{200} and $|V_{111}|$ found in this way, really mean what they purport to mean has been discussed in the last chapter. There is no doubt that with the analytical procedure used they do provide a precise specification of the Fermi surface, but there

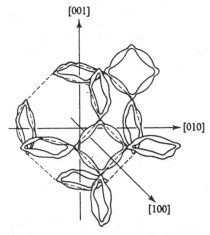

Fig. 2.16. For Al the free electron multiply-connected 'monster' of Fig. 2.14 breaks into 'ring of four' (after Volski, 1964).

are various approximations both in the basic assumptions of the simple pseudo-potential theory used and in the analysis, which modify the exact physical meaning of the Vs determined in this way. This shows up in Melz's (1966) experiments on the pressure dependence of the very low and medium de Haas–van Alphen frequencies, which he could not fit quantitatively to the changes expected on Ashcroft's scheme. For the medium frequencies, which arise from parts of the surface near the points U in the zone, Melz found he could improve the agreement with the observed pressure dependence by using 5 o.p.w.s, but for the very low frequencies (for which the 4 o.p.w.s used by Ashcroft should be adequate) the theoretically predicted pressure dependence was about twice the observed. Probably, such pressure studies, which reflects the differential properties of the pseudo-potentials, will eventually provide a useful guide in the refinement of the theory.

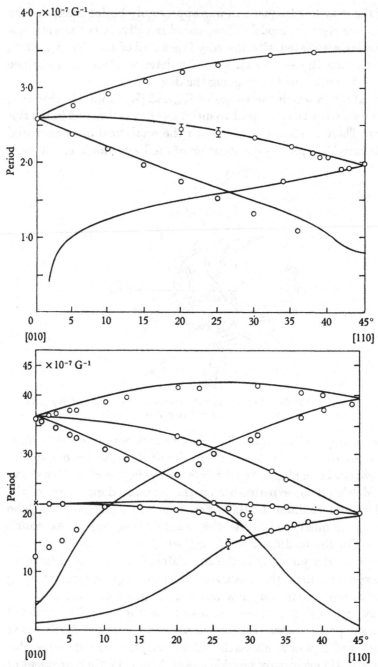

Fig. 2.17 a. Variations of medium- and low-frequency periods for Al in the (100) plane. The curves are computed from Ashcroft's model, the points are experimental (Larson & Gordon, 1967).

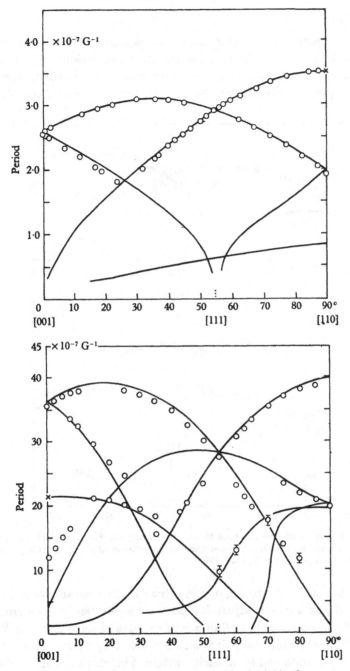

Fig. 2.17 b. As for Fig. 2.17 a, but in the (110) plane.

2.8.2 Lead

As already mentioned, Pb is more complicated than Al because spin-orbit coupling can no longer be neglected. Anderson & Gold (1965) ana-lysed their very detailed de Haas–van Alphen data in terms of 4 o.p.w.s involving pseudo-potential coefficients V_{200} and V_{111} as in Ashcroft's

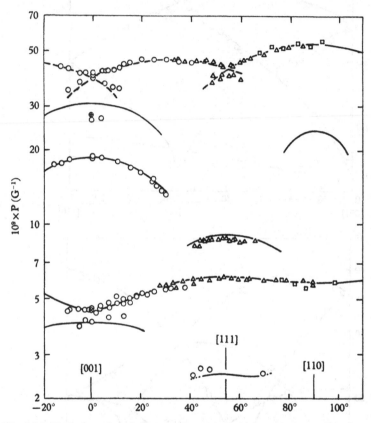

Fig. 2.18. Variation of periods (note logarithmic scale) for Pb in the (110) plane. The curves are computed from the 4-parameter model discussed in the text; the points are experimental (Anderson & Gold, 1965).

scheme, but also introducing a spin-orbit coupling parameter λ and treating E_F as another adjustable parameter. As can be seen from Fig. 2.18, they were able to find a set of values of the Vs, λ and E_F which fit the experimental points on the whole very well; only a poor overall fit was obtained in an earlier attempt in which the spin-orbit coupling was ignored. Because of the extra complications of the situation, inevitably even more simplifying assumptions have been introduced to make the calculation tractable, and the precise physical significance of the parameters is probably more in doubt than for Al. It may be mentioned, for instance, that the value found for λ is about 50 % greater than for the free atom and this excess is perhaps rather more than might be expected. Here too measurements of the pressure dependence (Anderson, O'Sullivan & Schirber, 1967) should provide useful clues to a more thorough theoretical understanding.

2.8.3   The divalent hexagonal metals

The series Be, Mg, Zn and Cd is of interest both in providing further illustrations of the use of the nearly free-electron model and in illustrating the complications caused by magnetic breakdown. The effect of magnetic breakdown is that the Fermi surface in magnetic experiments is effectively different according as the magnetic field is 'low' or 'high' (see §3.5). At low fields the 'single zone' picture (Fig. 2.19 (a) to (f)) must be used because of the spin-orbit splitting across the plane ALH, and the corresponding free-electron Fermi surface consists of:

(a) hole pockets in the 1st zone,

(b) a 'monster' of holes in the 2nd zone which is multiply connected at its corners both along the hexagonal axis and in directions perpendicular to it,

(c) 'needles' or 'cigars' along the edges of the 3rd zone,

(d) a 'lens' at the centre of the 3rd zone.

(e) partial discs, which remap to form 'four-winged butterflies' in the 3rd zone,

(f) electron segments, which remap to form 'cigars' in the 4th zone. When the field is high enough (10^2 to 10^3 Gs for Be and Mg, but 10^4 to 10^5 Gs for Zn and Cd) magnetic breakdown occurs and the electrons are able to cross the small energy gaps in the plane without feeling them, and the double-zone picture becomes appropriate. The effective free-electron Fermi surface is now considerably modified in the following respects:

(i) the 1st zone pockets remap on to the monster at the points H, and the monster is no longer connected along the hexagonal axis, though it still joins up laterally with the neighbouring monsters at these points,

(ii) the features 'e' and 'f' combine to give surfaces rather like partially opened clam shells (Fig. 2.19 (g)).

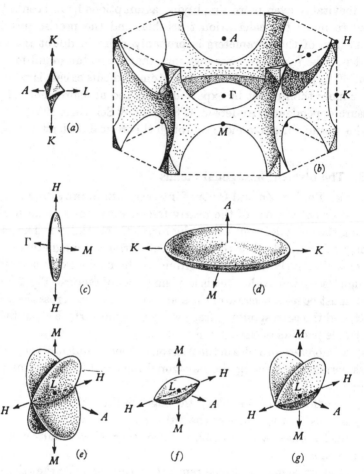

Fig. 2.19. Free-electron Fermi surface for a divalent hexagonal metal with the c/a ratio of Mg (after Ketterson & Stark, 1967). All the sketches except (g) refer to the single-zone picture: (a) 1st zone hole pocket centred on H (2 complete pockets per zone); (b) 2nd zone hole monster; (c) 3rd zone cigar centred on K (2 per zone); the cross-section of a cigar is shown schematically in Fig. 2.5; (d) 3rd zone lens centred on Γ; (e) 3rd zone butterfly centred on L (3 butterflies per zone); (f) 4th zone cigar centred on L (3 per zone). In the double-zone picture one-sixth of a pocket (a) is stuck on like a cap to each of the 12 arms of the monster in the top and bottom planes ALH in (b); the end faces of the double zone are at distances HK above and below the faces ALH of the single zone. The upper (or lower) half of each butterfly joins on to the lower (or upper) half of each 4th zone cigar to give the clam (g) of which there are 6 per zone. To understand the diagrams it is important to realize that only one typical zone point of each kind is shown in (b); thus there is a point M at the centre of *each* square face of the zone shown and adjoining zones, a point H at *each* hexagon corner and so on.

The needles and the lens are unaffected. For intermediate fields the situation is more complicated and features of both schemes appear.

The detailed dimensions of the features in either scheme depend on the c/a ratio and some are absent if c/a is too large or too small. Also, of course, the free-electron model is only an approximation to the truth and in the actual metals some of the dimensions are appreciably modified from the free-electron prediction, or even completely absent. With so many different sheets of complicated shape, there are many (typically 6 or so) extremal areas or calipers for any given field direction and correspondingly the pattern of orientation dependence of de Haas–van Alphen or magneto-acoustic frequencies is extremely complicated and difficult to interpret reliably.

In spite of what was said earlier about the relative disadvantages of the magneto-acoustic method, the recent beautiful measurements of Ketterson & Stark (1967) on Mg show that, handled with sufficient care, it can be as powerful a tool as the de Haas–van Alphen effect, and that the two methods used together are more powerful still. A particular advantage of the magneto-acoustic method in the present context is that the oscillations can be observed at very low fields (down to 100 Gs or so) because of the extreme purity of the samples, and so the complications of magnetic breakdown are completely avoided.

A great deal of work has gone into ever more accurate measurements of the de Haas–van Alphen and magneto-acoustic frequencies, and magneto-resistance measurements have been a considerable help in sorting out the topology of the surface and its modifications by magnetic breakdown. A fairly clear picture of the Fermi surfaces both at low and at high fields is now beginning to emerge, though there are still some ambiguities and the departures from the free-electron model have only recently begun to find an interpretation in terms of pseudopotential schemes.

We shall not attempt to describe all the observed features in detail, but mention just a few of the simpler results which have been well established. The 3rd zone cigars and lens are the easiest pieces of the Fermi surface to identify and their shapes and sizes have been accurately determined. It is found that their linear dimensions are usually fairly close to those predicted by the free-electron model. The case of Zn is particularly striking since the free-electron sphere radius is only $\frac{1}{2}$ % greater than the hexagon radius, so that only the slightest distortion would eliminate the overlap producing the needles. Yet in fact the 'needle' (a more appropriate name here than 'cigar' because of its thinness) has almost exactly† the diameter and length of the free-electron prediction. It is worth mentioning that the corresponding de Haas–van Alphen frequency $F \sim 1.5 \times 10^4$ Gs is about the lowest observed for any metal apart from Bi. Because of its extreme sensitivity to the c/a ratio, the diameter of the needle is very sensitive to pressure, increasing by about 40 % for a pressure of 3000 atmospheres, just about the expected amount for the corresponding change of c/a. For Mg the cigar has almost exactly the rather triangular section of the free-electron model, though it is slightly fatter and longer. For Be the cigar is considerably fatter and rounder and rather longer than the free-electron model, while for Cd the c/a ratio is sufficiently large to eliminate the cigar and none is indeed observed. The dimensions of the lens are also very nearly those of the free-electron model except for Be where it is entirely absent (Be has the smallest c/a ratio and should therefore have the smallest lens).

The study of the monster is considerably complicated by magnetic breakdown effects. Not only is there the complication of the changeover from the single- to the double-zone scheme which alters the possible extremal sections, but even in the double-zone scheme there is the further possibility of breakdown at a number of places where the band gap is small, so that all kinds of new orbits become possible for high enough fields. The classic case is that of the 'giant' orbit which was first observed in Mg for the field along the hexagonal axis; instead of following an orbit around the outside of the monster, the electron prefers at high fields to go round a large circular orbit made up of segments of the monster and the cigar (see Fig. 3.10).

Another interesting example is shown by Cd (Tsui & Stark, 1966) where the monster is broken at each of the 'waists' of Fig. 2.19. The 6 resulting pieces can then be remapped round the zone edge HK to form an undulating cylinder of trifoliate section which, in the periodically repeated scheme, continues indefinitely along the hexagonal axis (Fig. 2.20a). The 1st zone hole pockets are rather large and extend nearly from K to K; at very high fields where the double-zone picture is appropriate, half of each hole pocket 'caps' the cylinder at top and bottom giving the closed piece shown in Fig. 2.20b. At intermediate fields, breakdown occurs between the 1st and 2nd zone sur-

† A comparison to better than 20 or 30 % is difficult because the dimensions are extremely sensitive to the c/a ratio, which is not accurately enough known at low temperatures.

faces near K, leading to the self-intersecting orbits shown in Fig. 2.21 (see § 3.5).

From an experimental point of view, the simplest Fermi surface of the series is that of Be, paradoxically because it departs the most

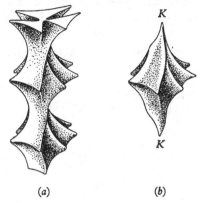

(a)   (b)

Fig. 2.20. Fermi surface of Cd (Tsui & Stark, 1966); (a) in the single-zone picture the broken monster remaps into the continuous cylinder shown, with each minimum section centred on H and each maximum trifoliate section centred on K; (b) in the double-zone picture the hole pocket 'caps' the top and bottom of that part of the cylinder between H and H, giving a closed surface centred on K and reaching nearly to K on either side (in the double zone, of course, the end-points K are no longer equivalent to the central point K).

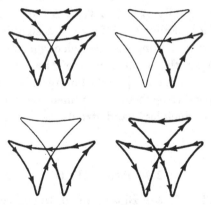

Fig. 2.21. The cross-section of the cylinder of Fig. 2.20 (a) in the plane ΓKM, showing various possible orbits arising from magnetic breakdown (Tsui & Stark, 1966).

seriously from the free-electron model. The relevant spin-orbit splitting is so small that the double-zone scheme applies at any fields big enough to give an appreciable de Haas–van Alphen effect, and moreover the lens and the clams are altogether suppressed. The

Fermi surface consists only of the 1st and 2nd zone monster (which is completely contained within the zone and has very thin waists) and the 3rd zone cigars; the monster is sometimes called a 'coronet' (Fig. 2.22). Because of this relative simplicity, Watts (1964a) was able to determine the leading dimensions of both cigars and the coronet without having to appeal to any detailed model; a gratifying feature of his analysis is that the volumes of the monster and the cigars are equal within experimental error as they should be if there are no other sheets of Fermi surface.

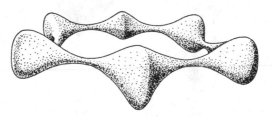

Fig. 2.22. Perspective sketch of the Be coronet
(after Loucks & Cutler, 1964).

So far the only attempt to interpret the departures from the free-electron model in terms of pseudo-potentials has been for Mg, and a fairly good fit to the departures was obtained with a 2-o.p.w. calculation involving only V_{0002} and V_{1011}. A 'first principles' band-structure calculation (Falicov, 1962) gives a significantly worse fit. For Be the departures from the free-electron model are a good deal greater than for the other metals of the series and they may well be too great to make the pseudo-potential approach profitable, but surprisingly good agreement is obtained with the Fermi surface indicated by 'first principles' band structures (e.g. Loucks & Cutler, 1964).

2.9 BISMUTH

Because the Fermi surface of Bi is so small, the de Haas–van Alphen effect can be observed in quite small fields (only a few kGs) and at fairly high temperature (up to 20 °K). It is in fact the metal in which the effect was first discovered and has probably been more studied ever since (by nearly all the methods discussed above) than any other metal. Another fortunate circumstance is that its Fermi surface consists entirely of ellipsoids so that it could be fairly completely determined at a time when the theory of the de Haas–van Alphen effect had been developed only for ellipsoidal energy surfaces.

The present position is that there are 3 constant-energy electron surfaces which are ellipsoids of the form

$$\alpha_1 k_x^2 + \alpha_2 k_y^2 + \alpha_3 k_z^2 + 2\alpha_4 k_y k_z = E(1 + E/E_g) \qquad (2.17)$$

and two others rotated by $\pm 120°$ about the k_z-axis, and also a single hole energy surface which is an ellipsoid of revolution

$$\beta_1 k_x^2 + \beta_2 (k_y^2 + k_z^2) = E, \qquad (2.18)$$

where energies are reckoned from the bottoms of the relevant bands and E_g is the small gap between the valence and conduction bands.

The form of (2.17), which implies that the band is non-parabolic, was suggested on theoretical grounds and confirmed by optical absorption and reflectivity measurements in a magnetic field, which reveal the structure of the Landau levels in the conduction and valence bands both below and above the Fermi level (Brown, Mavroides & Lax, 1963). The ellipsoidal form of the energy surfaces implies that the ratio $A/(dA/dE)$ or A/m at the Fermi surface should be the same for all extremal (i.e. central) sections, and this has been verified in the very careful measurements of Bhargava (1967) on the de Haas–van Alphen effect and of Kao (1963) on cyclotron resonance. This confirmation seems to disprove the suggestion of Cohen (1961) that fourth-order terms in k should be included in (2.17). It should be emphasized that the constancy of A/m for varying orientations shows only that the right-hand side of (2.17) is a function of E alone and does not prove that the function has the particular form of (2.17). If, however, this is assumed and E_g is taken as $0·015\,\mathrm{eV}$ (from the magneto-optical results), the value of E_F can be deduced from the observed value of A/m. It is found that

$$E_F = 0·025\,\mathrm{eV},$$

The Fermi energy for the hole ellipsoid is even smaller ($0·011\,\mathrm{eV}$).

As already mentioned, the dimensions of the various ellipsoids are very small compared with the size of the zone. The electron ellipsoids have principal axes of approximate lengths $0·005$, $0·085$ and $0·007$ in units of $10^8\,\mathrm{cm^{-1}}$ and the principal axes are tilted in the $k_y k_z$-plane by about $6°$ from the k_y- and k_z-axes; the principal axes of the hole ellipsoid in the same units are $0·014$, $0·014$ and $0·048$. There is agreement within experimental error between the volume of the hole ellipsoid and the total volume of the 3-electron ellipsoid, thus confirming that the number of electrons is equal to the number of holes as it should be; this number is about 3×10^{17} per c.c.

The peculiar nature of the Bi Fermi surface, and of the somewhat similar, but less extreme, Fermi surfaces of the other 'semimetals', Sb and As is interpreted theoretically in terms of their common crystal structures. Although they all have rhombohedral symmetry, their lattices are in fact only slightly distorted from the NaCl type characteristic of semi-conductors such as PbTe and it is not too surprising that they are only just metals. Both 'first-principles' band-structure calculations and semi-empirical pseudo-potential calculations give Fermi surfaces in fair agreement with those observed experimentally.

2.10 TRANSITION METALS

We have already discussed in some detail the noble metals, which because they are monovalent, have comparatively simple single sheet Fermi surfaces. The Fermi surfaces of the polyvalent transition metals are much more complicated (see, for instance, Fig. 2.23) basically because the Fermi level is right in the middle of the d-bands and no such simple approximation as the free-electron model can be used as a guide to the interpretation. No doubt with further development of the 'parameterization' approach discussed in §1.4, it will eventually be possible to interpret the experimental measurements of Fermi surfaces more quantitatively, but up to now the only interpretations attempted have been comparisons with band-structure calculations. A great deal of experimental work has already been done, mostly by the de Haas–van Alphen method, but also by the R.F. size effect (in an exceptionally pure crystal of W), and on the whole the Fermi surfaces found can be reconciled qualitatively and sometimes even fairly quantitatively with the calculated band structures. (The theory of transition metals is dealt with in Chapter 8; see particularly §8.1.7.)

One important general feature of the band structures, which is supported by the experiments, is that they are all very similar in character and it is possible to go from one transition metal to another of the same crystal structure merely by shifting the position of the Fermi level and making minor adjustments to some of the details of the individual bands. A good example of this 'rigid band' kind of approach is Coleridge's (1966) study of the Fermi surface of Rh, which he was able to reconcile in fair detail with a 'do-it-yourself' band structure based on that of Cu (at present no band structure for Rh has yet been calculated). However, as so often happens in Fermi surface studies, sheets of the Fermi surface indicated by the band structure have as yet found no observable counterpart in the de Haas–van Alphen effect, presumably because of high cyclotron mass, and until

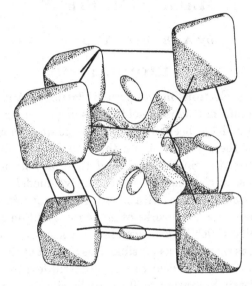

Fig. 2.23. Model of the Fermi surface of W (after Gold, unpublished).

experimental techniques are improved sufficiently to show up such predicted oscillations, even the most plausible conclusions based on the observed oscillations should be accepted with some caution.

Finally, it should be mentioned that rapid progress is being made in studying the ferromagnetic metals Ni and Fe, though the ferro-magnetism causes considerable experimental complications and the Fermi surface is also more complicated because there are separate sheets for spin-up and spin-down electrons.

CHAPTER 3

METALLIC ELECTRONS IN
A MAGNETIC FIELD

by A. B. PIPPARD†

3.1 INTRODUCTION

The behaviour of electrons in the periodic lattice of a metal, when subjected to a uniform magnetic field, raises very complex problems, and many papers have been written to elucidate various aspects. Examination of these papers shows them to fall usually into two contrasting categories; on the one hand there are straightforward expositions of semi-classical or semi-intuitive models, sometimes leading to important advances with startling economy of means, while on the other hand one finds works of immense erudition and mathematical ingenuity which seek to generalize, consolidate and extend what is known, as rigorously as possible. It is no part of the purpose of this chapter to survey the field thoroughly, still less to display the arguments in detail, but rather to illustrate by simple examples some of the principal results, and to indicate where reasonable certainty has been achieved and where there is still scope for deeper investigation. As far as possible the illustrations will employ real-space representation of the wave-function; this may not be the most economical way of reaching the answer to a problem, but it often has the advantage over more powerful techniques of revealing something of the physical processes at work.

One characteristic of the problem is the variety of representations possible even in real space. There are two reasons; first, the magnetic field enters the Schrödinger equation by way of the vector potential, $m\mathbf{v} + e\mathbf{A}$ being the momentum conjugate to $\mathbf{r}$, and any change of gauge of $\mathbf{A}$ alters the phase of ψ without changing its amplitude: and, secondly, as a result of a certain degree of translational invariance the eigenfunctions tend to have high degeneracy, so that different linear combinations of degenerate solutions can give equally valid alternative

† Dr Pippard is John Humphrey Plummer Professor of Physics at the Cavendish Laboratory, Free School Lane, Cambridge, and President of Clare Hall, Cambridge.

8 [113]

forms for ψ. We shall see an example of both points shortly when we discuss the free electron. Before turning to this, however, let us note the general rule for changes of gauge. If A is altered by the addition of $H\nabla\beta$, β being an arbitrary function $\beta(\mathbf{r})$ such that $\nabla^2\beta = 0$, $\psi(\mathbf{r})$ must be replaced by $\psi(\mathbf{r})\,e^{is\beta}$, where s is written for $eH/\hbar$.

It is sometimes convenient, especially if $|\psi|$ varies only slowly with position, to write $\psi(\mathbf{r})$ as $|\psi|e^{i\phi}$ and call $\nabla\phi$ the local wave-number $\mathbf{k}$. In a change of $\mathbf{A}$ by addition of $\delta\mathbf{A}$, as a result of which ϕ is replaced by $\phi + s\beta$, $\hbar\mathbf{k}$ becomes $\hbar\mathbf{k} + e\delta\mathbf{A}$, so that $\hbar\mathbf{k} - e\mathbf{A}$ is invariant in any change of gauge. If $|\psi|$ varies only slowly with $\mathbf{r}$, $\hbar\mathbf{k} - e\mathbf{A}$ is to be interpreted as $m\mathbf{v}$, and if in the absence of a field the energy varies with wave-number as $\mathscr{E}(\mathbf{k})$, it is often a good approximation to suppose that when a field is applied the energy, being essentially determined by $\mathbf{v}$, is given by $\mathscr{E}(\mathbf{k} - e\mathbf{A}/\hbar)$. In effect one treats $\mathscr{E}(\mathbf{k})$ as defining an equivalent Hamiltonian for Bloch electrons, and then inserts the magnetic field just as if they were ordinary particles. The justification and limitations of this device have been scrutinized in detail (Kohn, 1959 b; Blount, 1962), particularly in respect of its disregard of interband effects, and it has emerged triumphantly from this scrutiny as a method of reducing calculations of the utmost complexity to something more readily appreciated by the physical imagination. In particular the classical and semi-classical methods of treating particle dynamics, which play an important role in much that follows, are seen to be as applicable to Bloch electrons as to free particles, within fairly well delineated bounds.

Because $\mathbf{A}$ can be chosen to lie in the plane normal to $\mathbf{H}$, motion in the z-direction, parallel to $\mathbf{H}$, may usually be separated from the rest, and even when it cannot it tends to offer little of interest. We shall therefore illustrate the dynamical properties of individual particles in two dimensions only, the plane normal to $\mathbf{H}$, so that until we reach § 3.8 and have to treat complete assemblies we shall ignore the z-direction.

3.2 THE FREE ELECTRON

If electrons in the x–y-plane are entirely free the Hamiltonian takes the form $(p - e\mathbf{A})^2/2m$ and the resulting Schrödinger equation is exactly soluble. For example, in a 'linear' gauge, $\mathbf{A} = H(0, x)$, the variables separate and
$$\psi = F(x)\,e^{ik_y y},$$
where
$$F'' + [2m\mathscr{E}/\hbar^2 - (k_y - sx)^2]\,F = 0. \tag{3.1}$$

Since (3.1) is the oscillator equation, the x-variation of ψ is the same as for a harmonic oscillator centred on k_y/s and having an excursion equal to the orbit diameter for a particle of energy $\mathscr{E}$. The energy levels are $(n+\tfrac{1}{2})\hbar\omega_c$, ω_c being the cyclotron frequency eH/m, and each choice of k_y satisfying the boundary conditions leads to one of a set of degenerate functions differing only by their oscillator centre. Thus if the sample is rectangular, with dimensions X and Y, and we impose periodic boundary conditions in the Y-direction, k_y takes such values that $k_y Y = 2l\pi$, l being an integer. The oscillator centres are uniformly distributed along x at intervals of $2\pi/sY$, so that the total number that can be accommodated in the sample is $s/2\pi$ per unit area.

The energy level spectrum consists of a set of discrete levels, each of such degeneracy that the mean density of states is independent of H, and in consequence, as is well known, a metal exhibits only very weak diamagnetism due to orbiting electrons (classically, indeed, according to van Leeuwen's theorem, the total moment is zero). Yet each of the levels individually has a strong diamagnetic moment $\mu = -\partial\mathscr{E}/\partial H = -\mathscr{E}/H$. The moment of the orbits contributing to the sharp levels is almost exactly compensated by that due to electrons in incomplete orbits within one orbit diameter of the edge of the sample (Van Vleck, 1932, p. 100). In quantal terms one may think of electrons near the edge of a rectangular sample as described by oscillator-like wave-functions which are not altogether contained in the sample. These surface states are few in number, and their number is proportional to sample perimeter rather than area, yet because they produce circulating currents all along the edges they have a huge paramagnetic moment which increases with sample size. They form a virtually unquantized continuum of levels which are extremely sensitive to changes in H, for any such change induces a circulating electric field which acts powerfully upon the surface current.

The energy level diagram consists then of a set of highly degenerate levels on which is superposed a thin continuum of surface states. As the field is changed the sharp levels alter in spacing and degeneracy, and it is tempting to imagine that a given electron may leave its sharp level and climb up the continuum to the next level, in this way allowing a continuous evolution of the distribution function. But we may dispel this illusion by following an electron as the field decreases. Let us suppose that the boundary conditions for E are such that the induced field $\mathbf{E} = -\dot{\mathbf{A}}$, i.e. $\mathbf{E} = E_y$. Then under the influence of the crossed fields $\mathbf{H}$ and $\mathbf{E}$, the orbit centre moves at a speed E/H so as to stay at a point of constant A (k_y is unchanged in this process). The degenerate orbits move further apart by drifting towards the edge, so that the degeneracy is automatically reduced in proportion to H. At the same time the energy of each electron falls in proportion to H through the influence of the circulating electric field. So far, so good, but when the oscillator functions touch the edge their character alters; k_y is no longer conserved and their energy rises sharply because of the surface current interacting with E. According to a classical calculation the

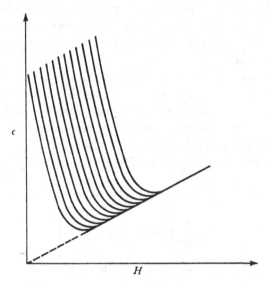

Fig. 3.1. Schematic diagram showing effect of surface
on energy levels.

orbit is never actually extruded from the sample, but $\mathscr{E}$ rises to infinity, as indicated in Fig. 3.1. For large Y the orbits arrive at the boundary in a steady steam, so that the rising paramagnetic levels form a continuum. From the quantal point of view the continuous distribution arises from the difference in boundary conditions at the two sides of the wave-function. If, for example, we attempt to fit an oscillator wave-function with energy $(n+\tfrac{1}{2})\hbar\omega_c$ there is no trouble on the inside, where ψ can be arranged to vanish at infinity; but at the boundary ψ will not automatically vanish, and an adjustment of $\mathscr{E}$ will be necessary to achieve this also. As k_y is virtually a continuous variable the necessary adjustment of $\mathscr{E}$ will also be virtually continuous.

It should be noted that there is no transfer of an electron from one degenerate level to another at the surface. The depletion of the lower degenerate levels proceeds according to expectation, but only by transferring electrons to surface states, not by filling upper levels. In fact, the distribution is not automatically adjusted so as to maintain equilibrium in changing H. This is not surprising, for we know that in the absence of collisions any attempt to alter H results in an eddy current screen, so that collisions are essential for maintenance of equilibrium.

At this point we may put the surface states to one side, until we come in §3.8 to discuss the magnetic properties of an assembly. But before then there is much to be done concerning electron behaviour uncomplicated by surfaces, beginning with an alternative description of the free electron.

The linear gauge we have chosen leads to a representation of the wave-function which does not closely resemble the classical model of an electron in a circular orbit. It is easy, however, to derive a different representation with this property, by choosing a circulating gauge, $\mathbf{A} = \frac{1}{2}\mathbf{H} \times \mathbf{r}$, and solving the Schrödinger equation in polar coordinates, in the form $\psi(r, \phi) = r^{-\frac{1}{2}}F(r)\,e^{il\phi}$, with F obeying the radial equation,

$$\left. \begin{aligned} &\frac{\hbar^2}{2m}F'' + (\mathscr{E} - V)F = 0, \\ \text{where} \qquad &V = \frac{\hbar^2}{2m}\left(\tfrac{1}{4}s^2r^2 + sl + \frac{l^2 - \frac{1}{4}}{r^2}\right). \end{aligned} \right\} \tag{3.2}$$

The effective radial potential has a minimum at

$$r_0 = \left(\frac{4l^2 - 1}{s^2}\right)^{\frac{1}{4}},$$

and rises quadratically from this minimum, as if the electron were bound in a very nearly harmonic well with characteristic frequency ω_c. The wave-function is most strongly confined when the oscillator is in its lowest state, corresponding to an energy $\frac{1}{2}\hbar\omega_c$ above the potential minimum of $ls\hbar^2/m$,† i.e. $l\hbar\omega_c$. As before, then, we arrive at quantized levels of $(l + \frac{1}{2})\hbar\omega_c$, but now represented by a narrow racetrack rather than an extended wave-function. The exact solution of (3.2) has the form

$$F = r^{l + \frac{1}{2}}e^{-\frac{1}{4}sr^2}, \tag{3.3}$$

from which one sees that the maximum occurs at a radius $[(2l + 1)/s]^{\frac{1}{2}}$, which is the classical radius for an electron of energy $(l + \frac{1}{2})\hbar\omega_c$, and

† This neglects terms of order $1/l^2$, which in fact should not be present, and only arise from our assumption of an exactly harmonic well.

the radial half-width of F^2 is $s^{-\frac{1}{2}}$. This is smaller than the radius by a factor $(2l+1)^{\frac{1}{4}}$; in metals where l may well be of order 10^3 upwards, the wave-function is closely confined to the vicinity of the classical orbit.

If we had attempted to derive this result by the w.k.b. approximation, we should perhaps have proceeded by noting that the phase path round the classical orbit must be an integral multiple of 2π, so that k_l, the tangential component of wave-number, must be l/r_c. Now an electron with such energy as would result in a classical orbit radius r_c would have, in zero field, a wave-number sr_c; because of the contribution of the vector potential, $eA/\hbar$, the wave-number is modified to $\frac{1}{2}sr_c$. We should thus have derived permitted values for r_c,

$$\frac{l}{r_c} = \tfrac{1}{2}sr_c, \quad \text{or} \quad r_c^2 = \frac{2l}{s},$$

instead of the correct form $r_c^2 = (2l+1)/s$. This is equivalent to energy levels of $l\hbar\omega_c$ rather than $(l+\frac{1}{2})\hbar\omega_c$. The error arises from the assumption that it is only the tangential variation of ψ that contributes to the kinetic energy, neglecting an extra $\frac{1}{2}\hbar\omega_c$ due to confinement of the wave-function, which must be included if the correct levels are to be deduced.

3.3 THE BLOCH ELECTRON

As soon as we attempt to solve the problem of an electron in a periodic lattice under the influence of a magnetic field, we find a mathematical problem of a quite different order of difficulty from that of a free electron. There are probably many ways of looking at the matter which illuminate the source of the difficulty. The point of view we shall adopt, which will be expounded in some detail later (§3.5), is that the field and the lattice potential, acting together, attempt to impose on the wave-function a periodicity in real space that is in general incommensurate with the lattice periodicity; the new period and the lattice period are related by a factor J, defined by

$$J = 2\pi/(s\Sigma), \tag{3.4}$$

where Σ is the area of the unit cell of the lattice. Only when the field is chosen so that J is a rational fraction do the periods have anything in common.

Ultimately this clash of periods is unavoidable, but considerable progress may be made by ignoring it, adopting the outlook that the principal effect of the lattice potential is to modify the gross dynamical properties of the electrons: in other words, we may insert the magnetic field into an equivalent Hamiltonian from which specific consequences of lattice periodicity have been eliminated (see §3.1).

This approach takes its most elementary form in the semi-classical treatment due originally to Onsager (1952) which treats the electron as a particle defined by $\mathscr{E}(\mathbf{k})$ and subject to a classical equation of motion $\hbar\dot{\mathbf{k}} = \mathbf{F}$. When $\mathbf{F}$ is the Lorentz force, $e\mathbf{v} \times \mathbf{H}$, the solutions are in the form of orbits of all energies, from which are chosen those that satisfy the Bohr–Sommerfeld rule $\oint \mathbf{p}.\mathbf{dq} = (n+\gamma)h$. From a simple geometrical analysis the following results emerge:

(a) In k-space an electron is driven by the Lorentz force round an energy contour in the plane normal to $\mathbf{H}$.

(b) In real space it executes an orbit similar in shape but scaled in dimensions by $1/s$ and turned through $\frac{1}{2}\pi$. It also has z-motion which we ignore.

(c) The permitted energies are those for which the energy contours have areas $\mathscr{A}_k$ such that

$$\mathscr{A}_k = (n+\gamma)\,2\pi s. \tag{3.5}$$

(d) Correspondingly in real space the orbit area obeys the relation

$$\mathscr{A}_r = (n+\gamma)\,2\pi/s. \tag{3.6}$$

The fractional correction γ is to allow for the fact, already discussed for the free electron, that the orbit integral neglects a radial contribution to the energy.

The result (3.5) determines the energy level diagram for the system. It differs from that for free electrons in that the levels are not necessarily evenly spaced in energy, though usually the spacing only varies slowly. As H is increased, the levels spread out, like a concertina, the movement being roughly proportional to $|\mathscr{E} - \mathscr{E}_0|$, $\mathscr{E}_0$ being the energy, which we shall call the vanishing point, at which $n+\gamma = 0$. For a free electron $\mathscr{E}_0 = 0$, but for electrons in hole orbits, where $\mathscr{A}_k$ decreases as $\mathscr{E}$ increases, $\mathscr{E}_0$ lies at the top end of the band.

The value of γ cannot be found by any simple argument, but entails a more rigorous analysis (Roth, 1966; Chambers, 1966b) which results in a justification of (3.5) and leads to the conclusion that although γ is always $\frac{1}{2}$ in weak fields, it is not constant for a given orbit but may vary with field,

$$\gamma = \tfrac{1}{2} + \gamma_1|H| + \dots. \tag{3.7}$$

Only if the effective mass of the electron is independent of energy is γ independent of H. By the same analysis it has been shown that the degeneracy of each level is $s/2\pi$ per unit area, just as for the free electron, and has no corrections of higher order in the field strength. This being so, if the leading term in γ were energy-dependent there would result the embarrassing consequence that the total number of states up to a given energy would not be constant but have a slight variation proportional to $|H|$; this would imply that even in zero field an assembly of spinless non-interacting electrons would possess a magnetic moment. The form of (3.7) in fact ensures that there is no permanent moment, though γ_1 contributes to a quadratic variation of energy and hence to the susceptibility (§ 3.8).

The most important aspect of (3.5) is that $\mathscr{A}_k$ is the sole feature determining the allowed energy levels. There are alternative ways of expressing this result. For example, from (3.6) we see that the flux contained within the real space orbit, i.e. $H\mathscr{A}_r$, is $(n+\gamma)h/e$. The quantity h/e is the flux 'quantum' which arises constantly in magnetic problems, for a change of flux by h/e in a closed circuit alters the phase integral by 2π; the flux 'quantum' that dominates certain aspects of superconductivity is of half this magnitude; hence the inverted commas, to indicate that too much significance should not be attached to quantization of flux. Another interpretation of (3.5) relates it to the cyclotron frequency of the electron in its orbit, which is also a geometrical property

$$\omega_c = \frac{2\pi s}{\hbar} \left(\frac{d\mathscr{A}_k}{d\mathscr{E}}\right)^{-1} = \frac{eH}{m_c}. \tag{3.8}$$

This equation defines the cyclotron mass m_c. Since from (3.5) neighbouring orbits differ in area by $2\pi s$, from (3.8) they differ in energy by $\hbar\omega_c$ in accordance with Planck's rule for oscillators.

The rule for quantizing orbits was derived by judicious neglect of the details of the lattice potential, and it is of value to analyse the problem from another point of view, in a limiting case where the lattice potential may be incorporated explicitly. If an electron is nearly free, so that the Fermi surface is distorted from spherical form only in the vicinity of the zone boundaries, we may imagine that in a magnetic field it moves in a circular arc until its direction of motion compels Bragg reflection, whereupon it is reflected into a new circular arc. The simplest example is provided by a two-dimensional metal having only one weak Fourier component of lattice potential, $V = V_0 \cos gx$, leading to energy contours as in Fig. 3.2 (a). In real space an electron of energy $\mathscr{E}_1$ can be represented as a race-track orbit as if V_0 were zero, except that at four points in the orbit its wave-vector satisfies the Bragg condition, $2k_x = g$, with the result shown in Fig. 3.2 (b) which is, as usual, similar to Fig. 3.2 (a) but turned through $\frac{1}{2}\pi$. The orbits involved have their centres separated by a distance $CC' = $ S such that (Pippard, 1964a)

$$\mathbf{g} = \mathbf{s} \times \mathbf{S}. \tag{3.9}$$

At first sight this diagram raises difficulties; for instance, we have seen that the wave-vector of the race-track orbit is not the zero field **k** but $\frac{1}{2}$**k**; again, the Bohr–Sommerfeld quantum condition for the lens-shaped orbit $\mathscr{L}$ would seem to involve the length L of the orbit rather

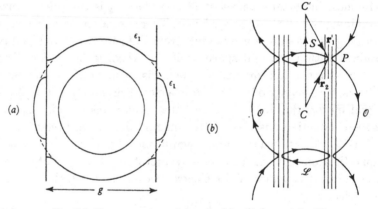

Fig. 3.2. Energy contours (a) and orbits (b) for metal with one weak lattice component.

than its area, through a relation such as $\frac{1}{2}kL = (n+\gamma)\,2\pi$. Both difficulties, however, vanish when we remember that the wave-vector **k**′ depends not only on **k** but on **A**, and the relation $\mathbf{k}' = \frac{1}{2}\mathbf{k}$ holds only when the gauge centre is the centre of the circular orbit. Consider, for example, the reflection at P. If we were to ignore the contribution of **A** we should represent the incident and reflected wave-vectors as in Fig. 3.3 (a), connected by a vector **g** to show the Bragg reflection. But because of **A** the real wave-vectors, **k**′, are $\frac{1}{2}\mathbf{k}_{\mathrm{inc}}$ and $\frac{1}{2}\mathbf{k}_{\mathrm{ref}}$, as in Fig. 3.3 (b), if C is the gauge centre for the incident, C' for the reflected wave. However, the Bragg condition applies only when the same gauge is used for both waves; when we transfer the gauge centre of the reflected wave from C' to C, we change **A** by $\frac{1}{2}\mathbf{H} \times (\mathbf{r}_2 - \mathbf{r}_1)$, i.e. $\frac{1}{2}\mathbf{H} \times \mathbf{S}$, and **k**′ by $\frac{1}{2}\mathbf{s} \times \mathbf{S}$ which from (3.9) is $\frac{1}{2}\mathbf{g}$. In the event, therefore, as Fig. 3.3(b) shows, the Bragg condition is satisfied, as if the field made no contribution to $\mathbf{k}'$; in fact the diagram 3.3(a) correctly represents the wave-vectors when both are referred to P as gauge centre.

We may now evaluate the phase length of $\mathscr{L}$. If the upper half has C as gauge centre, (see Fig. 3.4), its phase length is s times the area of the sector $CP'P$, as is obvious from (3.6), if the small phase correction γ is neglected. Similarly, if the lower half has C' as gauge centre, its phase length is s times $C'PP'$. But when we integrate round the orbit we must start and finish in the same gauge, and add any phase corrections incurred by gauge changes on the way. Thus at P, in

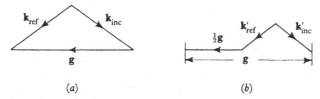

(a)   (b)

Fig. 3.3. Bragg reflection with different choices of gauge.

changing from C to C' a correction represented by s times the area of the quadrilateral $OCPC'$ must be added,† while at P' the correction in changing back is $OC'P'C$. When these four areas are added, taking clockwise areas as positive, the overall phase change is simply s times the (positive) area of $\mathscr{L}$, just as given by the Onsager method. A second example of this construction is shown in Fig. 3.5, with O chosen to confuse the diagram as little as possible. The phase length of the pin-cushion orbit is now represented by the sum of four (positive) sectors of the type $C_1P_4P_1$ and four (negative) quadrilaterals of the type $OC_4P_4C_1$, i.e. *minus* the area of the pin-cushion. The sign of the area indicates that it is a hole orbit.

† There are two separate phase corrections, arising from gauge changes and from the phase of the lattice wave at the point of reflection. In Fig. 3.4 take O for convenience as any point at which both lattice components $\pm\mathbf{g}$ have zero phase. Then when the gauge centre is moved from C to C' the vector potential suffers an increment $\frac{1}{2}\mathbf{S} \times \mathbf{H}$, which can be written as $\nabla[\mathbf{R}.(\frac{1}{2}\mathbf{S} \times \mathbf{H})]$ if the vector OP is designated as $\mathbf{R}$. In accordance with the result stated in section 1, the phase change accompanying this gauge change is $(s/H)\,\mathbf{R}.(\frac{1}{2}\mathbf{S} \times \mathbf{H})$, or $\frac{1}{2}s.(\mathbf{R} \times \mathbf{S})$, which is s times the area of the quadrilateral $OC'PC$. As drawn in the diagram the correction is positive, since $\mathbf{s}$ points out of the page.

Next we must add the phase of the lattice wave at P, and the component that is responsible for Bragg reflection is that travelling leftwards, with phase $-\mathbf{g}.\mathbf{R}$. From (3.9) this is seen to be $-\mathbf{s}.(\mathbf{R} \times \mathbf{S})$, so that the two corrections together amount to $-\frac{1}{2}\mathbf{s}.(\mathbf{R} \times \mathbf{S})$.

Although, as indicated in the footnote, the phase of the lattice at the point of Bragg reflexion affects the wave-function locally, in the examples just discussed the sum total round the orbit is zero; but this is not always the case. To discuss the point it is convenient to revert to the linear gauge, so that the problem is one-dimensional, the motion of an electron in a parabolic potential on which is superposed a weak sinusoidal potential. In the absence of the latter the wave-number k_x varies steadily, and there are regions (AB and $A'B'$ in Fig. 3.6) in which the kinetic energy of x-motion lies within the forbidden band

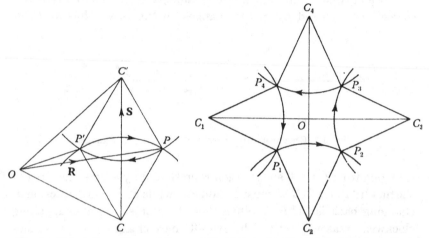

Fig. 3.4. Calculation of lens orbit integral in terms of areas.

Fig. 3.5. As for Fig. 3.4, but for four-sided hole orbit.

when the lattice potential is added. The wave-function contained within AA', except for exponential tails in the forbidden bands, represents the lens orbit bounded at both ends by Bragg reflections, while the wave-function contained between B and the classical limit C represents one of the open orbits $\mathcal{O}$ in Fig. 3.2(b). If the lattice potential is weak, over most of the range AA' the wave-function will be the free particle oscillator wave-function, but near A and A', where $2k_x \sim g$, it is deformed so that at A its nodes lie at the minima of lattice potential. Quantization of the energy arises of course out of the necessity of achieving the same disposition of the nodes at A' also. Consider now what happens when k_y is changed slightly, so that the parabola and forbidden zones are moved bodily by a distance rather less than the lattice period. The nodal pattern, being tied to the lattice rather than to the exact position of A, does not shift, and to a high approximation the new function, with a shifted envelope but unshifted nodes, has the same energy as the original function. Thus the lattice does not destroy the degeneracy of the states, or at most only very slightly.

The same cannot be said for the open orbits, whose quantization is determined by different conditions at B and C (Fig. 3.6). At B the nodes lie on potential maxima and remain fixed when k_y is changed; but at C the lattice is hardly effective and the tail of the wave-function is tied to the precise location of the classical limit, shifting as k_y alters.

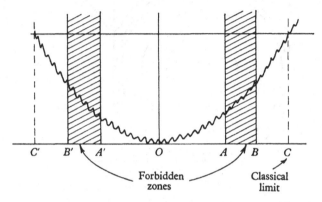

Fig. 3.6. Combined parabolic and periodic potential representing situation in Fig. 3.2.

Thus a change in k_y prevents the boundary conditions still being satisfied at both ends, and a change in energy is needed to give a valid solution. In consequence the different values of k_y give rise to a continuous spectrum of energy levels, and in effect the open orbits are unquantized. The reasoning underlying this result, and that for the lens orbit, is not dependent on the lattice potential being weak. There is a close resemblance between this argument and that given in §3.2 to show why the surface states form a continuum.

So far we have seen no reason to doubt the prescription provided by (3.5) and (3.7), together with a degeneracy of $s/2\pi$, but a simple example will make it clear that some problems still remain. Consider a square lattice, giving rise to energy contours as in Fig. 3.7(a), only the first zone being shown, with enough periodic replication to indicate the nature of the hole orbits. It is clear that the orbits are all closed, and will generate two sets of energy levels, as in Fig. 3.7(b); the electron set at energies $\mathscr{E}_0 + (n + \frac{1}{2})\hbar\omega_{c,el}$, and the hole set at

$\mathscr{E}_p - (n + \frac{1}{2}) \hbar \omega_{c,h}$, $\mathscr{E}_p$ being the energy at the corner P in zero field. The levels extend from $n = 0$ to the next integer before the energy $\mathscr{E}_1$ is reached. Now if the degeneracy of each level is $s/2\pi$ per unit area,

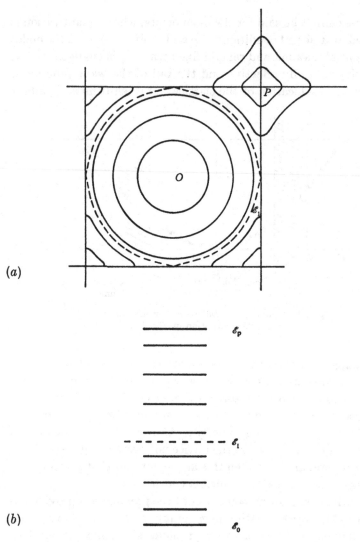

(a)

(b)

Fig. 3.7. (a) Schematic energy contours for square lattice, showing critical contour $\mathscr{E}_1$ separating electron and hole orbits. (b) Energy level diagram corresponding to (a).

the total number of states must be an integral multiple of $s/2\pi$ and so is bound to fluctuate as the field is changed. This, however, is impossible—states must be conserved, the band containing one state for

K

every lattice cell, i.e. $1/\Sigma$ per unit area if Σ is the area of the unit cell. These two statements are reconcilable only if H is such that the parameter J in (3.4) is integral, or

$$H\Sigma = \frac{1}{J} \cdot \frac{h}{e};\qquad(3.10)$$

the special field strengths are integral submultiples of the field which produces one flux quantum in the unit cell. Unless H takes one of these values we must doubt the universal applicability of the expression for the degeneracy and one may readily guess that it is round the energy $\mathscr{E}_1$, where the orbits change character abruptly, that the theory needs further refinement (§ 3.5).

3.4 SPIN-ORBIT INTERACTION

So far we have ignored the electron spin, but it may produce significant effects. Consider, for instance, the free electron whose quantized orbitals levels are $(n + \frac{1}{2})\hbar\omega_c$ or $(2n + 1)\mu_B H$, μ_B being the Bohr magneton. If we add to these energies the spin energy which may take either of the values $\pm \mu_B H$, we find the resultant levels are evenly spaced, as before, but at $2n\mu_B H$ rather than $(2n + 1)\mu_B H$. Levels for parallel and antiparallel spins coincide and are equally occupied in a metal, but below the lowest paired levels there is an extra parallel-spin level, which is responsible for the Pauli paramagnetism.

In general, for Bloch electrons, the spacing of the orbital levels does not coincide with the spin-splitting, and the diagram consists of two sets of levels with spacing $\hbar\omega_c$, shifted relatively by $2\mu_B H$ if the only interaction of the spins is with the applied field. In general, however, there is spin-orbit interaction (Yafet, 1963) which changes the spin-splitting to $g_{\mathrm{eff}}\mu_B H$. The effective g-factor may be very different from 2 for a reason which is easily understood in terms of the argument of the last section. For simplicity consider the situation represented in Fig. 3.2(a), where the potential is $V_0 \cos gx$ and by choice of a linear gauge the wave-function is separable in the form $e^{ik_y y}F(x)$. Now the motion of an electron with velocity $\mathbf{v}$ through an electric field $-\nabla V$ simulates for it a magnetic field $\mathbf{v} \times \nabla V$ which, interacting with the spin, contributes to the energy a term

$$\pm \tfrac{1}{2}\mu_B \mathbf{v} \times \nabla V$$

(the factor $\frac{1}{2}$ is the correction for 'Thomas precession'). In the simple case chosen, the only component of **v** which matters is v_y, which may be written $\hbar k_y/m$, and the spin-orbit interaction takes the form

$$\pm \frac{\mu_B^2 g k_y V_0}{e} \sin gx.\dagger$$

The electrostatic and spin-orbit interactions are in phase quadrature, so that they add up to a total interaction potential which may be displaced from the electrostatic potential by anything up to one-quarter wavelength, according to the relative strength of the two terms; moreover, the shifts are of opposite sign for the two spin directions.

In the example chosen, where the lattice potential is sinusoidal, the spin-orbit interaction is very weak compared with the electrostatic interaction. In a real metal, however, this need not be so. The spin-orbit interaction is dominated by regions well within the ions, where the field is strong and the velocity high. On the other hand these regions are largely ineffective for the electrostatic interaction, for which V_0 may be regarded as the strength of the pseudo-potential. It is quite possible for the pseudo-potential to be weak and the spin-orbit coupling strong, so that the latter is dominant. Thus the effective lattice potential may be written

$$V_{\text{eff}} = V_0' \cos{(gx \pm \theta)}, \tag{3.11}$$

in which the two signs refer to the two spin directions, and θ lies between 0 and $\frac{1}{2}\pi$, $\tan\theta$ being the ratio of spin-orbit to electrostatic interactions. It should be noted that a component of **v** parallel to the planes of lattice potential is needed to give spin-orbit interaction, and that the favourable condition for a strong effect is that the orbit should involve Bragg reflections at angles far removed from the normal.

Consider therefore an example of a situation in which this occurs, a triangular orbit of the type found in Mg, Zn and other hexagonal metals (Fig. 3.8). If the lattice potential is weak the orbit consists of three nearly free trajectories connected by Bragg reflections. The lines in 3.8(b) represent planes of lattice potential. The spin is normal to the plane and if there is spin-orbit coupling the effective potential is shifted from the positions shown, the lines being all moved inwards for one spin direction and outwards for the other. The phase path round the orbit is therefore different for the two spin directions, and the difference may amount to π for each Bragg reflection if θ takes its maximum value. This extreme situation is illustrated in Fig. 3.9

† Note that g here is a reciprocal lattice vector.

and in practice the spin-splitting of the levels may take any intermediate value. In terms of the cyclotron mass defined by (3.8), g_{eff} may take any value up to $3m/m_c$ (nm/m_c for an orbit involving n Bragg reflections). Since in zinc the triangular orbit has $m/m_c \sim 133$, g_{eff} can be as high as 400; experimentally (Stark, 1964; Myers & Bosnell, 1965) it is found to be about 90.

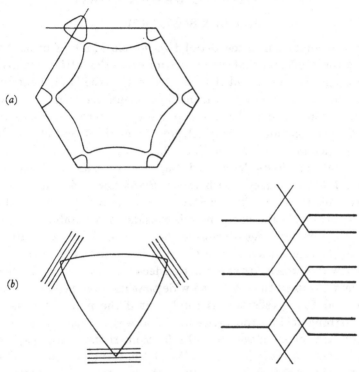

(a)

(b)

Fig. 3.8. (a) Energy contour as found in Mg and Zn, showing hexagonal hole orbit and triangular electron orbit. (b) Bragg reflections involved in producing the triangular orbit.

Fig. 3.9. Maximum spin-splitting for triangular orbits of Fig. 3.8.

The effect of spin-orbit interaction has been illustrated here by means of an example in which the Bragg reflections are clearly separated. Smaller orbits of the same family are less markedly triangular, and at the bottom of the band, where the energy is a parabolic function of wave-number (measured from the zone corner), it is no longer possible to distinguish the separate reflections. Nevertheless, it has been found by a different approach (Bennett & Falicov, 1964) that the same rule for the maximum value of g_{eff} holds. One may expect the ratio of spin-splitting to orbital level separation to vary slowly within a band. To the spin-orbit contribution there must be added the shifts of the bands $\pm \mu_B H$ because of the interaction of the spins with the external field; this of course is a constant energy shift over each band.

3.5 MAGNETIC BREAKDOWN

(Falicov & Stark, 1967)

We have examined in some detail two simple kinds of orbit, the free-electron orbit and that which results when the lattice potential causes Bragg reflections and changes the orbit shape fundamentally. Let us now consider how the transition occurs between these two forms as the strength of the lattice potential is increased from zero, the field being constant. We may ask how strong the potential must be for total reflection to occur, and then what happens if reflection is not total, a state of affairs designated Magnetic Breakdown (Cohen & Falicov, 1961) in analogy with Zener Breakdown (Ziman, 1964a, p. 163) to which it is formally almost identical. Figure 3.6 illustrates this point; if only one lattice component is considered, a suitable choice of linear gauge enables one to replace the effect of H by a parabolic potential. An electron starting at C is impelled towards the forbidden zone AB, within which its wave-function decays exponentially. Hitherto we have assumed that AB is so wide that no wave penetrates the barrier, and Bragg reflection is total. But if the parabola is steep enough (strong field) or the lattice potential weak enough, there may be partial transmission and partial reflection. If the energy gap between the two bands is $\Delta\mathscr{E}$, the width of the forbidden zone AB is proportional to $\Delta\mathscr{E}/H$ since, from (3.1), the constant of the parabola varies as H^2 and therefore the slope at a given energy varies as H. At the same time the extinction distance of the wave within the forbidden zone varies as $(\Delta\mathscr{E})^{-1}$, so that we expect the probability of penetration to be governed by the parameter $(\Delta\mathscr{E})^2/H$. The exact result can be taken over from the theory of Zener Breakdown, or derived by various alternative methods (Blount, 1962); P, the probability of an incident electron penetrating the barrier and Q, the probability of reflection, are given by the expression

$$P = 1 - Q = e^{-H_0/H}, \tag{3.12}$$

where $$H_0 = \frac{\pi(\Delta\mathscr{E})^2}{4\hbar e v_x v_y},$$

9

and v_x, v_y are the normal and tangential components of the velocity
which a free electron would have at the point where Bragg reflection
occurs. An alternative way of expressing (3.12), for a quasi-free par-
ticle of mass m^*, is in terms of its energy, $\mathscr{E} = \frac{1}{2}m^*v^2$, and the separation of Landau levels, $\delta\mathscr{E}_0 = eH_0\hbar/m^*$, in field H_0. Then

$$\mathscr{E}.\delta\mathscr{E}_0 = \tfrac{1}{4}\pi \operatorname{cosec} 2\theta.(\Delta\mathscr{E})^2, \qquad (3.13)$$

where θ is the Bragg angle; apart from a numerical constant $\Delta\mathscr{E}$ is
the geometric mean of $\mathscr{E}$ and $\delta\mathscr{E}_0$. In order to convey some idea of
the meaning of (3.13) in practice, if $\mathscr{E} = 10\,\text{eV}$ and $\Delta\mathscr{E} = 0\cdot1\,\text{eV}$,
$\delta\mathscr{E}_0 = 0\cdot001\,\text{eV}$, corresponding to $H_0 \sim 85\,\text{kGs}$ if $m^* = m$; penetration
is significant ($P = 0\cdot05$) at $30\,\text{kGs}$, and is almost complete ($P = 0\cdot95$)
at $1\cdot6\,\text{MGs}$. These are rather high field strengths, but if $\Delta\mathscr{E}$ is only four
times less, the range of the breakdown phenomenon, from 2 to 100 kGs,
is fairly readily accessible. It has in fact been observed in a number
of metals, especially Mg and Zn, and can be expected to occur in moderate fields whenever the band gap is largely due to spin-orbit coupling,
for then $\Delta\mathscr{E}$ is usually rather small.

The immediate consequence of magnetic breakdown is that an
electron is not confined to one orbit, as it is if either P or Q is zero,
but may switch from one orbit to another at such points that the
Bragg condition is satisfied (Pippard, 1962, 1964a). Free-electron
orbits which are so coupled have their centres separated by a distance
S related to the reciprocal lattice vector of the reflecting planes by
(3.9). Since for a two-dimensional lattice each $\mathbf{g}$ is related to an ionic
lattice vector $\mathbf{a}$ by the equation (in which $\mathbf{s}/s$ is used simply to define
the normal to the plane)

$$\mathbf{g} = \frac{2\pi}{s\Sigma}\mathbf{s} \times \mathbf{a}, \qquad (3.14)$$

it follows from (3.4), (3.9) and (3.14) that

$$\mathbf{S} = J\mathbf{a}. \qquad (3.15)$$

The free-electron orbits which are coupled by partial Bragg reflection
have centres lying on a lattice similar to the ionic lattice but scaled
by a factor J. It may be expected, though no general proof has been
given, that a representation may be found, generalizing the race-track
representation of the free electron, in which a typical wave-function
is confined to the arms of a regular network of circles, such as the
examples shown in Fig. 3.10. At higher energies, when the circles
are larger, the connections become more numerous as other than

nearest-neighbour orbits intersect, but no detailed analysis of such a case has been made, those shown having proved amply complex up to now.

In the belief that the network adequately schematizes the wave-function, we may imagine a wave running along the arms and dividing at each junction into a transmitted part of amplitude $|p| = P^{\frac{1}{2}}$ and

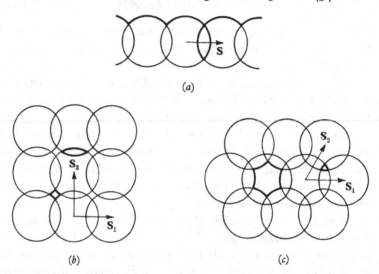

(a)

(b)   (c)

Fig. 3.10. Simple orbit networks produced by magnetic breakdown, showing orbit lattice vectors S and, in heavier lines, orbits present when no breakdown occurs (weak field).

a reflected part of amplitude $|q| = Q^{\frac{1}{2}}$. The phases of the emergent waves are represented by the phase angles of p and q. The problem of finding eigenfunctions of the system is now one of finding amplitudes for the waves on all arms which shall be self-consistent at the junctions. At first sight one might suppose that the network was strictly periodic, so that the eigenfunctions would have the wave-like character common to all periodic systems, but this view needs qualification. It can be seen from (3.15) that only when J is integral are all network junctions similarly located in the ionic lattice; in general, different junctions are not identical and the phase changes on Bragg reflection are not the same. In addition we must remember that the phase path along a given arm depends not only on its length but on its location relative to the gauge centre, so that identical-looking arms need not be the same in behaviour. This second difficulty is readily avoided by the device (§3.3) of choosing for each arm its own centre as gauge centre.

Then, as illustrated in Fig. 3.4, a phase correction must be applied when the electron switches from one orbit to another, whose magnitude is s times the area $OCPC'$. This correction includes the effect of the precise location of the junction on the ionic lattice. Only if J is integral are all phase corrections at corresponding points the same, apart from multiples of 2π, and only for such values of H that J is integral† can the network be regarded as strictly periodic with the vectors S_1 and S_2 (Fig. 3.10) defining the unit cell. If J is a rational fraction, M/N, a unit cell exists with basis vectors NS_1 and NS_2, but for irrational values of J no periodicity strictly exists. This peculiar property of the system has been noted by several authors (Azbel', 1964; Brown, 1964; Zak, 1964) using different approaches, and is clearly a general property, not just a consequence of the network model which illustrates it so directly.

If N is finite, solutions of the network can be found which are characterized by a wave-vector κ, in the sense that corresponding arms in two cells R apart ($R = N(\lambda S_1 + \mu S_2)$, λ and μ being integers) have the same amplitude of ψ and a phase difference of $\kappa \cdot R$. The range of κ is defined by a subzone of the same shape as the Brillouin zone of the lattice but smaller in linear dimensions by a factor NJ. The relation between $\mathscr{E}$ and κ is to be found by writing down the conditions to be satisfied by the amplitudes at each junction in the cell, so that, to take the square (Fig. 3.10(b)) as an example, there are $4N^2$ junctions each of which yields two equations, and a set of $8N^2$ linear algebraic equations results. By a change of gauge (W. Chambers, 1965) it is in fact possible to reduce the unit cell from a square net of N^2 orbits to a chain of N orbits, so that only $8N$ equations need be solved, but even so the labour is considerable even for small values of N. The solutions have the general form, as in the tight-binding model,

$$\cos (\pi \kappa_x/\kappa_0) + \cos (\pi \kappa_y/\kappa_0) = f_N(\mathscr{E}), \qquad (3.16)$$

in which κ_0 is the length of the side of the subzone, $2\pi/(NJa)$, and the form of $f_N(\mathscr{E})$ which emerges from the detailed analysis becomes progressively more complicated as N increases.

Corresponding to the decrease in size of the subzone, with increasing N the number of forbidden bands increases. The energy band structure is shown schematically in Fig. 3.11. On the left, where $Q = 0$, we get

† Pippard (1964a) found it necessary to suppose J to be an even integer when considering the hexagonal network, Fig. 3.10(c). It may be shown, however, by a transformation of the network that odd integers are just as good.

the expected sharp free-electron levels. On the right, where $Q = 1$ and Bragg reflection is complete, we have the lens orbits and 'square' hole orbits forming their separate series of sharp levels. In between these limits the levels broaden and reform. When J is integral ($N = 1$) the levels contact their neighbours, but when the fractional part of J, M'/N, is $\frac{1}{2}$ they split into narrower pairs which do not touch. When $M'/N = \frac{1}{3}$ each original band has split into three, and certain contact

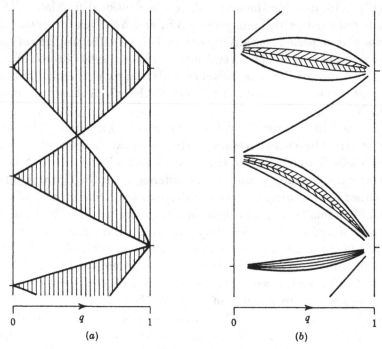

Fig. 3.11. Energy-level diagrams for the square network, Fig. 3.10(b), (a) when J is integral ($N = 1$); (b) when J is non-integral ($N = 5$) (W. Chambers, 1965).

points reappear. Further increase in N produces more and more splitting. Of course, as H is increased steadily, N fluctuates with infinite rapidity between finite and infinite values, according as J is rational or irrational, occasionally dipping down to small numbers. We must imagine the energy-level diagram as being in a steady state of rippling motion as H changes, the rapidity of the rippling being only limited by the finite total number of states imposed by the size of the sample, or by inhomogeneities of the field.† This extraordinary

† In practice collisions will also serve to smooth the behaviour, and it may well prove almost impossible to observe any significant consequences of this effect.

phenomenon, the consequence of two superimposed periodicities, makes it difficult to evaluate any measurable properties of the model, but some at least are independent of the exact value of J, as we shall see in the next section. It will be seen from the diagram that when $N = 1$ all states in the band deriving from a single free-electron level converge ultimately to a single level when $Q = 1$, but when $N > 1$ the sub-bands converge to different levels. In general the sub-bands split into two groups of M' and $N-M'$ respectively, converging to neighbouring levels.

As Fig. 3.11 shows, when J is integral and $N = 1$, each band derives from a single sharp free-electron level, whose degeneracy is $s/2\pi$ per unit area. Now in this case the subzone has side $2\pi/(Ja)$ and holds $s/(2\pi J)$ states per unit area. To conserve states we must suppose each individual state in the band to be J-fold degenerate, as if there were J independent networks needed to account for all the states. This need not be taken too seriously, since two networks displaced relative to one another are not quite orthogonal, but the numerical fact should be borne in mind. When J is non-integral, the subzone is smaller by a factor N^2, but each of the original ($N = 1$) bands is split into only N sub-bands, so that an extra N-fold degeneracy must be introduced to ensure conservation of states. Thus each state in the diagram should be regarded as having a degeneracy NJ, i.e. M.

We may now resolve the problem mentioned at the end of §3.3; what happens when J is not integral, and the degeneracy of each magnetic level is not a submultiple of the total number of states in the Brillouin Zone? If we note that the total number of states in the zone is just the number contained in J separate levels or bands each of normal multiplicity $s/2\pi$, we may expect, when J has a fractional part M'/N, to find somewhere in the level scheme a 'fractional level', or sub-band, and the fact that in the breakdown process the bands divide into groups of sub-bands in the proportion $M'/(N-M')$ provides just the mechanism required. Fig. 3.12 shows the conjectured solution of the problem; no detailed calculations have been carried out, and would be in fact rather troublesome. We imagine H to be kept constant while the strength of the lattice potential V is increased from zero, so that the free-electron energy contours divide into two sets, as in Fig. 3.7. The critical energy $\mathscr{E}_1$, at which division occurs, decreases as V rises, and so do the area and quantum number of the highest level below $\mathscr{E}_1$. Thus, a permitted orbit which lies just below $\mathscr{E}_1$ when V is small may find that as V increases it is drawn into contact with the zone boundaries; at the moment of contact, and in a small range of V around this, as the situation switches from none to complete Bragg reflection, there must be a transition region which may be loosely described as analogous to magnetic breakdown, even though the parameters may be unfavourable for breakdown on any but this critical orbit. It is in this region that the bands divide, broaden and reform as indicated in Fig. 12, and it will be seen that at any value of V there is always a sub-band of the required multiplicity to supplement the two series of normal levels.

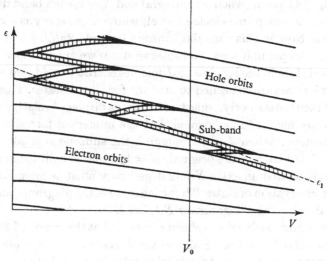

Fig. 3.12. Conjectured rearrangement of levels at change-over from electron to hole orbits, when J is non-integral. At the typical value V_0 of lattice potential, there is an extra sub-band, of multiplicity $M'/N \times s/2\pi$. On the left of the diagram break-down occurs at all energies above $\mathscr{E}_1$, but the diagram is not taken up to the point where the lens orbits appear and complicate the pattern considerably.

Let us now look briefly at some other networks that have been analysed in some detail, starting with those shown in Fig. 3.10(a) and (c). The linear chain (Pippard, 1962) which arises when only one Fourier component of lattice potential is operative, is the simplest extended network, and is genuinely periodic for all values of H. When $Q = 1$ the level structure consists of sharply quantized lens orbits and an unquantized continuum of open orbit states, as in Fig. 3.13(a). For any $Q < 1$ the continuum breaks into broad bands with narrow forbidden bands between. This may be visualized as exactly analogous to Bragg reflection causing gaps in the energy spectrum of a linear atomic chain. An electron on an open orbit is liable at certain points to pass through a lens orbit and be reflected back along a parallel open orbit, transferring itself from one side of the chain to the other; at certain energies all such reflections interfere constructively and create a band gap, whose width is governed by the probability of passage through a lens orbit.

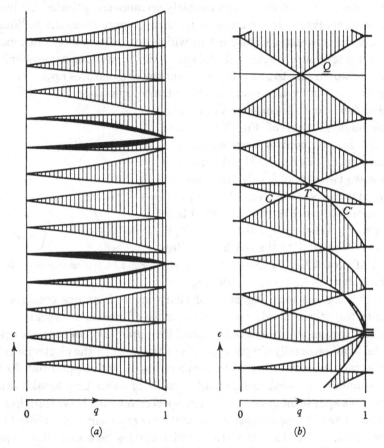

Fig. 3.13. Energy-level diagrams (a) for linear chain of Fig. 3.10(a); (b) for hexagonal network of Fig. 3.10(c) with integral J.

The hexagonal net (Pippard, 1964a) of Fig. 3.10(c) is more involved than the square net, having six junctions per unit cell rather than four, and gives a correspondingly more complicated band structure. When $Q = 1$ there are two sets of sharp levels, due to triangular electron orbits and hexagonal hole orbits, and for intermediate Q, when J is integral (the only case analysed), the bands expand to contact their neighbours, just as for the square net. There are, however, more singular contacts as well, as indicated in the diagram. At a point such as T a whole band shrinks to zero width at contact with a neighbour; sometimes T and its neighbouring contacts C and C' coalesce to form a quadruple contact Q, one band of zero width being squeezed between the contact of two others. Unfortunately no more complicated net has yet been analysed. It would be interesting, apart from the tedious algebra, to consider a square net in which not only nearest but next nearest neighbours intersected. This gives eight junctions per unit cell, and one would like to know if even stranger singularities appeared in the band structure as a result of the extra complication.

As a final example consider a finite network such as that of Fig. 3.14, which occurs in non-central sections of the hexagonal metals Zn and Mg. When $P = 0$ there are two sets of levels, those for the central triangular electron orbits, $\mathscr{T}_1$, and the threefold more degenerate set of hole orbits, $\mathscr{T}_2$. When $P = 1$ there is a single series of levels due to the self-intersecting orbit of area $3\mathscr{T}_2 - \mathscr{T}_1$. For intermediate values there is no level broadening, but

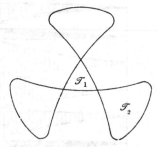

Fig. 3.14. Self-intersecting orbit.

merely a smooth rearrangement of sharp levels from one configuration to the other. Bands are a consequence of infinite networks.

The foregoing analysis has described the application of the network model in an especially simple type of situation, when the electrons are free except for a few very weak components of lattice potential. Not uncommonly the weak component capable of exhibiting breakdown effects is a spin-orbit gap in an energy spectrum which is far from free-electron-like because of other strong lattice components. It is probable (Chambers, 1966a) that the same model applies here as well, except that the coupled orbits are not circles, but race-track orbits of the same shape as the section of the energy surface by a plane normal to $\mathbf{H}$, the half-width of the track being $s^{-\frac{1}{2}}$ just as for free electrons. The same rule (3.15) determines the lattice of orbit centres, and the form of the theory remains the same as for free electrons. Where the model may be expected to fail, or at least to lose credibility, is when the disturbance of the energy by the lattice near one junction extends as far as the next, so that the junctions cannot be regarded as independent. The condition for this to occur is illustrated by the little triangular orbits of Fig. 3.10 (c). So long as these are triangular, in the sense of being largely formed of free-electron arcs, the lattice potential is clearly only sufficient to round the corners, not to distort the whole shape; the model is then applicable. But if $\Delta\mathscr{E}$ is large enough to convert the triangles into circular orbits, i.e. if $\Delta\mathscr{E}$ is at least comparable with the kinetic energy measured from the zone corner, the lattice components may be said to interact. From (3.13) it is seen that $\Delta\mathscr{E} \sim \mathscr{E}$ implies $\delta\mathscr{E} \sim \mathscr{E}$, that is, the quantum number of the orbit is of order unity. An alternative way of looking at this is to note that since the race-track is about $2s^{-\frac{1}{2}}$ wide, if the junctions are so close that the confused regions round each separate junction begin to overlap, the area of the orbit will be about $4/s$, and this again implies a quantum number of about unity. In Zn the triangles are so small that overlap is serious in a field of only $20\,\mathrm{kGs}$; breakdown is nearly complete and so it does not provide a really critical example. However, the good agreement between experiment and calculations based on the network (§ 3.10) encourages the belief that the model is sound up to, and perhaps beyond, the Quantum Limit, that is, the point of overlap.

The strength of the network model lies in its dissection of the theoretical problem into two parts, the behaviour in a region of imperfect Bragg reflection and the interaction of many such regions. The first is a straightforward problem which, nevertheless, needs delicate handling; for example, the form of (3.12) shows that it is not likely to succumb to perturbation treatment in powers of H, and the literature of Zener Breakdown offers some warnings against despising its subtleties. The second overcomes serious divergence problems always likely to arise with extended magnetic fields (since A is not bounded), and reduces the whole matter to finite algebraic form. In this way the network makes accessible the whole range of breakdown behaviour which can be reached by perturbation methods neither from above nor below, and reveals, as in Fig. 3.13 (b), complexities in the level structure that would be hard to discover by approximate methods. The model, however, is heuristic and could advantageously be investigated further from a fundamental point of view, as well as extended to more complicated situations and tested more fully against real metals.

3.6 GREEN'S FUNCTION ANALYSIS OF DENSITY OF STATES

(Falicov & Stachowiak, 1966)

The treatment in the last section aimed at explicit calculation of the complete energy-level diagram. An alternative analysis, which may provide certain information with less effort, derives the Fourier spectrum of the density of states as the transform of a Green's function. If a δ-function, $\Psi(\mathbf{r}, 0) = \delta(\mathbf{r} - \mathbf{r}_0)$, is established at time $t = 0$, it evolves according to the time-dependent Schrödinger equation so that at a subsequent moment

$$\Psi(\mathbf{r}, t) = G(\mathbf{r}, \mathbf{r}_0, t).$$

It is easy to show, by expanding $\delta(\mathbf{r} - \mathbf{r}_0)$ as a Fourier sum over all eigenstates, that the time variation of G at the point $\mathbf{r}_0$ determines $\mathcal{N}(\mathcal{E})$ the density of states

$$\mathcal{N}(\mathcal{E}) = \frac{1}{2\pi\hbar} \iint G(\mathbf{r}_0, \mathbf{r}_0, t)\, e^{i\mathcal{E}t/\hbar} d\mathbf{r}_0\, dt.$$

For a system showing complete translational invariance the integral over $d\mathbf{r}_0$ may be replaced by the volume of the sample, so that for unit volume

$$\mathcal{N}(\mathcal{E}) = \frac{1}{2\pi\hbar} \int_{-\infty}^{\infty} G(0, 0, t)\, e^{i\mathcal{E}t/\hbar} dt. \tag{3.17}$$

Let us interpret this result first for a two-dimensional free-electron metal, for which we know the energy levels to be evenly spaced. The initial wave-packet is created by superposition of all Landau functions for the system, in phase at the origin when $t = 0$. Since the functions have frequencies $(n + \frac{1}{2})\,\omega_c$, at all times $t = 2\nu\pi/\omega_c$ (ν integral) the functions recombine in phase at the origin and to a very good approximation $G(0, 0, t)$ consists of a series of evenly spaced pulses of alternating sign, suffering no distortion with time. It can be shown that the time-integral of each pulse is $m/\hbar$. This is of course very much like the behaviour of a classical particle, but it must be remembered that the wave-packet is something more like a superposition of all possible orbits passing through the origin; it spreads out in all directions and only reforms because all orbits have the same cyclotron frequency. We need not discuss here the precise form in time of the pulse, if we are concerned only with the structure of the energy levels, not the absolute values of $\mathcal{N}(\mathscr{E})$. We may therefore write, with sufficient accuracy

$$G(0,0,t) = \sum_{-\infty}^{\infty} (-1)^{\nu} m/\hbar \; \delta(t - 2\nu\pi/\omega_c), \qquad (3.18)$$

so that, from (3.17)

$$\mathcal{N}(\mathscr{E}) = \frac{m}{2\pi\hbar^2} \sum_{-\infty}^{\infty} (-1)^{\nu} e^{2\pi i \nu \mathscr{E}/\hbar\omega_c}, \qquad (3.19)$$

which represents a set of evenly spaced levels each of degeneracy $s/2\pi$, and which is only incorrect in failing to cut off the spectrum when $\mathscr{E} < 0$. For our purpose this is not a serious matter; it is easily rectified by a more careful treatment.

It will be observed that each successive pulse makes its separate contribution to the spectrum of $\mathcal{N}(\mathscr{E})$; a single pulse returning at time t', so that $G(0,0,t) = \delta(t-t')$, yields through (3.17) a sinusoidal variation of level density with period $2\pi\hbar/t'$. It is the regular succession of equal pulses that sharpens $\mathcal{N}(\mathscr{E})$ into a line spectrum. If through scattering the successive pulses are attenuated, the lines are broadened as a consequence of the absence of high harmonics (§3.7).

The argument has been developed for a free-electron orbit, but clearly it is capable of immediate extension to a closed orbit of any form, the sharp levels being spaced $\hbar\omega_c$ apart in conformity with the arrival of a regular succession of pulses at intervals of $2\pi/\omega_c$. It is the time interval that determines the period of the Fourier component of $\mathcal{N}(\mathscr{E})$, but its phase is determined by the phase integral of the orbit. Thus if the levels are at $(n+\gamma)\hbar\omega_c$, the νth pulse returns at time $2\pi\nu/\omega_c$ with phase $2\pi\gamma\nu$. In accordance with (3.5), therefore, the area of the orbit determines the phase of each Fourier component of $\mathcal{N}(\mathscr{E})$.

With these ideas in mind we may proceed to extend the treatment to a network. A wave-packet injected at some point does not now produce a regular sequence of pulses there, but something much more complex on account of the infinite variety of paths by which it may find its way back to the origin. Each of these paths gives rise to one Fourier component of $\mathcal{N}(\mathscr{E})$, whose period is determined by the time taken to execute the path, the phase by the area enclosed by the path, and the amplitude by the strength of the pulse on return. For any given network the parameters are easily found; e.g. if the network is composed of coupled free-electron orbits the flight-time is given by the total length of path, while the amplitude is $p^n q^m$ if during the path the pulse is transmitted through n junctions and reflected at m junctions. As for the phase, it is primarily determined by the area, but there may be a change of sign at some of the junctions, so that paths enclosing the same area may yield Fourier components that enhance or annul each other according to the precise values of m and n. When these points are correctly taken into account, limited tests of the method show very good agreement between the spectrum of $\mathcal{N}(\mathscr{E})$ calculated in this way and that derived from an exact solution of the network.

It must not be thought that the present method is a panacea. It has obvious attractions, not least that it avoids all consideration of interference between waves that have reached the same point by different routes. Moreover, if it is the spectrum of $\mathcal{N}(\mathscr{E})$ that is required, it goes straight to the heart of the problem. But the labour is not inconsiderable; the systematic enumeration of orbits is a major undertaking, and if one wished to find out the fine detail of $\mathcal{N}(\mathscr{E})$ it would involve literally thousands of orbits. On the other hand, being systematic it is ideally suited to machine computation, which is more than can be said for the direct network analysis, especially when J is non-integral.

It is interesting to note that the Green's function method takes no account of the precise value of J, and we conclude that the 'power spectrum' of $\mathcal{N}(\mathscr{E})$ is independent of J. Where the precise value enters is in the phase relationships of the components. In particular, when J is integral the phases are more simply related than usual. This may be illustrated by the square network Fig. 3.10(b). All paths on this network enclose areas that can be built up from the free-electron orbit ($\mathcal{O}$), the lens orbit ($\mathscr{L}$) and the square orbit ($\mathscr{S}$), and these orbits have areas related to the unit cell ($\mathscr{U}$) by the equation

$$\mathcal{O} + \mathscr{S} - 2\mathscr{L} = \mathscr{U}.$$

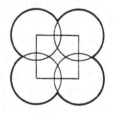

One may therefore express the phases of all components in terms of the phases contributed by three of these, say $\mathscr{U}, \mathscr{S}$ and $\mathscr{L}$. In general, there are three virtually independent phases, but if J is integral $\mathscr{U}$ has a phase which is a multiple of 2π and so ignorable, and this is true at all energies. In these special cases, then, there are only two independent phases, and there will be many more components of $\mathcal{N}(\mathscr{E})$ in phase with one another. For example, the two orbits shown in Fig. 3.15 have the same perimeter

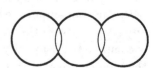

Fig. 3.15. Orbits differing in area by one unit cell.

L

and therefore contribute to the same component of $\mathcal{N}(\mathscr{E})$, but they differ in area by $\mathscr{U}$ and are in phase only when J is integral. Clearly integral values of J are a special case, but it must be admitted that the connection between this argument and the very precise rules governing the character of the energy bands has not been worked out.

3.7 LEVEL BROADENING

We have seen that a series of sharp levels corresponds in the Green's function treatment to a succession of pulses of undiminished amplitude, but in general we may expect the pulses to suffer attenuation from one cause or another, and we shall introduce the parameter β_ν to define the amplitude of the νth pulse, with $\beta_0 = 1$. Clearly β_ν is also the amplitude of the νth harmonic in $\mathcal{N}(\mathscr{E})$, relative to the amplitude for a series of sharp levels. Typical reasons for β_ν being less than unity are: (a) magnetic breakdown, (b) scattering, and (c) spatial inhomogeneity. We deal briefly with each in turn.

(a) *Magnetic breakdown* (Falicov & Stachowiak, 1966). An orbit which involves n transmissions and m reflections at junctions has

$$\beta_\nu = |p^n q^m|^{|\nu|}.$$

If all junctions have the same breakdown field H_0, the only orbits that can yield sharp levels are those involving only reflections or only transmissions; the former are responsible for sharp levels at low fields ($Q = 1$), the latter for the sharp free-electron levels when breakdown is complete at high fields ($Q = 0$). At intermediate values of Q we expect the density of states to be weak in short-wave components, which arise from multiple traverses of smaller orbits or single traverses of larger orbits, but which in any case involve many junctions. One should remember, however, that the larger orbits may have very high multiplicity, many different excursions contributing to the same component of $\mathcal{N}(\mathscr{E})$, whose character is therefore not so featureless as might be expected. Indeed, although the levels are broadened into bands, the bands retain sharp edges.

(b) *Scattering* (Falicov & Stachowiak, 1966; Dingle, 1952; Brailsford, 1966). Consider a simple orbit in a metal which through imperfections has isotropic scattering centres, whose action on a wave-packet is totally destructive, so that the chance of survival for time t is $e^{-t/\tau}$, τ

being the same relaxation time as determines the electrical resistivity, ρ. Then on the average the amplitude of the wave-packet decays as $e^{-t/2\tau}$. In a magnetic field the initial pulse returns from the νth transit after time $2\pi\nu/\omega_c$ with its amplitude diminished by $e^{-\pi\nu/\omega_c\tau}$. Care is needed with negative ν, that is, the pulses that precede the initial δ-function. To determine the function $G(\mathbf{r}, \mathbf{r}_0, -t)$ which will, after an interval t, evolve into a δ-function, we imagine time to be reversed. The evolution backwards in time then exactly mirrors the forward evolution, so that $G(\mathbf{r}_0, \mathbf{r}_0, t)$ must be an even function of t, and $\beta_\nu = e^{-\pi|\nu|/\omega_c\tau}$. The transform (3.17) of this sequence of pulses yields for the density of states a series of Lorentzian-broadened levels,

$$\mathcal{N}(\mathscr{E}) = \frac{1}{2\pi\hbar}\frac{\sinh(\pi/\omega_c\tau)}{\cosh(\pi/\omega_c\tau) - \cos 2\pi(\mathscr{E}/\hbar\omega_c + \tfrac{1}{2})}. \tag{3.20}$$

If $\omega_c\tau \gg 1$, each level has width at half-peak of $\hbar/\tau$, which is proportional to ρ.

If the electrons are scattered by phonons rather than impurities, so that at low temperatures each scattering is through only a small angle, many separate events must contribute to randomize the motion of a given electron. Thus the relaxation time τ which governs ρ is considerably larger than the mean time between electron–phonon collisions. On the other hand the deflection of a pulse moving in an orbit through even a minute angle is enough, when the orbit quantum number is large, to introduce such an error into its time of return to the origin that it might as well have been totally destroyed. Timing errors amounting to $1/n$ of the interval between transits cause the transform $\mathcal{N}(\mathscr{E})$ to be smoothed out for orbit quantum numbers greater than n, and this implies that an electron in an orbit of quantum number n is effectively dephased if scattered through an angle of $2\pi/n$ or more. Thus a tiny scattering angle may, through its effect on the phase, prove disastrous to the wave-packet. If then the orbit belongs to a Fermi surface of size comparable to the Brillouin zone, a phonon is effective for scattering if its energy is greater than $k\Theta_D/n$ or thereabouts. Except at extremely low temperatures, where phonon scattering is in any case very weak, it is safe to count every scattering event as destructive.

A rough estimate indicates that for a spherical Fermi surface, on which a phonon of energy kT scatters an electron through an angle, ϕ_0, of about $kT/(\hbar v_s k_0)$, v_s being the velocity of sound, the number of scattering events needed to randomize motion from the point of view of electrical resistivity is about $0 \cdot 2/\phi_0^2$, if $\phi_0 \ll 1$. Thus the level broadening given by (3.20) varies with temperature not as ρ but as ρ/T^2, i.e. roughly as T^3 if all scattering is by phonons. This is the same variation as shown by the thermal resistance in the ideal range, and for the same reason; vertical transitions caused by phonons of energy about kT are enough to destroy a thermal current, so that a very substantial fraction of all phonon collisions are effective as thermal scatterers. At low temperatures, if phonon scattering dominates, the Wiedemann–Franz Law is in error by a factor which is also $0 \cdot 2/\phi_0^2$ for this simple model; from this it appears that one should use the relaxation time derived from thermal, not electrical, resistance in (3.20) in order to obtain a realistic estimate of line broadening. This conclusion can doubtless hold also for non-spherical Fermi surfaces without too much error.

If the scattering mechanism is elastic, but predominantly small-angle, the same value of τ will be deduced from electrical and thermal resistivity, but it will be longer than that appropriate to level broadening.

(c) *Spatial inhomogeneity*, especially that due to lattice defects (Pippard, 1965 a). As discussed in § 3.3, if an orbit involves Bragg reflections the phase length between two points on either side of the point of reflection depends on the precise location of the lattice; but it was noted that differently centred orbits were still degenerate because the total phase path, involving more than one Bragg reflection, was not affected by a shift relative to the lattice. However, if the lattice is imperfect, particularly if it is distorted by dislocations, the lattice displacements at the various Bragg reflections are different, and the phase integral then depends on the nature of the distortion. A single edge dislocation running through the orbit is sufficient to create havoc even though it causes no appreciable scattering; for the lattice displacement may be anything between $\pm$ half a lattice spacing, and persists indefinitely far from the dislocation. In consequence phase shifts up to π can be introduced at any reflection, and the total shift depends on the position of the dislocation within the orbit. Different orbits are affected to various degrees, and in the event the degenerate levels may be spread into very nearly a continuum, if there are enough defects for each orbit to contain one or more. In a field of 10 kGs a typical orbit area might be $10^{-6}\,\mathrm{cm^2}$, so that a dislocation density of $10^6/\mathrm{cm^2}$ should be enough to smooth out $\mathcal{N}(\mathscr{E})$

considerably, if the orbits involved have Bragg reflections. This dislocation density is far less than would normally be needed to increase ρ to the point of appreciable level broadening according to (3.20). Evidence on this point is very meagre, but there have been some observations of loss of amplitude resulting from thermal cycling of a specimen between room temperature and helium temperature, which may be attributed to this effect. It may also be noted that orbits lying on a zone boundary, such as the neck orbits in the noble metals, may well be very sensitive to lattice defects, but no theory of this has been published: experimental evidence supports this view.

It is clear that the behaviour of the networks produced by magnetic breakdown is severely affected by dislocations, since the elements of the network are now largely random in their phase lengths, and all the Fourier components of $\mathcal{N}(\mathcal{E})$ are dephased. The only features that can be expected to survive are the free-electron periodicity and such long periods as are due to small orbits, most of which may be wholly situated in a dislocation-free neighbourhood and have their phase lengths undisturbed. We shall return to this point in §3.10, when discussing transport properties of the network.

3.8. MAGNETIC PROPERTIES OF THE CONDUCTION ELECTRON ASSEMBLY

3.8.1 Introduction

From individual particles we turn to an assembly such as the degenerate gas of conduction electrons in a metal, still regarding them as independent particles. If we are concerned solely to evaluate the thermodynamical properties of the assembly, including the magnetic moment, there are standard techniques which yield the correct answer. These techniques do not, however, give any insight into why the answer takes a particular form, and it is instructive to attempt a derivation by more direct physical arguments, for although these usually fail their very failure hints at the extremely subtle character of some of the problems. This section, then, is less concerned with the almost mechanical procedure for getting an answer than with the more formidable task of comprehending it once it is obtained. The general method of attack (noted briefly in §3.8.5) involves evaluating the Grand Potential Ω as a series expansion in H, the quadratic term, for example, yielding the steady susceptibility; and there are perturbation techniques which avoid any reference to the energy-level structure of the assembly, but instead operate directly on the Hamiltonian. Rather than proceeding in this powerful but obscure fashion we shall attempt in §3.8.2 to evaluate Ω in the elementary way as a sum over states, and shall see immediately that very great care and labour is needed to obtain the correct result for any system but the free-electron gas. Even for a free-electron gas an attempt at a yet more direct physical argument runs into trouble; as will be shown in §3.8.3, it is one thing to evaluate the magnetic moment correctly, but quite another to discover precisely which electrons are responsible. The difficulty arises over the weak temperature-independent Landau diamagnetism, not the oscillatory de Haas–van Alphen phenomenon which dominates the low temperature behaviour, and for which the direct attack proves adequate and instructive. The physical origin of the difficulty lies in the properties of electrons extremely close to the surface, and in §3.8.4 we look further into the matter and find a close connection with another well-known difficulty—the persistence of the Landau diamagnetism in the presence of electronic scattering. These problems still remain largely unsolved. The de Haas–van Alphen effect, however, which we discuss further in §3.8.6, is far easier to understand from a physical point of view; fortunately, since it is of practical interest to the experimental metal physicist, as the basis of the most powerful techniques available for studying the energy surfaces of real metals.

3.8.2   Elementary evaluation of Ω

The Grand Potential is defined as $E - TS - N\zeta$, N being the number of particles in the assembly and ζ the Fermi energy or chemical potential. It can be written as a sum over states,

$$\Omega = k - T \Sigma \ln\left(1 + e^{-\mathscr{E}'/kT}\right), \qquad (3.21)$$

in which $\mathscr{E}' = \mathscr{E} - \zeta$; from this the magnetic moment $M(\zeta, T)$ is derived as $-(\partial\Omega/\partial H)_{\zeta, T}$. It will be necessary to discuss what value is to be ascribed to ζ, but this can be postponed. The advantage of the use of Ω rather than free energy is that conservation of particles is not required, and each energy level is occupied independently of the others. We may therefore treat the three-dimensional metal as a stack of two-dimensional metals, by dissecting k-space into slices normal to **H** and considering separately each slice of thickness dk_z. In each the electron orbits are well defined, and if they produce sharp energy levels the degeneracy is $sdk_z/2\pi^2$, with both spin directions included.

Since in general the levels are not evenly spaced in energy it is more convenient to use the orbit area $\mathscr{A}_k$ as a variable, or its equivalent the quantum number n, defined as $\mathscr{A}_k/2\pi s$ in accordance with (3.5) and treated as a continuous variable. Then apart from the slow variation of γ_1, which for the moment we ignore, though it will be seen later to be important, the levels are equally spaced at half-integral values of n and may be written as a density of states, following (3.19),

$$\mathscr{N}(n) = \sum_{-\infty}^{\infty} (-1)^{\nu} a_{\nu} e^{2\pi i \nu n}, \qquad (3.22)$$

in which $a_{\nu} = sdk_z/2\pi^2$ for sharp levels or, more generally

$$a_{\nu} = \frac{\beta_{\nu} s dk_z}{2\pi^2}. \qquad (3.23)$$

Hence, from (3.21),

$$\Omega = -kT \int_0^{\infty} \sum_{-\infty}^{\infty} (-1)^{\nu} a_{\nu} e^{2\pi i \nu n} \ln(1 + e^{-\mathscr{E}'(n)/kT}) \, dn. \qquad (3.24)$$

This integral may be evaluated by integration by parts, followed by a simple contour integration, in the course of which it is found that if the gas is fully degenerate (so that $|\mathscr{E}'_0|$, the kinetic energy at the Fermi surface, is much greater than kT) the relationship between $\mathscr{E}'$ and n is only needed at the vanishing point ($n = 0$, $\mathscr{E}' = \mathscr{E}'_0$) and at those poles which lie close to the Fermi energy. In the former case we write $\mathscr{E}' - \mathscr{E}'_0 = n\hbar\omega_{c0}$, and in the latter case $\mathscr{E}' = (n - n_f)\hbar\omega_{cf}$, ω_{c0} and ω_{cf} being the cyclotron frequencies at the vanishing point and at the Fermi surface, and n_f the quantum number of the orbit at the Fermi surface. Then after evaluation we have

$$\Omega = \sum_{\nu=1}^{\infty} (-1)^{\nu} \frac{a_{\nu}\hbar}{2\pi^2\nu^2} \left(-\omega_{c0} + \frac{\omega_{cf}X}{\sinh X} \cos 2\pi\nu n_f \right), \qquad (3.25)$$

in which $\qquad\qquad X = 2\pi^2 \nu kT/\hbar\omega_{cf}. \qquad (3.26)$

We have omitted from Ω a constant term which is its value in zero field, and have made use of the fact that $\beta_{-\nu} = \beta_{\nu}$. On differentiating (3.25) to obtain M, it is clear that when n_f is large the second term is dominated by its oscillatory part; to a good approximation we may

148   A. B. PIPPARD

neglect the field variation of the rest. Then, by use of (3.5), (3.8) and (3.23),

$$M = M_s + M_{osc}$$

$$= \frac{e^2 H\, dk_z}{2\pi^4 m_{c0}} \sum_1^\infty (-1)^\nu \beta_\nu/\nu^2 - \frac{e\hbar \mathscr{A}_{kf}\, dk_z}{4\pi^4 m_{cf}} \sum_1^\infty (-1)^\nu \frac{\beta_\nu X}{\nu \sinh X} \sin\left(\frac{\nu \mathscr{A}_{kf}}{s}\right).$$

$$(3.27)$$

The first term, M_s, in (3.27) is the temperature-independent steady diamagnetism of the slice. For a free-electron gas we may put $m_{c0} = m$ and if there are no collisions so that the levels are sharp and $\beta_\nu = 1$ we arrive at the standard results for the Landau (1930) diamagnetism

$$\chi = -\frac{e^2 k_f}{12\pi^2 m}. \qquad (3.28)$$

Unfortunately this is about the only case in which the answer is correct. The dependence on m_{c0}, the mass at the bottom of the band, contradicts the result of Peierls (1933) who was able to express χ for a tight-binding model in terms of the mass at the Fermi surface.

In fact the appearance of ω_{c0} in (3.25) is rather misleading, as there are two separate contributions to the steady term. One of these arises at the Fermi surface and can be most clearly appreciated at $0\,°K$, when all levels below ζ are filled and all above ζ empty; moreover, if we measure electron energies from ζ so that filled states have negative energy, Ω is just the total energy of the assembly. Denoting by ζ' the energy half-way between the last filled and the first empty levels, we see that the energy of the assembly is the same as for a gas in zero field whose continuum of states is filled up to ζ'. When $\zeta' > \zeta$ there is an excess of electrons of positive energy, and when $\zeta' < \zeta$ there is a deficit of electrons of negative energy. In both cases Ω is increased above its zero field value, so that as well as being oscillatory, its mean value is increased, to an extent governed by ω_{cf}. There is a second contribution if ω_{c0} and ω_{cf}' are different, for then the energy band is not quadratic and the states in unit range of n that condense onto a sharp level suffer a slight change of mean energy. When this contribution, governed by $\omega_{c0} - \omega_{cf}$, is summed it has the effect of replacing ω_{cf} by ω_{c0} in the first term of (3.25).

One may see now why this is incorrect, for when $\omega_{c0} \neq \omega_{cf}$ the phase correction γ is not independent of energy, but varies in accordance with (3.7) in such a way as to provide yet another contribution to χ, and to throw the emphasis back onto the Fermi surface. In fact if interband effects are ignored Peierls's result is reproduced. It is very satisfactory to see that a rigorous determination (Roth, 1966) of γ_1, although elaborate, allows direct summation over energy levels to achieve what Peierls attained by a quite different approach, not involving individual levels, for it clears up one aspect of the rather long-standing puzzle of the precise relationship between orbit quantization and steady diamagnetism. As we shall see shortly, there are other aspects still only imperfectly understood.

3.8.3 The oscillatory diamagnetism

Let us, however, look first at the second oscillatory term M_{osc} in (3.27), whose origin in the relation of the quantized levels to the Fermi surface has already been pointed out, and which has great experimental significance since it is the most important example (de Haas-van Alphen effect) of a collection of oscillatory phenomena all having the same source. In contrast to M_s, M_{osc} has a measurable amplitude only at such low temperatures that X is small, the Fermi tail not spreading over more than a few levels. It is of interest to compare M_{osc} with the oscillations of the number of electrons in the slice, which can be found in a similar way by multiplying (3.22) by the Fermi function and integrating. Then we find

$$N = \frac{s dk_z}{2\pi^3} \sum_1^\infty (-1)^\nu \frac{\beta_\nu X}{\nu \sinh X} \sin (\nu \mathscr{A}_{kf}/s), \qquad (3.29)$$

and by comparison with (3.27) discover that the oscillatory moment is accounted for in terms of oscillations of N, if every electron is allowed a diamagnetic moment μ according to the expression

$$\mu = -\frac{e\omega_{cf}\mathscr{A}_{kf}}{2\pi s^2} = -\frac{e\omega_{cf}\mathscr{A}_{rf}}{2\pi}, \quad \text{from} \quad (3.6). \qquad (3.30)$$

Now, by Ampère's theorem, (3.30) is just the magnetic moment of an electron moving in an orbit of area $\mathscr{A}_{rf}$ with frequency ω_{cf}, so that the oscillatory part of M may be understood physically as due to the variation in the number of electrons in closed orbits. The electrons in surface states, whose number is so small in a large enough sample that they can be neglected in calculating Ω, form a continuum which contributes nothing to the oscillations.

The same does not apply to the steady diamagnetism. It is not obvious from (3.29) that there is any non-periodic variation of N, which only appears on integration over k_z. But M_s is not to be accounted for in these terms. If, for instance, we perform a direct summation of the moments of occupied orbital states in a free-electron gas, and compare it with the result for unquantized orbits, we might be tempted to believe that the difference represents the actual magnetic moment; the argument being that in an unquantized metal the orbital moments are exactly cancelled by the surface states discussed in §3.2 and the latter are sensibly unaffected by quantization. But this is too naïve; it gives the right value for the gross oscillatory effects but accounts for only half the more subtle steady diamagnetism, whose magnitude is smaller than the oscillatory amplitude at $0\,^{\circ}\mathrm{K}$ by a factor of about $n_f^{\frac{1}{3}}$. It seems that a direct computation of the steady effect would need to look carefully into the details of the surface states (Van Vleck, 1932, p. 356). Here we see the power of the method based on Ω, which can afford to ignore the surface states because they are weighted equally with the interior states, even though their individual moments may be so enormous. But this power is achieved at the cost of not understanding the detailed mechanism, which still remains somewhat obscure.

3.8.4. The steady diamagnetism

Even though we may not be able to describe in detail the contribution of each electron to the magnetic moment, it is interesting to proceed one step further, but at this stage the analysis becomes rather lengthy, and we shall only describe it schematically. Rather than work in terms of magnetic moments, we shall consider the system of currents due to the electrons. In the interior the motion of the particles of a free-electron gas is isotropic and there is no current, any magnetization of the sample being due to a surface current. For a classical gas this current vanishes, but the different quantal restrictions to orbital and surface states cause a surface current to appear. For the oscillatory moment, arising as we have seen from a change in number of orbiting electrons, the current is confined to a surface layer about as thick as the orbit diameter, and it may be noted that if the sample is thinner than this the magnetization is disturbed, and the oscillatory properties are lost. It is surprising to find that the same does not hold for the steady diamagnetism, which has its origin in a current sheath confined to a much thinner layer.

To show this we start paradoxically by eliminating the surface, and treat instead an infinite sample subjected to a spatially periodic field $H = H_0 \cos qx$, derived from a vector potential $A_y = H_0 \sin qx/q$. We are concerned to find the linear current response of the assembly to this perturbation, and we may do so even though the perturbation expansion for individual electron levels becomes divergent as we let q proceed to zero, keeping H_0 constant. The divergence is a consequence of the development of orbits, which cause the magnetization to become a highly non-linear function of H_0. However, with an assembly at a high enough temperature ($X \gg 1$) the oscillations due to individual electrons cancel one another out and when this happens it is found (Nakajima, 1955) that such statistical functions as Ω may be expressed as perturbation series even though the individual particle energies may not. It is then quite justifiable to pretend that the perturbation expansion of the individual energies converges and to calculate the first-order current due to each electron. One must remember, however, that this trick only gives, after summation, the first-order response of the assembly at a high enough temperature—it does not give correctly the contribution of each individual electron. Thus we are still no nearer to understanding the detailed mechanism, but may extend the result (3.28) to spatially-varying fields, and calculate the susceptibility χ_q as a function of wave-number. The only result of this analysis (Bardeen, 1956) that need be quoted is that χ_q has (3.28) for leading term, but falls monotonically to zero as q increases. It can be expressed as a power series in $(q/k_f)^2$

$$\chi_q = -\frac{e^2 k_f}{12\pi^2 m}\left(1 - \frac{q^2}{20 k_f^2} + \dots\right). \tag{3.31}$$

We may proceed from this result to discuss the current sheath at the surface of a bounded sample in a uniform field, using exactly the same methods as prove valuable in the analysis of the skin effect (see §4.5). Without entering into details the qualitative result is easily appreciated; if χ_q falls to zero in a range of q around k_f, the depth within which the surface currents flow is of the order of k_f^{-1}, i.e. of atomic dimensions in most metals. It appears then that the surface states are able to balance the currents due to the orbiting electrons everywhere except very close to the surface, and that the orbit size is irrelevant to the character of the steady diamagnetism. This conclusion, which does not depend on whether the surface is rough or smooth, is consistent with what is known about the diamagnetism, which survives collisions numerous enough for $\omega_c \tau$ to be much less than unity, so that the orbits are effectively destroyed. It has indeed been conjectured (Peierls, 1955, p. 155) that only when the free path is as short as k_f^{-1} will the diamagnetism disappear. This is made plausible by considering a thin film in a parallel field. If the thickness is

152 A. B. PIPPARD

more than a few times k_f^{-1} there is no interaction between the surface currents on either side, and the film should behave as a slice of bulk metal. This should hold even when the surfaces scatter electrons diffusively, so that we must picture the diamagnetism as surviving the extremely frequent collisions suffered by the electrons as they bounce from side to side, and by extension surviving also equally frequent collisions in the body of the sample.

There seems to be no rigorous proof of this result. It is indeed possible to show that collision-broadened levels, whose structure is independent of H, reproduce the diamagnetism unaltered (Mott & Jones, 1936). But when the collision-broadening is larger than the separation of magnetic levels it seems a little cavalier to treat the magnetic field as the primary perturbation, and the collisions as secondary, rather than the other way round. One would like to see a full treatment of the effect of a magnetic field on collision-perturbed electrons, possibly along the lines suggested by Bardeen (1956). But at present we must leave this as an incompletely solved problem, like others concerned with the detailed physical picture of the steady diamagnetism.

3.8.5 Exact formal methods

If, however, we are only concerned with obtaining a cerreet expression for χ, there are powerful methods available (Hebborn *et al.* 1964) for computing Ω in real metals as a power series H, at temperatures high enough to destroy the oscillations, provided we are prepared to ignore collisions so that the zero field state of the metal has the translational symmetry of a perfect lattice. The calculations are extremely laborious, and the general result complicated to a degree that stupefies the imagination. It is enough to say here that χ has contributions from three principal causes, the conduction electrons (Peierls diamagnetism), the bound electrons of the ionic lattice, and interband terms resulting from the coupling of the first two. These last are liable to be important when band gaps are small, and one may expect similar complexities when different sheets of the Fermi surface are imperfectly decoupled, as in magnetic breakdown.

3.8.6 The de Haas–van Alphen effect

Let us then leave these subtle and complex matters and return to the coarser oscillatory effects. The second term of (3.27) gives the contribution of one slice of k-space, and the oscillatory moment of the whole metal is found by integration. As is well known (Pippard, 1965b, p. 21), the integral is dominated by those sections for which $\mathscr{A}_{kf}$ is extremal, where the oscillatory phase is momentarily stationary with respect to k_z. In fact everything may be thrown away except for an effective zone of thickness $\delta k_z = (2\pi s/\nu \mathscr{A}_k'')^{\frac{1}{2}}$ at each extremum, A_k'' being the value of $d^2\mathscr{A}_{kf}/dk_z^2$ at the extremum. The phase of the resultant differs from that of the extremum, by $+\frac{1}{4}\pi$ for a minimum and $-\frac{1}{4}\pi$ for a maximum. The result of the integration is as follows:

$$M_{osc} = -\left(\frac{e}{\hbar H}\right)^{\frac{1}{2}} \frac{kT}{(2\pi)^{\frac{3}{2}}} \sum_{\substack{\text{all} \\ \text{extrema}}} \frac{\mathscr{A}_{kf}}{(\mathscr{A}_k'')^{\frac{1}{2}}} \sum_{\nu=1}^{\infty} \frac{(-1)^{\nu}}{\nu^{\frac{1}{2}}}$$

$$\times \frac{\beta_{\nu}}{\sinh X} \sin\left(\frac{\nu \mathscr{A}_{kf}}{s} \pm \frac{\pi}{4}\right) \cos \frac{\nu g_{\text{eff}} \pi m_c}{2m}. \quad (3.32)$$

The final cosine factor allows for the spin splitting of the levels, g_{eff} being the parameter discussed in §3.4. This expression for M_{osc} lies at the heart of the use of oscillatory phenomena such as the de Haas-van Alphen effect for determining the shape of the Fermi surface in real metals. The success of such determinations, coupled with much incidental information about the dependence of oscillatory amplitudes on spin, temperature and scattering, as well as the variation of amplitude when magnetic breakdown occurs, leaves one in little doubt of the essential correctness of the result. This topic is discussed in greater detail in §2.2.1.

One final point should be made. We have assumed throughout this section that the chemical potential ζ is constant, and (3.32) is strictly $M_{osc}(H, \zeta)$. In fact, neutrality of the metal ensures that the number density of electrons is constant, and ζ adjusts itself accordingly. Integration of (3.29) over k_z shows that at constant ζ the number density oscillates, each extremum making such a contribution N_{osc} as accords with (3.30),

$$M_{osc} = \sum_{\text{all extrema}} N_{osc}\mu.$$

Because only a small fraction of the Fermi surface belongs to the effective zones the oscillations of ζ are not very great, since most of the Fermi surface acts as a reservoir to smooth out the oscillations at the extrema. It is not difficult to see that, to conserve number, ζ must oscillate at $0 °$K with an amplitude which is less than the separation of levels by something like the fraction of the Fermi surface contained within the effective zones. For a free-electron metal the oscillation of Fermi level is less than the level spacing by a factor $(8n)^{\frac{1}{2}}$, n being the quantum number of the equatorial orbit. There is thus a slight, but usually negligible, modification of the oscillatory pattern. The small oscillations in ζ required to preserve neutrality in the interior are not necessarily such as will also preserve neutrality in the surface layer. The kinetic energy at the Fermi level may therefore need to vary close to the surface, and in consequence there may be an electric field in the surface layer, normal to the surface, so that the total chemical potential is constant throughout the sample. Under these conditions the potential of the surface is different at points where H is normal and parallel, and an external electric field is set up whose strength and direction oscillate with H; this can in principle be picked up by a suitable capacitive arrangement, just as if it were a change in contact potential. For potassium at 0 °K in a field of 50 kGs the oscillation should have an amplitude of about $1\,\mu V$. The effect has been observed in a number of metals (e.g. Caplin & Shoenberg, 1965).

3.9   THE SHOENBERG (1962) EFFECT

In all that has been said up to now the electrons have been regarded as independent particles subject to a time-dependent single-particle Schrödinger equation, scattering being introduced only as a perturbation. This is, of course, a considerable over-simplification, since there are electron–phonon and screened electron–electron interactions which lead to significant dynamical modifications near the Fermi surface. These have been examined most carefully (Pines, 1961) in the absence of a field, and rather less exhaustively (Luttinger, 1961; Kohn, 1961) with a magnetic field present, and on the whole the conclusion is that the independent particle approach is remarkably close to the mark, provided one recognizes that what we have thought of as an electron near the Fermi surface is better imagined as a long-lived excitation of a many-particle ground state, with all the characteristics of a charged particle. It is surprising how few experiments can reveal unambiguously the magnitude of the interactions (comparison of electron specific heat and spin susceptibility is one). The general conclusion of most theoretical investigations has been that experiments which yield geometrical information about the dynamics of conduction electrons may be interpreted in independent particle terms without need of correction. Thus if we use the de Haas–van Alphen effect and the anomalous skin effect to determine the shape of the Fermi surface, they both refer to the same shape, and it is a surface which contains the correct number of states to account for the total number of electrons present; moreover, the same cyclotron mass is involved in interpreting Azbel'–Kaner resonance and the temperature variation of the de Haas–van Alphen amplitude. In other words, however much the interactions may contribute to the actual properties of the system, including the precise character of the particle-like excitations, when all is said and done it is possible to construct a single-particle energy function $\mathscr{E}(\mathbf{k})$ which reproduces exactly the majority of the observable effects. This conclusion is important, and if we dismiss the topic summarily it is not because the absence of significant effects implies triviality, but because there is little of value to be said, short of a detailed discussion for which this is not the place.

There is, however, one interaction which leads to observable effects, and which fortunately is macroscopic in character and capable of analysis without probing too deeply into fundamentals. This is the magnetic interaction which becomes important when the de Haas–van Alphen effect is strong enough to make the internal field in the sample significantly different from the applied field $\mathbf{H}_0$ (Shoenberg Effect). The internal field responsible for determining the energy levels inside the sample is the macroscopic $\mathbf{B}$, and in the highly oscillatory de Haas–van Alphen effect it does not take much to make $\mathbf{B}$ differ from $\mathbf{H}_0$ by a substantial fraction of a cycle of the oscillation. For an ellipsoidal sample having demagnetization coefficient D,

$$H = H_0 - 4\pi D I, \tag{3.33}$$

and
$$B = H_0 + 4\pi(1 - D)I. \tag{3.34}$$

There is no reasonable doubt about the magnetizing field experienced by the conduction electrons in the body of the metal being $\mathbf{B}$, even in a ferromagnetic metal (Kittel, 1963) where the answer is physically far less obvious than in the present case. We might expect, if the field varied in magnitude within the unit cell of the lattice, that the appropriate average would be weighted according to the probability of finding the electron at any point. In a sense this is true, but we must remember that what enters the Hamiltonian is not $\mathbf{H}$ but $\mathbf{A}$. If we divide the field into its average, $\mathbf{B}$, and the deviations therefrom, $\mathbf{h}$, the vector potential generating $\mathbf{h}$ fluctuates from cell to cell, with typical amplitude $a|\mathbf{h}|$, where a is a lattice dimension. But the vector potential generating $\mathbf{B}$, imagined as centred on the orbit centre, has typical amplitude Br at the orbit radius r. Thus B is weighted relative to h by the large factor r/a, and even substantial local fluctuations of field are in practice rather insignificant. When, as here, the oscillatory magnetization results from a disturbance of balance between complete orbits and surface states, the internal field differs from the applied field only by virtue of currents near the surface, and there is no reason to suppose that the additional field possesses any unusual microstructure in the interior. The internal state of the metal is in fact exactly as if it were subject to a uniform applied field of magnitude **B**. We may assume, then, that the quantized orbits and their individual moments are determined by **B**. We must look carefully, however, before jumping to the conclusion that the magnetization of the whole sample is determined by **B**, since the extremely important surface states exist in a field which varies from $\mathbf{H_0}$ outside to **B** inside the surface layer. Nevertheless, the conclusion is correct, for reasons analogous to those just mentioned that allow neglect of local field fluctuations. Although the field variation near the surface must have some influence on the precise quantum treatment of surface states, the most important factor is the mean value of **A** round the surface and this is determined by the total flux in the sample, i.e. essentially by **B**. As we allow the sample to increase in size, the local surface effects remain constant in magnitude while the influence on **A** of $(\mathbf{B} - \mathbf{H_0})$ in the interior increases without limit. We conclude, then, that for large samples all eigenstates are determined by **B**, and that the magnetization is to be evaluated by substituting **B** for **H** in the expression derived for non-interacting electrons.† We may further conjecture that in a non-uniform field $\mathbf{I(r)}$ is locally determined by $\mathbf{B(r)}$, provided the variations of **B** are on a scale considerably larger than the orbit radius.

If the temperature of the sample is not too low, so that only the leading term of M_{osc} in (3.32) survives, we may approximate $I(B)$ in a uniform field by the simple function

$$I = I_0 \sin \lambda B,$$

in which λ may be taken as constant over a single cycle, but really has a slow variation with H, since $\lambda = \hbar \mathscr{A}_{kf}/(eH^2)$. Then the relation

---

† This argument does not apply to the evaluation of anything involving the energy, since the interaction energy is not equivalent to each magnetic moment being subjected to a field **B** rather than **H**; such a procedure leads to twice the correct value of the interaction energy. It may be shown (Condon, 1966) that the free energy is obtained by substituting **B** for **H** in the non-interacting expression, and supplementing the answer with $2\pi V I^2$.

between I and the applied field can be seen, by use of (3.34), to take the form

$$y = a \sin \{x_0 + (1 - D) y\}, \qquad (3.35)$$

in which $y = 4\pi\lambda I$, $x_0 = \lambda H_0$ and $a = 4\pi\lambda I_0$. This, however, is not the whole story, for when $D \neq 0$ there is the possibility of instability (Condon, 1966), leading to domain formation and non-uniform magnetization. We shall therefore discuss first the case of a long rod

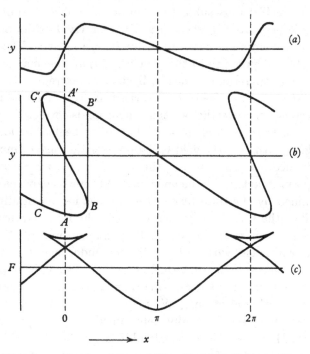

3.16. The Shoenberg effect in a long rod ($D = 0$): (a) oscillations of magnetization when $a = 0.8$, (b) the same when $a = 2$, (c) free-energy oscillations when $a = 2$; x represents the applied field.

parallel to H, for which $D = 0$ and (3.35) is applicable without correction. Plotting out $y(x_0)$ for various values of a, as in Fig. 3.16, shows how the oscillations become distorted as a approaches unity, and for $a > 1$ I is a many-valued function of H_0. The parameter a, which determines the degree of non-linearity, is 4π times the amplitude of the differential susceptibility $\partial I/\partial B$. Even though the oscillations of I itself may be very small in amplitude, a high de Haas–van Alphen frequency may lead to a becoming large enough to matter. Since λ

M

varies as H^{-2}, it is at low fields that one expects to see the Shoenberg effect, provided that the temperature is also low enough and the material pure enough not to reduce the amplitude excessively. The effect has been studied in the noble metals (Shoenberg, 1962; Joseph & Thorsen, 1965; Joseph et al. 1965) and particularly in Be (Condon, 1966; Plummer & Gordon, 1964; Broshar et al. 1966; Le Page et al. 1964) where it is very strongly marked, and the observations reveal clearly that a can be made to exceed unity.

The resolution of the ambiguity resulting from a multi-valued $\Gamma(H)$ has been the subject of some discussion, and the most plausible answer is that the speed with which the applied field is changed determines the behaviour. There is no doubt from the free-energy curve that the extreme ends of the line AA' represent two states in equilibrium, and that in a slightly non-uniform field stable domain boundaries can exist on those planes parallel to $\mathbf{B}$ where $\lambda \mathbf{B} = 2n\pi$. Neighbouring domains are magnetized parallel to the boundary but in opposite senses and the magnetization gradually reverses as one crosses a domain. In the limit of large a, B is virtually constant at $2n\pi/\lambda$ in each domain, with n changing by unity from one to the next. This is the equilibrium state, but creation of a nucleus from which a new domain may grow is inhibited by eddy currents, and if the field is increased rapidly the jump in magnetization may be deferred until the line BB' is reached, beyond which resolution is automatic; similarly in a sufficiently rapidly decreasing field the jump may occur along CC'.

Once a domain boundary exists it will migrate until it reaches condition A, the speed of migration being limited by eddy currents; the currents induced by the magnetization changes accompanying the boundary movement are kept at such a level by the movement that the local situation at the boundary corresponds to A. This aspect of the problem is exactly analogous to the motion of a superconducting-normal boundary (or indeed any phase boundary, e.g. liquid–solid, in the presence of a temperature gradient). If a spatially non-uniform field be caused to grow, there is a critical rate of growth at which the eddy current skin depth is comparable to the width of the domains. For slower rates the boundaries only need to move so slowly that they maintain local equilibrium in state A, for faster rates they move as well as they can subject to eddy current limitations. The mathematical problem of determining their motion is rather hard except in the simplest cases. It is still more difficult if the field lines are significantly not parallel, so that the surfaces of constant B do not coincide with field lines. The problem of domain configuration involved here has not been investigated, but obviously resembles the problem in ferromagnetics, though the very rapid variation of oscillatory phase with angle in most metals, because of Fermi surface anisotropy, adds an extra complication here. It should be noted, however, that in most experiments involving the de Haas–van Alphen effect particular care is taken to obtain a uniform field, so that domain complications should in practice be minimal.

It is clear from Fig. 3.16 that the Shoenberg effect leads to an increase in harmonic content of the oscillations, as well as to a reduction of amplitude, particularly at lower temperatures when a is larger. It was observation of these phenomena that led to the discovery of the effect, and subsequent work has confirmed the sharpened profile of the oscillations, which in a rod of Be were found to have very nearly the limiting saw-tooth form. But even clearer evidence has appeared somewhat indirectly. When two rather close frequencies are present, in the absence of interactions they beat together to produce an amplitude-modulated oscillation at the mean frequency, but under the right conditions interaction may lead to rectification (Condon, 1966) so that the difference frequency appears as a real oscillation in addition to the higher frequencies. One would not expect this to happen with the magnetization directly, for the interactions do not alter the mean level; positive and negative parts of the cycle continue to match. Thus if the amplitude is modulated the harmonic content will fluctuate at the beat frequency, but no variation of mean level can be expected at this frequency and therefore no real difference tone should appear. If, however, the phenomenon investigated depends on a property whose oscillations are in phase quadrature with I, the situation is quite different. Figure 3.16(b) shows how the discontinuity at A prevents the material from traversing a substantial part of the sine wave, the range between something less than $-\frac{1}{2}\pi$ and something over $+\frac{1}{2}\pi$ being cut out. Any property, such as the free energy F, which is in quadrature with I loses in this process the whole of its positive half-cycle and a little of its negative half-cycle, as Fig. 3.16(c) shows. If only the part below the intersection is realized, the mean value is forced down, by an amount depending on the value of a. One may then expect an amplitude-modulated signal to be distorted by interaction so that F consists of a signal at the original frequency superimposed on a signal at the modulation frequency (difference tone).

An example of this phenomenon is provided by magneto-thermal oscillations. An isolated sample oscillates in temperature to keep its entropy constant, according to the relation,

$$\left(\frac{\partial T}{\partial H_0}\right)_S = -\frac{T}{C_H}\left(\frac{\partial S}{\partial H_0}\right)_T.$$

Since S and F oscillate in phase, rectification can take place as illustrated by Fig. 3.17. The high frequency, which in the absence of interaction would reveal a degree of amplitude modulation, is almost overwhelmed by the difference tone; some of this is instrumental, reflecting the inability of the thermometers to respond fast enough, but there is no reason in principle why if a is sufficiently large, the original signal should not be reduced to an almost insignificant ripple superposed on the difference tone.

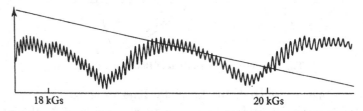

18 kGs   20 kGs

Fig. 3.17. Magneto-thermal oscillations in Be (Le Page *et al.* 1964). The rapid oscillations have the de Haas–van Alphen frequency of the 'needle' orbits, with an amplitude considerably reduced by instrumental limitations; the slow oscillations are the difference tone created by the Shoenberg effect. The sloping line records the applied field.

The foregoing arguments have been applied specifically to a sample of negligible demagnetization factor. If D is not negligible the sample splits into domains (Condon, 1966) for part of the cycle. The argument is quite simple and is exactly parallel to the argument for intermediate state formation in type 1 superconductors. The magnetic field inside the sample, H, is not the same as H_0, but we may assume that locally the relation between I and H is the same as that between I and H_0 when $D = 0$, i.e.

$$y = Y(x), \tag{3.36}$$

in which, from (3.33), $x = \lambda H = x_0 - Dy,$ (3.37)

and $Y(x)$ is the discontinuous function in Fig. 3.16(b). Now although the line AA' does not represent along its length any physically realizable states, we may attain any point on AA' as a macroscopic average by dividing the sample into domains, whose individual properties correspond to the extreme ends of AA' and whose proportion is adjusted as required by the circumstances. We may therefore regard $Y(x)$ as a continuous function and determine the magnetization as a function of external field by solving (3.36) and (3.37) graphically, as in Fig. 3.18. To convert $Y(x)$ into $y(x_0)$, choose any point x_0 on the x-axis and draw a line of gradient $-1/D$ to cut $Y(x)$ at a value y, which is the required $y(x_0)$. The straight line portions of $y(x_0)$ are where domains are formed. In the limit of large a $Y(x)$ consists of straight lines of unit negative slope joined by vertical lines, and in $y(x)_0$ the former are replaced by

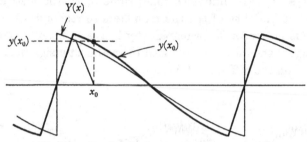

Fig. 3.18. Graphical solution of (3.36) and (3.37).

lines of slope $-(1-D)^{-1}$, the latter by lines of slope D^{-1}. Thus in a flat plate normal to $H(D = 1)$ the saw-tooth is reversed. Nothing has been said about the scale of the domain structure, which, as in the superconductor, must be determined by the surface energy of a domain boundary. The boundary must have a thickness comparable with an orbit radius, but there is as yet no analysis of energy levels in a spatially varying field to enable the details to be worked out. As in superconductors and ferromagnetics the behaviour near the surface of the sample is a complicated problem, and may be still more complicated here on account of the great sensitivity of the magnetic properties to angular variations.

3.10 TRANSPORT (see Chapter 4)

Up to this point we have considered only such properties as may be derived from the energy-level structure of the particles. The dynamical properties of the assembly raise a whole new set of problems, adequate analysis of which would take us too far afield. We shall therefore concentrate on one topic alone, electrical conductivity in a magnetic field, and even here do no more than classify and illustrate the varieties of behaviour that would need detailed discussion in any complete survey. One reason why we can afford to dismiss the matter briefly is that peculiarly quantal effects, which are the principal topic of this chapter, may usually be included in transport theory by an elementary extension of the usual semi-classical method (e.g. Boltzmann equation), while the rest demand such powerful methods of analysis as to rule out all but brief mention here. Thus in a reasonably pure metal the transport problem may be dissected into two parts, the motion of charge under the influence of an electric field in a perfect lattice, and the scattering which ultimately determines the state of dynamical equilibrium. So far as the first part is concerned, the fact that quantized orbits can be represented as so nearly conforming to classical orbits, with the energy levels separated by $\hbar\omega_c$, ensures that wave-packets move very nearly in classical trajectories at the classical speed. As for the scattering, the adequacy of the w.k.b. approximation under most circumstances, when the orbital quantum number is large, guarantees that scattering centres which are small in dimension will interact with a wave-packet without noticing that the basis functions are not Bloch functions. The same holds with phonon scattering, and it is normally true that the electron may be treated as if passing steadily from one Bloch state to another under the influence of the magnetic field, and responding to other perturbations in the manner appropriate to its momentary condition. There are of course exceptions, notably when the field is so strong that all the electrons are in the lowest orbital state; this Quantum Limit, which has been fully discussed elsewhere (Kubo *et al.* 1965) is normally unattainable in metals and we shall say no more of it. Other exceptions will be treated in due course, particularly the phenomena due to magnetic breakdown which are only imperfectly susceptible to classical methods.

As far as possible, then, we shall describe the effects classically, and for this purpose the idea of Effective Path (Pippard, 1965a) serves as a unifying concept. This is no more than one way of writing the solution of the transport problem without use of the Boltzmann equation, and is especially valuable in situations where a relaxation time cannot be defined. It is a time-reversed modification of the Chambers (1952a) path-integral method. A steady uniform electric field, by accelerating the electrons and thereby bodily shifting the distribution in k-space, may be thought to create new particles (or holes) all round the Fermi surface; thus the rate of change of occupation number due to **E** may be written

$$\frac{\partial f(k)}{\partial t} = -\dot{\mathbf{k}} \cdot \nabla_k f = -\frac{e}{\hbar}\mathbf{E} \cdot \nabla_k f, \qquad (3.38)$$

METALLIC ELECTRONS AND MAGNETISM 163

and $\partial f/\partial t$ may be interpreted as the local creation rate. Consider now the new electrons created at a certain point in real space, moving away under the influence of a magnetic field and being scattered. After a while their motion will have become randomized in the sense that their centre of mass no longer moves, but has reached its ultimate position at some distance $\bar{L}$ from the point of creation. At this stage the current in the metal due to these particles is nil, and they may be removed from the assembly without disturbance.

We therefore conceive of a steady state in which the particles are removed as fast as they are created, but on the average a distance $\bar{L}$ away from the point of creation. If the rate of creation and removal is N per unit volume, the fact that this operation is necessary in order to maintain the steady state implies that the momentum density of the assembly is $Nm\bar{L}$, and hence $Ne\bar{L}$ is the current density; the holes make an identical contribution. Now it is readily seen from (3.38) that an element $d\mathbf{S}$ of the Fermi surface creates particles at a rate $dN = e\mathbf{E}.d\mathbf{S}/(4\pi^3\hbar)$, and if these particles have an effective path $\mathbf{L}$, varying round the Fermi surface, we may write

$$ \mathbf{J} = e \int \mathbf{L}\, dN = \frac{e^2}{4\pi^3\hbar} \oint \mathbf{L}(\mathbf{E}.d\mathbf{S}). \tag{3.39} $$

We now analyse the influence of a magnetic field on the conductivity in terms of its effect on $\mathbf{L}$, and it is convenient to distinguish the following categories, depending on different principal effects of $\mathbf{H}$:

Classical effects:

(a) Curvature of trajectories (Pippard, 1965b, p. 90; Fawcett, 1964).

(b) Variations of scattering probability consequent upon motion round the Fermi surface (Pippard, 1965b, p. 99).

(c) A mixture of (a) and (b) (Pippard, 1964b).

Quantal effects:

(d) Oscillatory variations of scattering probability, linked to de Haas–van Alphen oscillations (Pippard, 1960).

(e) Topological changes in trajectories due to magnetic breakdown (Falicov et al. 1966); added to which certain transport phenomena in the presence of breakdown which are not susceptible to classical analysis in terms of effective path (Pippard, 1964a).

Let us take these categories in turn, giving typical examples rather

164　　　　　　A. B. PIPPARD

than a complete discussion. The references given above should lead the interested reader into the extensive literature.

(a) In a free-electron gas, having constant relaxation time, the trajectories are changed from straight lines (up to the moment of collision) to circular helices. The z-component of velocity, parallel to H, is unchanged, and so therefore are $\bar{L}_z$ and the longitudinal component σ_{zz} of the conductivity tensor. In the plane normal to H the trajectories are curved, and $\bar{L}$ is readily shown to lie on a semicircle as in Fig. 3.19, with $\tan\phi = \omega_c \tau$.

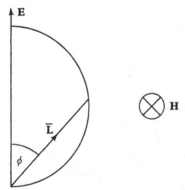

Fig. 3.19. Variation of mean effective path $\bar{L}$ with magnetic field for free-electron metal.

What is usually measured is not the conductivity σ_{ij} but the resistivity ρ_{ij}; for example, the potential developed along a rod when current is passed along it is determined by the diagonal components of ρ_{ij} (ρ_{zz} for longitudinal magneto-resistance, ρ_{xx} and ρ_{yy} for transverse magneto-resistance). In a metal having more than twofold symmetry about the axis of H, ρ_{ij} is isotropic in the plane normal to H and the relation between σ_{ij} and ρ_{ij} is particularly clear. Thus Fig. 3.19 shows that the component of $\bar{L}$ parallel to E steadily decreases as H and $\omega_c \tau$ increase, so that σ_{xx} varies as $\cos^2\phi$ or $(1+\omega_c^2\tau^2)^{-1}$; but the component of E parallel to $\bar{L}$ remains exactly proportional to $\bar{L}$ at all fields, and ρ_{xx} is thus field-independent. Where the field shows its influence is on the non-diagonal components, such as ρ_{xy} which describes the Hall effect.

It is useful to remember that when the x–y-plane is isotropic, circles with diameters on the E-vector, and which pass through the origin, represent trajectories of constant transverse magneto-resistance. In metals whose Fermi surfaces are closed but not spherical, the trajectory of $\bar{L}$ describes a curve which is not a semicircle, but which finishes, as $\omega_c\tau \to \infty$, normal to E at the origin. Its curvature defines the circle to which it is asymptotic, and so the saturation value of ρ_{xx} at high fields. If scattering is independent of H, the transverse magneto-resistance is always positive, i.e. the trajectory ends on a circle of smaller radius than that from which it started, as in Fig. 3.20 which illustrates a case where the saturation resistance is three times the zero field resistance.

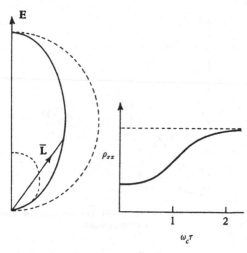

Fig. 3.20. Variation of mean effective path $\bar{L}$ with magnetic field for a metal with closed non-spherical Fermi surface, and corresponding magneto-resistance curve.

If the Fermi surface is not closed, and the direction of H is such as to permit open orbits, the electron assembly may be divided into two parts, the closed-orbit assembly which behaves as in Fig. 3.20, and the open-orbit assembly which at high fields can carry current only in one particular direction in the transverse plane, the direction in which the mean velocity of the open orbit lies. The open-orbit conductivity becomes field-independent as $H \to \infty$. The transverse plane is no longer isotropic, but a simple diagram shows what can be expected of the resistance in the high-field limit. In Fig. 3.21 the required current direction, determined by the crystallographic orientation of the specimen rod, is vertical and does not coincide with the open-orbit direction which is at an angle θ to it. Since in a strong field the open-orbit conductivity does not fall to zero like the closed-orbit conductivity, the former becomes dominant, and to get a current to pass in the required direction it is necessary to suppress the open orbit almost completely. Let us start by putting **E** normal to the open orbit, so that it sets up a small current $\mathbf{J}_c$ in the closed-orbit assembly. In order to bring **J** vertical, a slight tilt of **E** is required to bring in some open-orbit current $\mathbf{J}_0$, but it is clear that the resultant current **J** is now very small indeed, so that the open orbit has the effect of greatly increasing the resistivity. If the closed orbits, acting alone, had a saturation resistivity r times the zero-field resistivity ρ_0, $\mathbf{J}_c$ has a component along E of magnitude $E\cos^2\phi/(r\rho_0)$, and the resultant current

$$J = E\cos^2\phi/(r\rho_0\sin\theta).$$

Hence

$$\frac{\rho_{xx}}{\rho_0} = r\sin\theta\sec^2\phi = r(1+\omega_c^2\tau^2)\sin\theta.$$

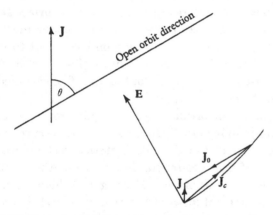

Fig. 3.21. To illustrate magneto-resistance in the presence of open orbits.

The resistance tends to infinity as H^2 unless the open-orbit direction coincides with the specimen axis.

Another important case in which the resistance tends to infinity occurs when there are equal numbers of electrons and holes, and no open orbits. This is a not uncommon situation in metals of even valency which in these circumstances are called 'compensated'. The effective paths for electrons and holes are swung away from the direction of **E** in opposite senses, and in such fashion that when the numbers are equal the resultant **L** has a trajectory which is not asymptotically normal to **E**, and therefore ρ_{xx} does not saturate. If there is any difference between numbers of electrons and holes, however small,

ρ_{xx} should ultimately saturate, but it may be only in very strong fields and at a very high value of resistivity.

(b) If τ varies considerably from one point to another round the Fermi surface, the effect of H is to allow electrons to sample different scattering probabilities during their paths. For a spherical Fermi surface this may provide the only mechanism for magneto-resistance, and may indeed be the origin of the weak magneto-resistance in Na and K (MacDonald, 1956; Justi & Auch, 1963), whose Fermi surface is very nearly spherical. In this case, however, much doubt remains, and it is possible that the scattering probabilities are themselves field-dependent, or some as yet unthought-of process is at work.

(c) If scattering is isotropic, so that each event terminates a free path, processes (a) and (b) should describe the situation fairly well. But with small-angle scattering there may appear new effects resulting from the need for many scattering processes before randomization is complete. An electron wandering over the Fermi surface as a result of small-angle scattering may move from one orbit to another, and its final destination may depend strongly on the extent to which magnetic field and scattering combine to determine its path. The process has been analysed to some extent for the case of longitudinal magneto-resistance in the noble metals, whose Fermi surfaces, being multiply connected, afford opportunity for changing from electron to hole orbits, and for jumping rapidly from one part of the surface to another by Bragg reflection and small-angle scattering in close succession. The whole pattern of magneto-resistance may be markedly altered by a change in the character of the scattering, and there is some evidence to support the general ideas of the theory (Pippard, 1964b) which has, however, not yet been worked out in sufficient detail, but which may account for the slower-than-expected approach to saturation in many metals.

(d) Even though the coupling of electrons to scattering agents may be field-independent, the probability of scattering is proportional to the density of states available for scattering into, and as this oscillates at the de Haas–van Alphen frequency, so the relaxation time oscillates, and $\overline{L}$ with it. This is one of the principal mechanisms of resistance oscillations (de Haas–Schubnikov effect). The magnitude of the effect may be estimated, perhaps not quite precisely, by a very simple argument. The relative magnitude of the oscillations of τ is the same as that of the density of states $\mathcal{N}_0$ and of the entropy (since $S \propto \mathcal{N}_0$ at a given temperature); the last can be related thermodynamically to oscillations of moment, so that with very little manipulation we can write for $\Delta\tau$, the amplitude of oscillations of τ,

$$\frac{\Delta\tau}{\tau} = \frac{2\pi m_c^2}{e^2 h^2 \mathcal{N}_0} \Delta I \, \delta(1/H) \, F(T),$$

in which $\mathcal{N}_0$ is the mean density of states per unit volume, ΔI is the amplitude of de Haas–van Alphen oscillations, $\delta(1/H)$ is the periodicity of the oscillations in $1/H$, and $F(T)$ is the temperature factor $-(3/X)(d/dX)(X/\sinh X)$, X being defined by (3.26); $F(0) = 1$. This formula is perhaps not very transparent, but it may be noted that when applied to a free-electron metal at $0\,°\text{K}$ it yields oscillations of relative amplitude $(2n)^{-\frac{1}{2}}$. For metals with large Fermi surfaces, this means one cannot expect oscillations of more than a few percent even at very high field strengths, but in semimetals the effect may be strong, since n may be quite low.

When the de Haas–van Alphen amplitude is large enough to bring in the Shoenberg effect, the beating together of high-frequency oscillations may be rectified to produce a lower frequency. We have seen (§ 3.9) that S is one of the functions that may behave in this way, and it follows that $\mathcal{N}_0$ and τ will also do so. This may be the explanation of the low-frequency magneto-resistance oscillations (Reed, 1964) in Be which also exhibits a strong Shoenberg effect with beats.

(e) When magnetic breakdown occurs a particle may switch from one orbit to another, and in this way execute a path quite different from what would have happened without breakdown. A full analysis of this problem is extremely difficult, but certain special cases have been treated with success. Experimentally the effect of breakdown is most dramatic when the metal is compensated, so that in the absence of breakdown ρ_{xx} would rise without limit. In Mg and Zn, with H along the hexad axis, this is what happens in low fields (< 2 kGs), as can be seen from Fig. 3.22. But at higher fields the quadratic increase ceases, and the resistance oscillates vigorously about a steady saturation level. This is readily understood qualitatively if one remembers that in the central regions of the Fermi surface the electron orbits form a hexagonal net as in Fig. 3.10(c). In low fields they traverse hexagonal hole orbits and small triangular electron orbits, but in higher fields breakdown compels them all into circular free-electron orbits. The number of electrons in hole orbits is reduced at the expense of those in electron orbits, and the low-field compensation is broken, with the consequence that ρ_{xx} saturates, as described in (a).

The non-oscillatory part of the phenomenon can be accounted for (Falicov & Sievert, 1964) in a straightforward way if we allow the possibility of choice of path for a particle at each junction of the network, and regard it as a classical particle undergoing random walk on the net. Since this problem brings out the strength of the effective

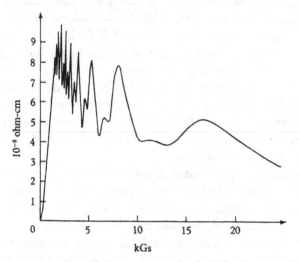

Fig. 3.22. Transverse magneto-resistance of Zn (Stark, 1964) with current along the hexad axis.

path analysis we shall work through the simplest example as an illustration. The quantity to be determined is the mean distance travelled by a particle starting from a given point on the net, and we shall consider only a linear chain, as in Fig. 3.10(a), with no collisions. Because of this last, all particles starting on AB (Fig. 3.23) end up on the average at the same point X, OX being represented by a complex number X, which is to be determined. Now as a result of division at the junction, a particle entering the lens at B has a chance R of leaving along BC and $(1 - R)$ of leaving along DE, R being determined by the junction conditions. In the former case it must have a mean end-point $X + S$, in the latter $-X$. Hence, for consistency,

$$X = R(X+S) - (1-R)X,$$

or $$X = \tfrac{1}{2}RS/(1-R).$$

(3.40)

In this elementary way the effective path may be found for particles created at any point on the net, and hence L and the conductivity are

determined; the method can be readily extended to more complicated nets. We have not yet discussed the value of R, but in the strictly
classical model this follows immediately from the junction coefficients
P and Q. By following the particle round the lens we may deduce its
chances of leaving by the alternative routes, and find $R = 2Q/(1+Q)$,
so that, from (3.40)

$$X = QS/P. \qquad (3.41)$$

This clearly has the right sort of behaviour. When breakdown is
complete ($Q = 0$, $P = 1$), $X = 0$ and the particle moves in a circular
orbit whose centre is O; when there is no breakdown $X \to \infty$, the particle moving unchecked along an open orbit in the absence of collisions.

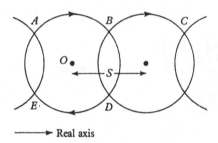

Fig. 3.23. To illustrate calculation of effective path on a
linear chain of orbits.

When this argument is applied to the hexagonal net and the parameters chosen as well as possible, the general lie of the curve shown
in Fig. 3.24 clearly corresponds well with Fig. 3.22.

To understand the oscillations one must recognize that their
period is the same as that of the slow de Haas–van Alphen oscillations
in zinc, which are attributed to the 'needle' part of the Fermi surface,
i.e. that part whose section is one of the small triangles in Fig. 3.10(c).
Clearly then, it must be necessary to take account of the quantization
of these small orbits by incorporating into the theory the phase
coherence of waves on the network (Pippard, 1965a). To return to the
linear chain, we might as a compromise between classical and quantal
theories suppose that the lens orbits are all identical but that the larger
orbits are totally randomized by lattice defects; it is not unreasonable
then to calculate R for the lens orbit on the assumption of phase
coherence, but otherwise to treat the net completely classically. But
it should be emphasized that this procedure has no rigorous justifica-

tion. Nevertheless, if we proceed hopefully we readily find that R is now an oscillatory function of energy

$$R = \frac{2Q(1-\cos\phi)}{1+Q^2-2Q\cos\phi},$$

and, from (3.40),

$$X = \frac{2QS}{P^2}\sin^2\tfrac{1}{2}\phi,$$

in which ϕ is the phase length round the lens orbit. At not too high fields, when P is small, the conductivity oscillates with very large amplitude, between zero and $2/P$ times the 'classical' value (3.41). The same analysis may be applied to the hexagonal net, triangles being treated as phase-coherent, but all else as random. As the energy

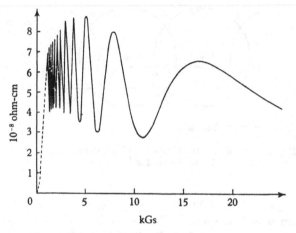

Fig. 3.24. Theoretical curve (Falicov *et al.* 1966) corresponding to Fig. 3.22.

changes, the probabilities of the electron leaving a triangular orbit by different paths oscillate in such a way as to favour now electron orbits, now hole orbits, and in this manner the degree of compensation of electrons and holes fluctuates periodically. Figure 3.24 shows how this reflects itself in the resistivity, and comparison with Fig. 3.22 indicates a very fair measure of agreement. Much of the discrepancy is removed by incorporating the effects of electron spin (§3.4); in this orbit $g_{\text{eff}} \sim 90$.

The success of the theory given here is rather fortuitous, depending on a clear distinction between small and large orbits in Zn, with the latter large enough to be seriously disturbed by a moderate concentration of dislocations. A less random network presents a formidable problem, since it represents a transition situation between two clear-cut limits, pure classical on the one hand and pure quantal on the other. Let us therefore look briefly at the quantal extreme, exemplified by a network in a perfect crystal. When the magnetic field takes one of the special values that make J integral the energy bands are at their broadest, and resemble the bands of Bloch electrons. The resemblance is indeed close, for it can be proved (Harper, 1955; Pippard, 1964a), that each eigenstate represents a 'quasi-particle' moving with velocity $\hbar^{-1}\nabla_{\kappa}\mathscr{E}$ (κ is defined in §3.5), and in the broadest bands this velocity is comparable with the electronic velocity in zero magnetic field. The result of magnetic breakdown is to create a situation in which electrons can pass through the metal in straight lines in any direction, in spite of the presence of the field. Clearly this can have a profound effect on the conductivity which is otherwise severely limited by the curling-up of paths into closed orbits.

However, the quasi-particle conductivity is obviously extraordinarily sensitive to lattice defects (Pippard, 1965a). A single dislocation disrupts the phase relationship of the orbits close at hand, and must act as a scattering centre whose dimension is of the order of the orbit lattice spacing rather than atomic spacing. A sparse distribution of defects may well be supposed to act independently, just like impurities in normal conduction processes, but a distribution almost dense enough to destroy all phase coherence produces a situation more analogous to an alloy or even a liquid metal, and very deep analysis will be needed to interpolate between the two limits in this regime of the behaviour.

An additional complication arises when J is not integral, for then the energy bands of a network split and become much narrower, so that $\nabla_{\kappa}\mathscr{E}$ and the quasi-particle velocity are much reduced. This behaviour is very closely parallel to the condensation of the continuum of an electron gas into sharp degenerate levels under the influence of a magnetic field, as the following argument shows. If the magnetic field is such that J is slightly different from an integer, $J = J_0 - 1/J_1$, J_0 and J_1 being integers, according to (3.4) we have that

$$s = s_0 + \Delta s, \left.\vphantom{\begin{array}{c}a\\b\end{array}}\right\}$$

$$\text{where} \quad s_0 = 2\pi/(J_0\Sigma) \quad \text{and} \quad \Delta s \doteqdot 2\pi/(J_0^2 J_1\Sigma). \tag{3.42}$$

Let us imagine a gas of quasi-particles established by the main field s_0, and perturbed by the weak additional field Δs, which should condense the energy bands in the usual way into narrow levels, each of degeneracy $\Delta s/2\pi$ per unit area. Now the Brillouin zone in κ-space contains $1/\Sigma J_0^2$ states per unit area, so that from (3.42) the band within one zone is divided into J_1 sub-bands. This is just the conclusion

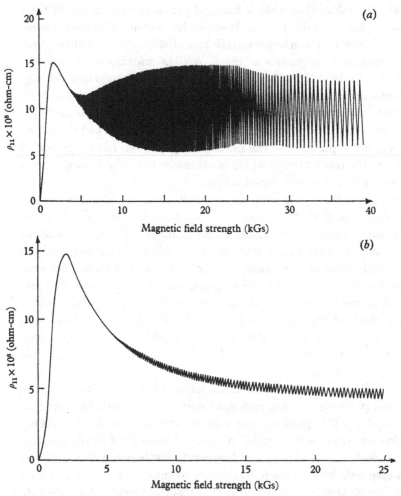

Fig. 3.25. Transverse magneto-resistance (Falicov & Stark, 1967) of one specimen of Mg to show loss of oscillatory amplitude due to thermal cycling: (a) a new sample, (b) after one thermal cycle.

reached in § 3.5 The sub-bands are not perfectly sharp levels, as might possibly have been expected, so one must not take the argument too far, but it does at least give a physically intelligible picture of the quasi-particles (or wave-packets constructed from network functions)

N

bent into large hyper-orbits and suffering additional quantum restrictions thereby. If the lattice is so regular that the hyper-orbits can be traversed without much scattering ($\omega_c \tau > J_1$, ω_c being the normal cyclotron frequency and τ the relaxation time for quasi-particles), the conductivity is reduced by the curvature of the path in the same way as shown in Fig. 3.19. But if $\omega_c \tau$ is not greater than J_1, the additional field Δs hardly influences the conductivity; there is then a band of field values around each s within which the quasi-particle conductivity is close to the maximum value. In any case if, as is likely, the quasi-particles dominate the transport process, the argument of (a) suggests that the resistivity will not depend on J, whatever may happen to the Hall coefficient.

Experimentally there is very little systematic evidence for the existence of network quasi-particles, but Fig. 3.25 is well worth looking at, for it shows how the oscillatory component of magneto-resistance in Mg may be much larger than can be accounted for without quasi-particles, when the sample is prepared unusually free of defects; however, a single warming to room temperature and re-cooling seems to be enough to remove all phase coherence in the network apart from that within the smallest orbits. This example should be enough to indicate that there is scope here, as elsewhere in the topics covered by this chapter, for careful experiment and still deeper theoretical analysis.

CHAPTER 4

TRANSPORT PROPERTIES: SURFACE AND SIZE EFFECTS

by R. G. CHAMBERS†

4.1 INTRODUCTION

The d.c. transport properties of a bulk sample of metal are very simple to measure experimentally, but the results tell us rather little about the detailed behaviour of the conduction electrons in the metal. For example, if we assume that collisions can be represented by a relaxation time τ_k, the d.c. conductivity σ_0 of a cubic metal in zero magnetic field can be written

$$\sigma_0 = \frac{e^2}{12\pi^3\hbar} \int_{FS} l_k dS_F = \frac{e^2}{12\pi^3\hbar} S_F \bar{l} \qquad (4.1)$$

(cf. Ziman, 1964a, (7.25)), where $l_k = \tau_k v_k$ is the mean free path of an electron of velocity v_k, and the integration extends over the area S_F of the Fermi surface (FS) in k-space. The value of σ_0 at any temperature is thus determined by the total FS area S_F and by the average mean free path $\bar{l}$ over the FS. It is not possible to deduce the value of either of these quantities separately; still less is it possible to deduce the detailed shape of the FS, or the detailed variation of τ_k and v_k over it.

To learn more it is necessary to modify the experimental conditions so that (4.1) is no longer valid. This can be done in several ways; for instance, by applying a magnetic field, or by using a thin film or wire instead of a bulk sample, or by working at a high frequency instead of at d.c. In each case, the effect is to introduce a new length into the problem, in addition to the mean free path $\bar{l}$—either the radius r of a typical electron orbit in the magnetic field, or the thickness d of the wire or film, or the skin depth δ of high-frequency field penetration—and the departure from (4.1) will become marked when r, d or δ becomes comparable with or less than $\bar{l}$. The effect of a magnetic field on the d.c. conductivity of bulk samples is discussed in §3.10;

† Dr Chambers is Professor of Physics at the H. H. Wills Physics Laboratory, University of Bristol, Tyndall Avenue, Bristol, 2.

here we shall be concerned with size effects and high-frequency effects, and in particular with the theory of these effects. Their application to FS determination is discussed in §2.2. Much of the theory has been excellently reviewed by Pippard (1960, 1965 b), and we shall try to avoid too great an overlap with his reviews.

Size-effect studies have been carried out for many years: the first paper on the subject was published by a student of Michelson's, Miss Isabelle Stone, who reported in 1898 that very thin metal films showed a higher apparent resistivity than the bulk metal. Patterson (1901) reported similar observations, and at the same time an approximate theory of the effect was given by Thomson (1901). But the early experimental work, carried out on thin films evaporated or sputtered under ill-controlled conditions, was of doubtful validity, and perhaps the first reliable observations were those of Lovell (1936, 1938), carried out in Mott's laboratory in Bristol. These observations led Fuchs (1938), working in the same laboratory, to produce the first detailed theory of the effect. A year or two later London (1940), also working in Bristol, first reported the apparent breakdown of equation (4.1) at low temperatures and high frequencies, and correctly ascribed the resultant 'anomalous skin effect' to the mean free path becoming large compared with the skin depth.

The experimental situation has been transformed in the past five or ten years by the development of new methods of purifying materials. It is now possible to produce some metals in such pure form that at low temperatures the mean free path rises to 1 mm or more. This means that size effects can be detected in samples of macroscopic dimensions, so that instead of working with polycrystalline thin films of uncertain structure, the experimentalist can use macroscopic single crystals grown under well-controlled conditions, and so learn a great deal more about the details of the electronic behaviour.

4.2 INTERNAL SCATTERING AND SURFACE SCATTERING

Under either size-effect or anomalous skin-effect conditions, the current (d.c. or r.f.) is confined to a region very close to the metal surface, so that in discussing the theory of these effects we must take into account collisions of electrons with the surface as well as collisions with phonons, imperfections, etc., in the interior of the metal. We start by reviewing briefly the theory of internal scattering, as far as it concerns us here. We are not concerned with evaluating scattering probabilities *ab initio*—this problem is fully discussed in Chapter 5— but with the formal description of scattering effects in transport theory. From this point of view, the quantities which describe the scattering—relaxation times, mean free paths, etc.—are regarded as parameters in the theory, to be determined either by experiment or by more detailed calculation.

In the simple case of a bulk sample of metal (i.e. large enough for surface effects to be negligible), subjected to a steady uniform electric field E and magnetic field H, the Boltzmann equation reduces to

$$e\mathbf{E}\cdot\mathbf{v_k}\frac{\partial f_k^0}{\partial\mathscr{E}}+\frac{e}{\hbar}(\mathbf{v_k}\times\mathbf{H})\cdot\frac{\partial f_k}{\partial\mathbf{k}}=\left(\frac{\partial f_k}{\partial t}\right)_{\text{coll}},\tag{4.2}$$

where

$$\left(\frac{\partial f_k}{\partial t}\right)_{\text{coll}}=\int\{f_{k'}(1-f_k)Q(\mathbf{k'},\mathbf{k})-f_k(1-f_{k'})Q(\mathbf{k},\mathbf{k'})\}\,d\mathbf{k'}.\tag{4.3}$$

Here $Q(\mathbf{k'},\mathbf{k})$ represents, as usual, the probability per unit time that collisions will cause a transition from state $\mathbf{k'}$ to state $\mathbf{k}$ if $\mathbf{k'}$ is full and $\mathbf{k}$ is empty. It is possible to solve this equation by iteration, and to express the result compactly in terms of a vector mean free path $\mathbf{L}_{E,\mathbf{k}}$:

$$g_k=f_k-f_k^0=-e\mathbf{E}\cdot\mathbf{L}_{E,\mathbf{k}}\frac{\partial f_k^0}{\partial\mathscr{E}}\tag{4.4}$$

(Price, 1957, 1958; Taylor, 1963 b). The vector $\mathbf{L}_{E,\mathbf{k}}$ has a direct physical significance: it is the average distance travelled by an electron before passing through a given point in the metal with velocity $\mathbf{v_k}$ (Price, 1957). Thus if $\langle\mathbf{v}(t)\rangle_\mathbf{k}$ is the average velocity at time t of an assembly of electrons all of velocity $\mathbf{v_k}$ at $t=0$, then

$$\mathbf{L}_{E,\mathbf{k}}=\int_{-\infty}^{0}\langle\mathbf{v}(t)\rangle_\mathbf{k}\,dt.\tag{4.5}$$

In computing $\langle\mathbf{v}(t)\rangle_\mathbf{k}$, the effect of the applied field E on the velocity of the electrons can be neglected; only the effect of collisions, and of applied magnetic fields, if any, need be taken into account.

Just as collisions occurring after time $t=0$ will tend to randomize the velocities of the electrons, so that $\langle\mathbf{v}(t)\rangle_\mathbf{k}\to 0$ as $t\to\infty$, so their velocities will become increasingly randomized if we trace their history backwards in time before $t=0$, so that $\langle\mathbf{v}(t)\rangle_\mathbf{k}\to 0$ as $t\to-\infty$ also. In fact the vector describing their subsequent motion,

$$\mathbf{L}_{E,\mathbf{k}}^{\pm}=\int_{0}^{\infty}\langle\mathbf{v}(t)\rangle_\mathbf{k}\,dt,\tag{4.6}$$

will be equal to $L_{E,\mathbf{k}}$ in the absence of a magnetic field, and if the electrons are moving in a steady field $\mathbf{H}$, it is not difficult to show that $L_{E,\mathbf{k}}(\mathbf{H}) = L_{E,\mathbf{k}}^{+}(-\mathbf{H})$.

The quantity $e\mathbf{E}.L_{E,\mathbf{k}}$ in (4.4) is thus equal to the average energy $\Delta\mathscr{E}_{\mathbf{k}}$ acquired by an electron from the field $\mathbf{E}$ before the time $t = 0$, so that (4.4) can be rewritten in the form

$$f_{\mathbf{k}} = f_{\mathbf{k}}^{0} - \Delta\mathscr{E}_{\mathbf{k}}\,\partial f_{\mathbf{k}}^{0}/\partial\mathscr{E} = f^{0}(\mathscr{E}_{\mathbf{k}} - \Delta\mathscr{E}_{\mathbf{k}}). \tag{4.7}$$

In other words, the applied field effectively perturbs the distribution function f by supplying an average energy $\Delta\mathscr{E}$ to the electrons in the unperturbed system. Note that in general there is no reason why $L_{E,\mathbf{k}}$ should be parallel to $\mathbf{v}_{\mathbf{k}}$ (even in the absence of a magnetic field), so that in general we cannot define a simple scalar relaxation time by writing $L_{E,\mathbf{k}} = \tau_{\mathbf{k}}\mathbf{v}_{\mathbf{k}}$.

Since the electric current density is given by

$$\mathbf{J} = \frac{e}{4\pi^{3}}\int g_{\mathbf{k}}\mathbf{v}_{\mathbf{k}}d\mathbf{k}, \tag{4.8}$$

it follows from (4.4) that the conductivity tensor σ_{ij} is given by

$$\sigma_{ij} = -\frac{e^{2}}{4\pi^{3}}\int v_{i}L_{E,j}\frac{\partial f^{0}}{\partial\mathscr{E}}d\mathbf{k} \simeq \frac{e^{2}}{4\pi^{3}\hbar}\int_{FS} v_{i}L_{E,j}\,dS_{F}/v. \tag{4.9}$$

In the second form of (4.9), we have used the approximation

$$-\int F(\mathbf{k})\frac{\partial f^{0}}{\partial\mathscr{E}}d\mathbf{k} = -\int_{FS} F(\mathbf{k})\frac{\partial f^{0}}{\partial\mathscr{E}}\frac{1}{\hbar v_{\mathbf{k}}}d\mathscr{E}\,dS_{F} \simeq \int_{FS} F(\mathbf{k})\,dS_{F}/\hbar v_{\mathbf{k}}$$

to transform the integral over all $\mathbf{k}$ to an integral over the Fermi surface S_{F}.

A slightly different approach (Lax, 1958; Pippard, 1964 b), described in §3.10, yields an alternative expression for σ_{ij} in terms of the mean free path $L_{E,\mathbf{k}}^{+}$ of (4.6):

$$\sigma_{ij} = -\frac{e^{2}}{4\pi^{3}}\int L_{E,i}^{+}v_{j}\frac{\partial f^{0}}{\partial\mathscr{E}}d\mathbf{k} \simeq \frac{e^{2}}{4\pi^{3}\hbar}\int_{FS} L_{E,i}^{+}v_{j}\,dS_{F}/v. \tag{4.10}$$

These two expressions for σ_{ij} are in fact equivalent, since Onsager thermodynamics tells us that $\sigma_{ij}(\mathbf{H}) = \sigma_{ji}(-\mathbf{H})$, and we have seen that $L_{E}(\mathbf{H}) = L_{E}^{+}(-\mathbf{H})$. It is worth noting that equations (4.9) and (4.10), with L_{E} and L_{E}^{+} defined by (4.5) and (4.6), are simply special cases of the very general Kubo formula for the conductivity: see, for example, Chester & Thellung (1959).

If the metal is subjected to a uniform temperature gradient instead of a uniform electric field, equation (4.4) is replaced by

$$g_{\mathbf{k}} = \eta_{\mathbf{k}} k_B (\nabla T) . \mathbf{L}_{T,\mathbf{k}} \frac{\partial f^0}{\partial \mathscr{E}}, \tag{4.11}$$

where k_B is Boltzmann's constant and $\eta_{\mathbf{k}} = (\mathscr{E}_{\mathbf{k}} - \mathscr{E}_F)/k_B T$. The vector mean free path $\mathbf{L}_{T,\mathbf{k}}$ is now defined by

$$\eta_{\mathbf{k}} \mathbf{L}_{T,\mathbf{k}} = \int_{-\infty}^0 \langle \eta(t) \, \mathbf{v}(t) \rangle_{\mathbf{k}} dt, \tag{4.12}$$

and thus depends on the rate at which the energy of the electron is randomized by collisions—that is, the rate at which $\langle \eta(t) \rangle$ tends to zero—as well as the rate at which the velocity is randomized. In general, therefore, we expect L_T to be smaller than L_E. In particular, phonon scattering at low temperatures rapidly randomizes η, because the electron energy changes by an amount of order $k_B T$ at each collision, but only slowly randomizes $\mathbf{v}$, because each collision can change $\mathbf{k}$ and hence $\mathbf{v}$ by only a small amount. The distance L_E may therefore be many times the distance between individual scattering events, whereas L_T will not be much greater than this distance.

Now the mean d.c. electrical conductivity of any metal (neglecting size effects) can be written

$$\sigma_0 = \tfrac{1}{3}(\sigma_{xx} + \sigma_{yy} + \sigma_{zz}) = \frac{e^2}{12\pi^3\hbar} S_F \overline{L_{E,\parallel}},$$

and the mean thermal conductivity can likewise be written

$$\kappa_0 = \tfrac{1}{3}(\kappa_{xx} + \kappa_{yy} + \kappa_{zz}) = \frac{\pi^2 k_B^2 T}{3} \frac{1}{12\pi^3\hbar} S_F \overline{L_{T,\parallel}},$$

where $\overline{L_{\parallel}} = \overline{\mathbf{v}.\mathbf{L}/v}$, averaged over the whole FS. Under these conditions, therefore, the Lorentz ratio $\mathscr{L}(T) = \kappa_0/\sigma_0 T$ may become much less than the 'ideal' value $\mathscr{L}_0 = \pi^2 k_B^2/3e^2$ predicted by the Wiedemann–Franz law, which holds only if $\overline{L_{E,\parallel}} = \overline{L_{T,\parallel}}$. On the other hand, scattering by static imperfections is essentially an elastic process, which leaves η unchanged, and at high temperatures phonon scattering can also be treated as approximately elastic, in the sense that $\langle \eta \rangle$ decays much more slowly than $\langle \mathbf{v} \rangle$. Thus both at high temperatures $(T > \theta_D)$ and in the residual resistance region, where phonon scattering is negligible, we shall expect $L_E \simeq L_T$, and the Wiedemann–Franz law should hold.

The concept of a vector mean free path thus makes it possible to discuss in a very compact fashion the exact solution of the Boltzmann equation for the simple case of bulk metal subjected to steady uniform applied fields or temperature gradients. Unfortunately it is difficult to extend the concept usefully to problems involving surface scattering, or to problems involving non-uniform fields. It is possible to write down a formal expression for $\Delta\mathscr{E}$, but the expression is too complicated to be very useful. If the field is non-uniform, for instance, the average energy $\Delta\mathscr{E}$ acquired from the field is no longer $e\mathbf{E}.\mathbf{L}_{E,\mathbf{k}}$, so that $g_{\mathbf{k}}$ is no longer given by (4.4). One might assume that for electrons passing through point $\mathbf{r}$ at time t, $\Delta\mathscr{E}$ would be given by

$$\Delta\mathscr{E}_{\mathbf{k}}(\mathbf{r},t) = e\int_{-\infty}^{t} \mathbf{E}(\mathbf{r}',t').\langle\mathbf{v}(t')\rangle_{\mathbf{k}}dt', \qquad (4.13)$$

where $\mathbf{E}(\mathbf{r}',t')$ is the field existing at time t' at the point

$$\mathbf{r}' = \mathbf{r} - \int_{t'}^{t} \langle\mathbf{v}(u)\rangle_{\mathbf{k}}du.$$

But this is incorrect, because the electrons which pass through $\mathbf{r}$ at time t with velocity $\mathbf{v}_{\mathbf{k}}$ have not all passed through $\mathbf{r}'$ at time t': this is merely the position of their centre of gravity at that time, and individual electrons may have experienced fields very different from $\mathbf{E}(\mathbf{r}',t')$. If the probability of one of these electrons being in a particular volume element $d\mathbf{r}''$ at time t' is $p_{\mathbf{k}}(\mathbf{r}'',t')\,d\mathbf{r}''$, and if its probable velocity is then $\langle\mathbf{v}(\mathbf{r}'',t')\rangle_{\mathbf{k}}$, $\Delta\mathscr{E}$ will be given by the expression

$$\Delta\mathscr{E}_{\mathbf{k}}(\mathbf{r},t) = e\int_{-\infty}^{t} dt' \int d\mathbf{r}''\, p_{\mathbf{k}}(\mathbf{r}'',t')\, \mathbf{E}(\mathbf{r}'',t').\langle\mathbf{v}(\mathbf{r}'',t')\rangle_{\mathbf{k}}, \qquad (4.14)$$

but this expression is too cumbersome to be of much practical value.

But there is one particular situation in which $\Delta\mathscr{E}$ is given correctly by an expression of the form (4.13), and that is when every individual scattering process randomizes the velocity of the electron completely, so that it is equally likely to be scattered into any point on the FS. Each collision then effectively restores the electron to the unperturbed distribution function f^0; in other words, each collision effectively destroys any memory of the excess energy previously acquired from the field $\mathbf{E}$. Thus the excess energy of an electron at $\mathbf{r}$, t is now limited to the energy acquired since its last collision, and moreover all electrons which pass through $\mathbf{r}$ at time t with velocity $\mathbf{v}(t) = \mathbf{v}_{\mathbf{k}}$ will have followed a common path since their previous collisions, and will therefore have been subjected to the same field $E(r', t')$. But r' is now the position at time t' of these *unscattered* electrons.

If a magnetic field is applied to the metal, the wave-vector and hence the velocity of each electron will vary as it moves around its orbit, and we shall denote by $v(u)_k$ the velocity at time u of an (unscattered) electron which at time t is in state k, with velocity $v_k = v(t)_k$. With this notation, we have

$$r'(t') = r - \int_{t'}^{t} v(u)_k \, du. \qquad (4.15)$$

If $H = 0$, we can put $v(u)_k = v_k$, since in the linear approximation the effect of the electric field E on the trajectory of the electron can be neglected.

Similarly, we shall denote by $\tau(u)_k$ the relaxation time of the electron at time u, so that the probability of collision in time interval du is $du/\tau(u)$. Like $v(u)$, $\tau(u)$ will only vary with time if $H \neq 0$; if $H = 0$, $\tau(u)_k = \tau_k$. The probability that the electron will survive from time t' to time t without collision is then given by

$$\exp\left[-\int_{t'}^{t} du/\tau(u)_k \right],$$

which becomes simply $\exp - (t - t')/\tau_k$ if $H = 0$.

Since each collision is assumed to randomize the electron velocity completely, it follows that only unscattered electrons contribute to $\langle v(t') \rangle_k$ in (4.13), so that

$$\langle v(t') \rangle_k = v(t')_k \exp\left[-\int_{t'}^{t} du/\tau(u)_k \right],$$

and the perturbation $g_k(r, t)$ to the distribution function becomes

$$g_k(r, t) = -\Delta \mathscr{E}_k(r, t) \frac{\partial f^0}{\partial \mathscr{E}} = -e \frac{\partial f^0}{\partial \mathscr{E}} \int_{-\infty}^{t} dt' \, E(r', t') \cdot v(t')_k$$

$$\exp\left[-\int_{t'}^{t} du/\tau(u)_k \right]. \qquad (4.16)$$

This path-integral expression for g_k (Chambers, 1952a) forms a convenient starting-point for the discussion of size-effect and skin-effect problems. In deriving it, we have implicitly assumed that $f_k^0(r') = f_k^0(r)$; this will not be true if temperature gradients or volume charges exist in the metal, but the derivation can readily be generalized to allow for these. The major assumption in deriving (4.16) is that every scattering process is 'catastrophic', in the sense that the electron is equally likely to be scattered to any point on the *FS* at each collision. In terms of the Boltzmann equation, this is equivalent to assuming that the collision integral can be written in the form

$$(\partial f_k/\partial t)_{coll} = -g_k/\tau_k, \qquad (4.17)$$

with a universal relaxation time τ_k which is independent of the form of g_k, and it is not difficult to show that (4.16) is simply a convenient integrated form of the resultant simplified Boltzmann equation.

In fact this assumption is rather more restrictive than appears at first sight. By considering the form of the exact collision integral, equation (4.3), one can show that if a universal relaxation time τ_k is to exist, independent of the form of g_k, it cannot in fact be an arbitrary function of **k**. At most it can depend on $\mathscr{E}(\mathbf{k})$, and then only if the scattering is completely elastic; if the scattering is inelastic, it cannot even depend on $\mathscr{E}(\mathbf{k})$, but must be the same for all electrons. The physical reason for this is easy enough to see. Suppose that some external disturbance has produced a positive perturbation $g(\mathbf{k}_1)$ near $\mathbf{k}_1$, and a negative perturbation $g(\mathbf{k}_2)$ near $\mathbf{k}_2$, leaving $g(\mathbf{k}) = 0$ elsewhere. If the external disturbance is removed, these perturbations will relax exponentially to zero, according to (4.17), with time constants $\tau(\mathbf{k}_1)$ and $\tau(\mathbf{k}_2)$. But if the total electron density is not to change in the process, these two time constants must clearly be equal: the net effect of collisions is simply to transfer electrons from $\mathbf{k}_1$ to $\mathbf{k}_2$; and since the positions of $\mathbf{k}_1$ and $\mathbf{k}_2$ are arbitrary, this must clearly be a two-stage process: the electrons near $\mathbf{k}_1$ must be scattered randomly over the *FS*, and at the same time electrons from random points on the *FS* are scattered into $\mathbf{k}_2$.

The conditions for the existence of a universal relaxation time, and for the validity of (4.16), are therefore highly restrictive, and it is very unlikely that they are ever satisfied in a real metal. Despite this, one almost invariably uses (4.16), or some equivalent approximation, as the starting-point in any but the simplest transport problems, simply because a more exact treatment—based for instance on (4.14)—would introduce too many unknown parameters to be practically useful. Moreover, one almost invariably assumes for generality that τ can vary with **k**, instead of being the same for all electrons. Although these assumptions may be logically inconsistent, they usually form the best available approximation to the complicated truth. The parameters appearing in (4.14), for instance, will certainly be **k**-dependent, and if we are to idealize the problem by assuming that they may be replaced by a single parameter τ it is more realistic to allow τ likewise to be k-dependent. But we shall not be too surprised if different experiments, interpreted using such a theory, yield different values for the parameter $\tau(\mathbf{k})$.

Similar problems arise in the treatment of surface scattering. When an electron in state $\mathbf{k}$ arrives at the surface of the metal, it will be scattered back into the metal in some different state $\mathbf{k}'$, and the probability of scattering from $\mathbf{k}$ to $\mathbf{k}'$ at the surface will be governed by some function $Q_s(\mathbf{k}, \mathbf{k}')$ analogous to the $Q(\mathbf{k}, \mathbf{k}')$ of (4.3). (cf. Greene, 1966). But very little is known about the probable form of $Q_s(\mathbf{k}, \mathbf{k}')$ for real metal surfaces, and in any case a formally exact treatment of surface scattering would be just as difficult as an exact treatment of internal scattering. As there, the problem is almost always simplified drastically by introducing a single parameter p (Fuchs, 1938) to describe the surface scattering. A fraction p of the incident electrons are supposed to suffer specular reflection, as from a mirror; the remainder $1 - p$ are diffusely reflected, so that they emerge in random directions. In other words, diffuse reflection acts like 'catastrophic' internal scattering, to restore the electrons leaving the surface to the unperturbed distribution function f^0 with no memory of the energy previously acquired from the applied field. The concept of specular reflection is not so readily defined, in general. For free electrons, it offers no difficulty: if the surface of the metal lies in the plane $z = 0$, we can assume that an incoming electron with wave-vector components k_x, k_y, k_z is specularly reflected into the state $k_x, k_y, -k_z$. But if the energy surfaces are non-spherical, this state may have a very different energy from the incoming state, and since surface scattering will certainly be elastic, such a boundary condition would be quite unacceptable. Price (1960) has suggested, as a plausible alternative assumption, that k_x and k_y are conserved, and that k_z adjusts itself so that the outgoing state has the same energy as the incident one. But this assumption leads to serious ambiguities, because in the periodic zone scheme there will be many different emergent states with the same values of k_x, k_y and $\mathscr{E}$ (cf. Pippard, 1965 b, p. 47). Thus the attempt to simplify the surface-scattering problem by introducing the parameter p does not in general produce much real simplification. Luckily, most of the available evidence suggests that scattering is usually effectively diffuse ($p = 0$), and in this limit no ambiguities remain: each surface collision restores the electron to the distribution f^0. There is indeed evidence that specular reflection, or something equivalent to it, does occur under special conditions, notably in very thin films of gold (cf. Bennett & Bennett, 1966), but diffuse reflection seems to be the general rule, and for simplicity we shall assume it in most of what follows. (See, however, pp. 222–3.)

Equation (4.16) is readily generalized to include the effect of diffuse surface scattering. An electron which passes through point $\mathbf{r}$ at time t with velocity $\mathbf{v_k}$ will have followed the path defined by (4.15) since its last collision. If this path intersects the surface at time $t - t_s$, an electron which has suffered no intervening collisions must have collided with the surface at that time, and if surface scattering is diffuse it will thus have started at time $t - t_s$ with $\Delta\mathscr{E} = 0$. The excess energy at time t is therefore equal to the energy acquired since time $t - t_s$, and to take account of this we merely have to replace the lower limit of integration in (4.16) by $t - t_s$.

In high-frequency problems, where $\mathbf{E(r)}$ is non-uniform, this diffuse boundary condition introduces considerable mathematical complications. In order to avoid these, we shall sometimes use a very artificial definition of 'specular reflection' as our boundary condition: we assume that an electron coming up to the surface in state $\mathbf{k}$ is reflected into state $-\mathbf{k}$, with reversal of its excess energy, so that

$$\Delta\mathscr{E}(-\mathbf{k}, z = 0) = -\Delta\mathscr{E}(\mathbf{k}, z = 0).$$

In terms of the perturbation $g(\mathbf{k}, \mathbf{r})$ to the distribution function, this means that
$$g(-\mathbf{k}, z = 0) = -g(\mathbf{k}, z = 0). \tag{4.18}$$

This boundary condition is obviously highly implausible, but it is at least free from the difficulties and ambiguities of the previous definitions, because the central symmetry of the FS ensures that

$$\mathscr{E}(\mathbf{k}) = \mathscr{E}(-\mathbf{k}).$$

And it follows from (4.8) and (4.18) that since $\mathbf{v(k)} = -\mathbf{v}(-\mathbf{k})$, the total current density at $z = 0$ is the sum of two equal terms: the electrons leaving the surface $(v_z > 0)$ contribute exactly as much as those approaching it. Thus the overall effect of this boundary condition is to produce a reasonably 'specular' current pattern, and in fact it is not difficult to show that for free electrons in zero magnetic field (4.18) is equivalent to the usual specular boundary condition,

$$g(k_x, k_y, k_z, z = 0) = g(k_x, k_y, -k_z, z = 0).$$

As an alternative boundary condition, equally convenient mathematically and equally implausible physically, we may assume that an electron is reflected at the surface from $\mathbf{k}$ to $-\mathbf{k}$ without change of $\Delta\mathscr{E}$, so that

$$g(-\mathbf{k}, z = 0) = g(\mathbf{k}, z = 0). \tag{4.19}$$

We can call this 'antispecular' reflection, with $p = -1$; the total current density at $z = 0$ is now zero, because the electrons leaving the surface carry a current equal and opposite to those arriving. For the physically interesting case of diffuse reflection, where the electrons leave the surface with $\Delta\mathscr{E} = g = 0$ and carry no current, we might expect the properties of the metal to be intermediate between those predicted using (4.18) and (4.19); as we shall see, this is sometimes but not always so.

4.3   D.C. SIZE EFFECTS, $H = 0$

Under d.c. conditions, (4.16) takes the form

$$g_\mathbf{k}(\mathbf{r}) = -e\frac{\partial f^0}{\partial\mathscr{E}} \int_{-t_s}^{0} dt\, \mathbf{E}(\mathbf{r}')\cdot\mathbf{v}(t)_\mathbf{k}\exp\left[-\int_{t}^{0} du/\tau(u)_\mathbf{k}\right], \tag{4.20}$$

where the effect of diffuse surface scattering has been taken into account by writing $-t_s$ as the lower limit of integration; $t_s = t_s(\mathbf{k}, \mathbf{r})$ is the time taken for an electron to travel from the surface to point $\mathbf{r}$ along the path defined by (4.15). Knowing $g_\mathbf{k}(\mathbf{r})$, the current density $\mathbf{J}(\mathbf{r})$ can at once be found from (4.8). In the case of a thin wire or thin film, $\mathbf{J}(\mathbf{r})$ will depend on the position of $\mathbf{r}$ within the metal, and the apparent resistivity will be determined by the average current density over the cross-sectional area A

$$\bar{\mathbf{J}} = A^{-1}\int \mathbf{J}(\mathbf{r})\, dA.$$

If $H = 0$, we can put $\mathbf{v}(t)_\mathbf{k} = \mathbf{v}_\mathbf{k}, \tau(u)_\mathbf{k} = \tau_\mathbf{k}$, independent of time, and if in addition the electric field $\mathbf{E}$ is uniform throughout the sample, (4.20) can be integrated at once to yield

$$g_\mathbf{k}(\mathbf{r}) = -e\frac{\partial f^0}{\partial\mathscr{E}}\mathbf{E}\cdot\mathbf{v}_\mathbf{k}\tau_\mathbf{k}(1 - \exp(-d_s/l_\mathbf{k})),$$

where $d_s(\mathbf{k}, \mathbf{r}) = v_\mathbf{k} t_s(\mathbf{k}, \mathbf{r})$ is the distance an electron in state $\mathbf{k}$ travels in moving from the surface to point $\mathbf{r}$, and $l_\mathbf{k} = v_\mathbf{k}\tau_\mathbf{k}$ is the mean free path. Using (4.8), we thus have

$$\mathbf{J}(\mathbf{r}) = -\frac{e^2}{4\pi^3}\int \frac{\partial f^0}{\partial\mathscr{E}}\mathbf{v}_\mathbf{k}(\mathbf{E}\cdot\mathbf{v}_\mathbf{k})\,\tau_\mathbf{k}(1 - \exp(-d_s/l_\mathbf{k}))\, d\mathbf{k}$$

$$= \frac{e^2}{4\pi^3\hbar}\int_{FS} dS_F\, \mathbf{n}_\mathbf{k}(\mathbf{E}\cdot\mathbf{n}_\mathbf{k})\, l_\mathbf{k}(1 - \exp(-d_s/l_\mathbf{k})), \tag{4.21}$$

where $\mathbf{n_k} = \mathbf{v_k}/v_k$. Thus if we write $J_i(\mathbf{r}) = \sigma_{ij}(\mathbf{r})\,E_j$, the position-dependent conductivity $\sigma_{ij}(\mathbf{r})$ is given by

$$\sigma_{ij}(\mathbf{r}) = \frac{e^2}{4\pi^3\hbar} \int_{FS} dS_F (n_i n_j l)_\mathbf{k}\,(1 - \exp(-d_s/l_\mathbf{k})), \qquad (4.22)$$

where n_i, $n_j = v_i/v, v_j/v$. The position-dependence of $\sigma_{ij}(\mathbf{r})$ arises from the position-dependence of d_s. In a bulk sample, of dimensions large compared with $l_\mathbf{k}$, the term $\exp(-d_s/l_\mathbf{k})$ is negligibly small almost everywhere within the sample, and (4.22) then yields the usual expression (4.1) for the mean conductivity $\sigma_0 = \frac{1}{3}(\sigma_{xx} + \sigma_{yy} + \sigma_{zz})$. In a thin sample, the term $\exp(-d_s/l_\mathbf{k})$ will lead to a reduction in the apparent conductivity, and an enhancement of the apparent resistivity. In effect, the electron travelling through point $\mathbf{r}$ in the direction $\mathbf{n}$ has now travelled on the average only a distance

$$l_\mathbf{k}(1 - \exp(-d_s/l_\mathbf{k}))$$

since its last collision (either with an internal scatterer or with the surface, a distance d_s away) and picked up a correspondingly reduced energy from the applied field $\mathbf{E}$, so that it contributes less to σ_{ij}.

For a free-electron metal, with l independent of $\mathbf{k}$, it is a straightforward matter to evaluate $\sigma_{ij}(\mathbf{r})$, and hence $\bar{\sigma}_{ij} = A^{-1}\int \sigma_{ij}(\mathbf{r})\,dA$, for a long thin wire or film of thickness d. If the sample axis is in the x-direction, the off-diagonal components $\sigma_{yx}(\mathbf{r})$ and $\sigma_{zx}(\mathbf{r})$ vanish by symmetry, so that the apparent resistivity is simply $\rho(d) = 1/\bar{\sigma}_{xx}$. The variation of $\rho(d)/\rho(\infty)$ with d/l for a thin film (Fuchs, 1938) and for a thin wire of circular cross-section (Dingle, 1950; Graham, 1958) is shown in Fig. 4.1. For $d/l \gg 1$, the theory predicts

$$\rho(\infty)/\rho(d) = 1 - 3l/8d\ldots \text{ (film); } = 1 - 3l/4d\ldots \text{ (wire)}, \qquad (4.23)$$

and for $d/l \ll 1$,

$$\rho(\infty)/\rho(d) = (3d/4l)(\ln(l/d) + 0\cdot423\ldots) \quad \text{(film);}$$

$$= d/l\ldots \quad \text{(wire)}. \qquad (4.24)$$

For a thin wire, the exact free-electron result departs by only 5 % at most from the simple Nordheim approximation

$$\rho(d)/\rho(\infty) = 1 + l/d \qquad (4.25)$$

over the whole range of l/d, and for a thin film, the approximation $\rho(d)/\rho(\infty) = 1 + 0\cdot4l/d$ is good to 5 % for l/d less than 10.

Thus for a free-electron metal with l independent of $\mathbf{k}$, one could deduce the Fermi surface area S_F by measuring $\rho(d)$ for a series of samples of different thicknesses (but all having the same purity and all measured at the same temperature, so that all had the same bulk mean free path l), and deducing l by comparison with the theoretical curves of Fig. 4.1. The value of S_F would then follow at once from (4.1). Alternatively, as shown by Cotti (1964), l could be deduced from measurements on a single sample, by comparing the apparent d.c. resistivity with the resistivity deduced from eddy current decay measurements. Unfortunately, few metals except Na and K approximate to the spherical FS of the free-electron model, but the curves of

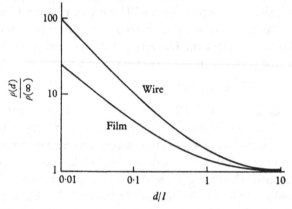

Fig. 4.1. The enhancement of apparent resistivity of a thin film and of a thin wire, as a function of d/l, for the free-electron model.

Fig. 4.1 have often been used in this way to interpret size-effect data on polycrystalline thin films and wires, on the hopeful assumption that the apparent mean free path deduced in this way for a polycrystal would be the appropriate average to insert in (4.1). But the values of S_F deduced in this way are usually about half the true values (Olsen, 1958, 1962), for a number of reasons.

First, the mean conductivity $\bar{\sigma}_{xx}$ of a real metal, with non-spherical FS and anisotropic mean free path, will depend on the crystal orientation relative to the x-axis. For a thick wire, in the limit $d/l \gg 1$, the *average* of $\bar{\sigma}_{xx}$ over all crystal orientations is given by $\langle \bar{\sigma}_{xx}\rangle/\sigma_0 = 1 - 3\overline{l^2}/4d\bar{l}$ (Bate, Martin & Hille, 1963). Thus if measurements of $\langle \bar{\sigma}_{xx}\rangle$ are interpreted using (4.23), the apparent mean free path will exceed $\bar{l}$ in the ratio $\overline{l^2}/\bar{l}^2$, and the apparent value of S_F will be smaller in the same ratio. Secondly, the apparent resistivity of a long thin polycrystalline wire will in fact be more nearly given by $\langle(\bar{\sigma}_{xx})^{-1}\rangle$ than by $\langle\bar{\sigma}_{xx}\rangle^{-1}$, since the crystallites will be effectively in series. Unless this is allowed for, S_F will be still further underestimated, because for any variable X, $\langle X \rangle \langle X^{-1} \rangle \geqslant 1$. Thirdly, the effective resistivity $\bar{\rho}_{xx}$ of a thin single crystal will in general exceed $\bar{\sigma}_{xx}^{-1}$, unless the off-diagonal components σ_{xy}, σ_{xz} of (4.22) vanish; that is, unless the crystal has a symmetry plane normal to the sample axis. What is worse, the derivation of (4.22) will in fact be invalid unless these components vanish, because an applied field E_x acting alone will generate transverse currents J_y, J_z, and to cancel these out transverse fields E_y, E_z will be set up within the sample. Since $\sigma_{ij}(\mathbf{r})$ is position-dependent, these transverse fields will likewise be position-dependent, whereas (4.22) was deduced on the assumption that the field E was uniform throughout the sample. To obtain an exact solution, we should now have to return to (4.20), and seek a form of $E(\mathbf{r})$ which satisfied the condition $J_y = J_z = 0$ for all $\mathbf{r}$. This would be a formidable and somewhat unrewarding task.

Finally, there is the complication that a simple mean free path l_k may not adequately describe the electron-scattering process. It is usually assumed that scattering by static imperfections is essentially isotropic, so that a well defined τ_k and l_k should exist in the residual resistance region, though Pippard (1964b) has shown that even in this region small-angle scattering may be important. Small-angle phonon scattering will certainly be important at somewhat higher temperatures, and Olsen (1958) first pointed out the relevance of this to size-effect measurements. According to the approximation (4.25), we have $\rho(d) = \rho(\infty) + \Delta\rho(d)$, where $\Delta\rho(d) = \rho(\infty)\,l/d$. $\Delta\rho(d)$ is thus independent of temperature, so that the resistivity of a thin wire should appear to obey Matthiessen's rule, with a residual resistivity greater than the bulk value by $\Delta\rho(d)$. Olsen found that thin wires showed marked deviations from Matthiessen's rule, and attributed this to the effect of small-angle scattering. The problem has been examined theoretically by Lüthi & Wyder (1960), Blatt & Satz (1960) and Azbel' & Gurzhi (1962), who have shown that Olsen's results can be accounted for semi-quantitatively in this way.

Wyder (1965) has demonstrated the importance of small-angle scattering in a different and rather direct way, by measuring the size-dependence of both $\rho(d)$ and the thermal resistivity $W(d)$, using the same samples. If it is assumed that scattering can be described by a simple mean free path, the ratio $W(d)/W(\infty)$ for a free-electron metal should vary with d/l in precisely the same way as $\rho(d)/\rho(\infty)$, as shown in Fig. 4.1. Using these curves to analyse his results, Wyder found that the ratio of the apparent mean free paths l_T and l_E deduced from thermal and electrical measurements varied with temperature as shown in Fig. 4.2. This figure also shows the variation of the Lorentz ratio $\mathscr{L}(T) = \rho/WT$, measured on bulk samples. As we saw in §4.2, $\mathscr{L}(T)/\mathscr{L}_0 = \overline{L_{T,\parallel}}/\overline{L_{E,\parallel}}$, and it will be seen that the two curves are indeed very similar. Admittedly it is a little inconsistent to deduce the ratio l_T/l_E by using the simplified theory of Fig. 4.1, because the very fact that $l_T \neq l_E$ shows that the simplified theory is incorrect; nevertheless, this experiment does show rather directly the difference between the effective thermal and electrical mean free paths.

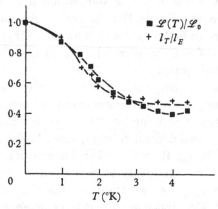

Fig. 4.2. The ratio l_T/l_E of thermal to electrical mean free path in In and the ratio $\mathscr{L}(T)/\mathscr{L}_0$ of the Lorentz ratio to its 'ideal' value, as a function of temperature (from Wyder, 1965).

4.4 D.C. SIZE EFFECTS, $H \neq 0$

In the presence of a magnetic field, the trajectories of the electrons will become helices rather than straight lines, so that the distance d_s measured along the trajectory from the surface to any point $\mathbf{r}$ is altered, and in general increased. In high enough fields, the majority of the helical orbits may miss the surface altogether; if they do, the resistivity of a thin film or wire in this limit will tend to the bulk value. Early theoretical work on these effects and on size-effects generally, for the free-electron model, has been well reviewed by Sondheimer (1952). Even on this simple model, it is difficult to evaluate (4.20) precisely unless the Hall field is either zero or uniform across the sample. In a thin film, for example, with its normal in the y-direction and $\mathbf{J} = J_x$, $\mathbf{H} = H_z$, a non-uniform Hall field E_y will be set up across the film to reduce the transverse current J_y to zero at all points in the film,

o

and the self-consistent determination of this field and of the consequent apparent resistivity is a difficult task (Macdonald & Sarginson,
1950; Ditlefsen & Lothe, 1966). The corresponding problem for a thin
wire is still more difficult, and has not been attempted. For the thin
film, the resistivity $\rho(H)$ falls towards the bulk value, as one would
expect, when the field is strong enough for the orbit diameter $2r$ to
become small compared with the thickness of the film. From the shape
of the $\rho(H)$ curve one can deduce the field at which $2r_{max} = d$, and
hence deduce the Fermi momentum, since for free electrons
$r_{max} = \hbar k_F/eH$.

For $H \| J$, no Hall field will be set up in a free-electron metal, and
the problem becomes relatively simple to solve, either for a thin wire
(Chambers, 1950) or for a thin film (Koenigsberg, 1953; Azbel', 1954;
MacDonald & Barron, 1958; Kao, 1965). Again, $\rho(H)$ falls to its
bulk value for $2r_{max} \ll d$, and again k_F can in principle be deduced
from the shape of the curve.

The remaining configuration of interest is the thin film with H
normal to the plane of the film. Here the Hall field lies in the plane of
the film, and is therefore uniform across its thickness, and this problem
can be solved fairly simply (Sondheimer, 1950). This configuration is
of particular interest, because of the 'Sondheimer oscillations' which
the resistivity then shows as a function of H, and we discuss these more
fully below.

For real metals, the complications discussed in §4.3 are now com-
pounded by the complications of distorted helical trajectories, and
an exact treatment is virtually out of the question. Azbel' (1963b) has
given a qualitative discussion of the effects to be expected in a variety
of situations, including thin films and thin wires of compensated
($n_e = n_h$) and uncompensated ($n_e \neq n_h$) metals, with H at various
angles to J. The general features are qualitatively much the same as
for the free-electron model, with fairly obvious modifications for compensated metals, in which the bulk resistivity rises as H^2 in large fields
instead of saturating.

Azbel' (1963a) has also drawn attention to a curious 'static skin-
effect' which may occur even in bulk samples at high magnetic fields.
The ordinary theory of bulk magneto-resistance (§3.10) neglects
collisions of the electrons with the sample surface, and for a bulk
sample this is of course valid for the great majority of electrons.
But those within a distance $\sim l$ of the surface will 'see' the surface,
and their transport properties will be correspondingly modified. In

effect, the sample consists of two conductors in parallel: the core region, whose transport properties are described by the bulk conductivity tensor $\sigma_{ij}(H)$, and the surface sheath surrounding it, with effectively a modified conductivity tensor $\sigma^s_{ij}(H)$. According to Azbel's original analysis, the form of $\sigma^s_{ij}(H)$ was such that at very high fields the current tended to flow almost entirely in the surface sheath, and the apparent resistivity of the sample grew linearly with H, for both compensated and uncompensated metals. Experimentally, however, no such behaviour has been observed, and more recently Azbel' & Peschanskii (1965) have produced a revised theory, in which the static skin effect occurs only for compensated metals, and then in such a way as to leave the form of the magneto-resistance unaffected ($\rho \propto H^2$). Azbel's (1963 b) predictions concerning the behaviour of thin samples at very high fields correspondingly need some revision.

We return now to the thin film with **H** normal to its surface, and to the Sondheimer oscillations which then arise. Alone among the phenomena so far discussed, these oscillations arise from localized regions of the FS, and because of this it is possible to derive a relatively simple expression for the period of the oscillations, not only for free electrons but also for real metals (Gurevich, 1958), in terms of the properties of these localized regions.

It is easy enough to see physically how the oscillations arise. We first recall the rules for finding the form of the electron orbit, in k-space and in real space (Ziman, 1964a, p. 250). From the Lorentz force equation $\hbar\dot{\mathbf{k}} = e\mathbf{v} \times \mathbf{H}$, it follows that **k** moves around an orbit defined by the intersection of a constant-energy surface with a plane normal to **H**. If the orbit is closed, the electron completes one orbit in the time τ_c, related to the cyclotron frequency ω_c and the cyclotron mass m_c^* by

$$\omega_c = 2\pi/\tau_c = eH/m_c^*. \qquad (4.26)$$

Here
$$m_c^* = (\hbar^2/2\pi)\,(\partial\mathscr{A}/\partial\mathscr{E})_{k_H}, \qquad (4.27)$$

where $\mathscr{A}$ is the area of the orbit. Further, by integration of the Lorentz force equation it follows that

$$\mathbf{k} - \mathbf{k}_0 = (e/\hbar)\,(\mathbf{r} - \mathbf{r}_0) \times \mathbf{H}, \qquad (4.28)$$

where $\mathbf{k}_0$ and $\mathbf{r}_0$ are integration constants. Thus the real-space orbit, projected on the plane normal to **H**, has the same shape as the k-space orbit, but rotated through $\frac{1}{2}\pi$. The drift motion along the field direction is less easily evaluated. As **k** moves around its orbit, the electron velocity in the field direction v_H will vary with period τ_c, and may even change sign, so that the orbit is spread out into a distorted helix. If v_H does change sign around the orbit, the helix will be correspondingly 'sheared', as shown in Fig. 4.3, so that the distance moved in the field direction,

$$r_H = r_H(0) + \int_0^t v_H(t')\, dt' \qquad (4.29)$$

will no longer be a single-valued function of t. And the time t_d taken by an electron to cross a film of thickness d in the z-direction, with $H\|z$, will correspondingly be an awkward non-analytic function of d and of the magnetic field strength. But the *mean* velocity

$$\bar{v}_H = \frac{1}{\tau_c} \int_0^{\tau_c} v_H(t')\, dt'$$

is given by the simple expression (Harrison, 1960)

$$\bar{v}_H = -\frac{\hbar}{2\pi m_c^*} \left(\frac{\partial \mathscr{A}}{\partial k_H} \right)_{\mathscr{E}} \qquad (4.30)$$

and to a first approximation we can therefore write $t_d = d/\bar{v}_H = d/\bar{v}_z$. Now if $d \ll l$, most collisions will occur at the surfaces, rather than in the body of the film, and if $t_d = n\tau_c$, the electron will just complete n orbits in crossing the film, so that (if the k-space orbits are closed, as we are assuming) its net displacement in the xy-plane is zero, and the energy $\Delta\mathscr{E}$ picked up from an electric field E_x or E_y is therefore also zero. On the other hand, if $t_d = (n+\frac{1}{2})\tau_c$, the electron will complete $n+\frac{1}{2}$ orbits in crossing the film, and $\Delta\mathscr{E}$ will be relatively large.

Thus $\Delta\mathscr{E}$ will fluctuate with a period given by $t_d\,\Delta(\tau_c^{-1}) = 1$, i.e. by

$$\Delta H = \pm\, 2\pi m_c^* \bar{v}_z/ed = \mp\, (\hbar/ed)\,(\partial\mathscr{A}/\partial k_z)_{\mathscr{E}}. \qquad (4.31)$$

Clearly, the oscillations in $\Delta\mathscr{E}$ will still have the same period if we evaluate t_d from the more accurate expression (4.29); they arise simply because the distorted helix shrinks uniformly in size as $1/H$, so that more and more turns of it can be fitted into the thickness d as H increases. And the oscillations will still have roughly the same amplitude; the only effect of using the more accurate expression will be to introduce some distortion into the detailed form of the oscillations. Since we are concerned primarily with the period of the oscillations, we can therefore replace $v_z(t)$ by $\bar{v}_z$, which greatly simplifies the theory.

In general, $(\partial\mathscr{A}/\partial k_z)_{\mathscr{E}}$ will vary from one slice-plane k_z to the next, so that electrons on different slice-planes will give rise to oscillations in $\Delta\mathscr{E}$, and hence in σ_{ij} ($i,j = x,y$), of different periods in H. As so often in FS studies, the oscillations actually observed will come from the regions on the FS where the period is extremal, or where there is an abrupt cut-off in the range of periods. On a spherical FS, for instance, $|\partial\mathscr{A}/\partial k_z|$ varies continuously from zero to $2\pi k_F$ when $k_z = \pm k_F$, the radius of the sphere. At these limiting points, the slice-plane $k_z =$ constant just touches the FS, and the orbit diameter has shrunk to zero. Though there is no extremum of $|\partial\mathscr{A}/\partial k_z|$ at the limiting point, the abrupt cut-off here in the range of integration gives rise to observable

Fig. 4.3. The sheared helical path that an electron may follow, if the velocity component v_H in the field direction changes sign periodically.

oscillations of the corresponding period $\Delta H = 2\pi\hbar k_F/ed$. More generally, on any FS observable oscillations will arise from the 'elliptic limiting points' at which the slice-plane $k_z =$ constant is just tangent to the FS. (Such points may not always exist if the FS is open: for instance on a cylindrical FS with H parallel to the cylinder axis.) If the two principal radii of curvature of the FS at this point are ρ_1 and ρ_2, so that $K = (\rho_1\rho_2)^{-1}$ is the Gaussian curvature, it is easy to show that $|d\mathscr{A}/dk_z| = 2\pi/\sqrt{K}$ at the limiting point, so that

$$|m_c^* v_H| = \hbar/K^{\frac{1}{2}} \quad \text{and} \quad \Delta H = 2\pi\hbar/edK^{\frac{1}{2}}. \qquad (4.32)$$

Thus a measurement of ΔH yields the Gaussian curvature K at the limiting point directly.

But the limiting-point oscillations in σ_{xx}, σ_{yy} are rather weak, and fall off in amplitude as H^{-4}, because the electrons actually at the limiting point have $\mathbf{v}\|\mathbf{H}$, so that in a field H_z, they have $v_x = v_y = 0$ and can carry no current in the xy-plane. The oscillations in fact come from a small region of the FS around the limiting point, where v_x and v_y, though non-zero, are still very small; hence their weakness. As shown by Gurevich, much stronger oscillations can occur if $|\partial\mathscr{A}/\partial k_z|$ passes through an extremal value for some intermediate value of k_z, where v_x and v_y are no longer small, and these oscillations fall off only as $H^{-2\cdot5}$. Still stronger oscillations, falling off only as H^{-2}, will occur if

$|\partial\mathscr{A}/\partial k_z|$ happens to be sensibly constant over a finite range of k_z (Bloomfield, 1966b); for example, if some part of the FS has the paraboloidal form $k_\alpha^2 + b^2 k_\beta^2 + ck_\gamma = d$. If θ is the angle between the γ and z-axes, we then have $|\partial\mathscr{A}/\partial k_z| = \pi c/b \cos^2\theta$, independent of k_z.

We now sketch briefly the derivation of these results, starting from (4.20). We assume that the FS is rotationally symmetric about the z-axis, so that the orbits are circles in **k**-space and undistorted helices in real space, and we assume that the relaxation time τ is constant around each orbit. These assumptions simplify the problem somewhat, without significantly restricting the validity of the results. We describe the position of the electron around its **k**-space orbit by an orbit angle α (Ziman, 1964a, p. 252), such that $\dot{\alpha} = \omega_c$, and for a circular orbit we can then write $v_x(t) - iv_y(t) = v_r e^{i(\alpha_0 + \omega_c t)}$, where $v_r = v_r(k_z)$. For a uniform applied field $\mathbf{E} = E_x$, we can then integrate (4.20) at once to find

$$g = -e\frac{\partial f^0}{\partial\mathscr{E}} E_x v_r \tau \mathscr{R} \frac{e^{i\alpha_0}}{1 + i\omega_c\tau}[1 - \exp\{-(1 + i\omega_c\tau)t_s/\tau\}]. \quad (4.33)$$

We now use (4.8) to find J_x and J_y, and hence σ_{xx} and σ_{yx}. In doing so, we write $d\mathbf{k} = dk_z dk_n dk_l$, where dk_n is normal to the orbit at the point k_z, α_0, and dk_l is parallel to the orbit at that point, so that

$$d\mathbf{k} = dk_z dk_n dk_l = dk_z \frac{1}{\hbar v_r} d\mathscr{E} \frac{|ev_r H|}{\hbar} dt = \frac{|m_c^*|}{\hbar^2} dk_z d\mathscr{E} d\alpha_0. \quad (4.34)$$

This useful transformation is not confined to circular orbits, but is applicable to any system of closed orbits in which α on each orbit is defined to satisfy $\dot{\alpha} = \omega_c$. Using $v_x = v_r \cos\alpha_0, v_y = -v_r \sin\alpha_0$ at $t = 0$, we find from (4.8) that σ_{xx} and σ_{yx} can conveniently be combined in the form

$$\sigma_c = \sigma_{xx} + i\sigma_{yx} = \frac{e^2}{4\pi^2\hbar^2}\int dk_z \frac{|m_c^*| v_r^2 \tau}{1 + i\omega_c\tau}[1 - \exp\{-(1 + i\omega_c\tau)t_s/\tau\}]. \quad (4\cdot35)$$

Finally, to find the apparent conductivity we have to average σ_c over z from 0 to d. For electrons with $\bar{v}_z < 0$, $t_s = (z - d)/\bar{v}_z$, and for those with $\bar{v}_z > 0$, $t_s = z/\bar{v}_z$. From the central symmetry of the FS, these two groups of electrons contribute equally to $\bar{\sigma}_c$, and we find

$$\bar{\sigma}_c = \frac{e^2}{4\pi^2\hbar^2}\int dk_z \frac{|m_c^*| v_r^2 \tau}{1 + i\omega_c\tau}\left[1 - \frac{|\bar{v}_z|\tau}{(1 + i\omega_c\tau)d}\{1 - \exp[-(1 + i\omega_c\tau)d/|\bar{v}_z|\tau]\}\right]. \quad (4.36)$$

It is convenient to split this expression into a non-oscillatory part

$\bar{\sigma}_c^N$ and an oscillatory part $\bar{\sigma}_c^{\text{osc}}$ arising from the exponential term. At high fields $(\omega_c^2 \tau^2 \gg 1)$ we find

$$\bar{\sigma}_{xx}^N = \frac{e^2}{4\pi^2\hbar^2} \int dk_z \frac{|m_c^*||v_r^2|\tau}{\omega_c^2\tau^2} \left[1 + \frac{|\bar{v}_z|\tau}{d} \right] \qquad (4.37)$$

and $\qquad \bar{\sigma}_{yx}^N = -\frac{e^2}{4\pi^2\hbar^2} \int dk_z \frac{|m_c^*||v_r^2|}{\omega_c} \left[1 - \frac{2|\bar{v}_z|\tau}{\omega_c^2\tau^2 d} \right]. \qquad (4.38)$

The first term in $\bar{\sigma}_{xx}^N$ is simply the ordinary bulk conductivity, falling at high fields as ω_c^{-2}, that is, as H^{-2}; in thin films, this is enhanced by the factor $[1 + \langle|\bar{v}_z|\tau\rangle/d]$, where $\langle|\bar{v}_z|\tau\rangle$ is the appropriate weighted average over the FS. For free electrons, $\langle|\bar{v}_z|\tau\rangle = 3l/8$. In $\bar{\sigma}_{yx}^N$ the first term is again simply the bulk Hall term $\pm ne/H$, where n is the number of electrons or holes per cm^3 within the FS, and the sign depends on the sign of m_c^*, positive for electrons and negative for holes. The second term in $\bar{\sigma}_{yx}^N$ is negligible at high fields, unless the metal is compensated. In a compensated metal, containing equal numbers of electrons and holes, the electrons and holes contribute equal and opposite amounts to the bulk term, which therefore vanishes, and we are left with the very small second term, proportional to H^{-3}.

For the oscillatory term, we have

$$\bar{\sigma}_c^{\text{osc}} = \frac{e^2}{4\pi^2\hbar^2 d} \int dk_z \frac{|m_c^*\bar{v}_z|v_r^2\tau^2}{(1+i\omega_c\tau)^2} \exp\left(-d/l_z\right) \exp\left(-i\omega_c d/|\bar{v}_z|\right) \qquad (4.39)$$

where $l_z = |\bar{v}_z|\tau$ and is independent of H. Thus at high fields, $\bar{\sigma}_c^{\text{osc}}$ depends on τ only through the factor e^{-d/l_z}. Writing

$$\omega_c d/|\bar{v}_z| = \phi = \pm 2\pi e H d/\hbar |\partial\mathscr{A}/\partial k_z|,$$

we see that the period of the oscillations from each orbit will indeed be given by (4.31), as expected. If $|\partial\mathscr{A}/\partial k_z|$ remains essentially constant over a finite range Δk_z, so that all the orbits within Δk_z have the same phase ϕ_e, this range will dominate the integral, and we shall have $\bar{\sigma}_c^{\text{osc}} \propto H^{-2} e^{-i\phi_e}$. Alternatively, if $|\partial\mathscr{A}/\partial k_z|$ simply passes through an extremum at $k_z = k_e$ say, we can write $\phi = \phi_e[1 + \beta(k_z - k_e)^2...]$, and the effective range Δk_z over which ϕ remains essentially constant will be of order $(\beta\phi_e)^{-\frac{1}{2}}$, $\propto H^{-\frac{1}{2}}$, so that $\bar{\sigma}_c^{\text{osc}} \propto H^{-2\cdot5} e^{-i\phi_e}$. To derive this result a little more carefully, we note that if $|\beta|$ is large enough, the oscillations will rapidly get out of phase and interfere destructively except for $k_z \sim k_e$, and over this small range we can regard the integrand of (4.39) as approximately constant, apart from the $e^{-i\phi}$ factor. Thus (at high fields)

$$\overline{\sigma}_c^{\text{osc}} = -\frac{e^2}{4\pi^2\hbar^2 d}\left[\frac{|m_c^*\overline{v}_z|v_r^2}{\omega_c^2}e^{-d/l_z}\right]_{k_z=k_e}e^{-i\phi_e}\int dk_z\, e^{-i\beta\phi_e(k_z-k_e)^2}$$

$$= -\frac{e^2}{4\pi^2\hbar^2 d}\left[\frac{|m_c^*\overline{v}_z|v_r^2}{\omega_c^2}e^{-d/l_z}\right]_{k_z=k_s}e^{-i\phi_e}\left(\frac{\pi}{i\beta\phi_e}\right)^{\frac{1}{2}}$$

where we have used the fact that

$$\int_{-L}^{L}e^{-x^2}dx \to \pi^{\frac{1}{2}} \quad \text{for} \quad L \gg 1.$$

This 'Cornu spiral' type of integration through a stationary-phase region recurs constantly in FS studies: we shall meet it again several times in the theory of cyclotron resonance, and it enters in a similar way into the theory of the de Haas–van Alphen effect (§ 3.8).

Finally, if $|\partial\mathscr{A}/\partial k_z|$ has no extrema in the range $-k_L \leqslant k_z \leqslant k_L$, where $\pm k_L$ are the limiting points, the dominant oscillations will be those of phase ϕ_L, from the limiting points themselves. Near k_L we can put $\phi = \phi_L[1+\beta k'...]$, where $\phi_L = eHdK^{\frac{1}{2}}/\hbar$ (cf. (4.32)), and $k' = k_L - k_z$. We can also put $|m_c^*v_z| = \hbar K^{-\frac{1}{2}}$, and

$$v_r^2 = \hbar^2 k_r^2/m_c^{*2} = 2\hbar^2 k'/m_c^{*2}K^{\frac{1}{2}}.$$

Then (4.39) becomes (inserting an extra factor 2 for the similar contribution from $k_z \sim -k_L$)

$$\overline{\sigma}_c^{\text{osc}} = -\frac{2e^2}{4\pi^2\hbar^2 d}\left[\frac{2\hbar^3}{m_c^{*2}\omega_c^2 K}e^{-d/l}\right]_{k_z=k_L}e^{-i\phi_L}\int_0^{\sim k_L}dk'\, k'\, e^{-i\beta\phi_L k'}. \quad (4.40)$$

Clearly, we can neglect the contribution from the upper limit of this integral, because in the exact expression (4.39), to which this is an approximation, the phase of the integrand tends to infinity and its amplitude tends to zero for $|v_z| \to 0$, i.e. for $k' \to k_L$. We thus find

$$\overline{\sigma}_c^{\text{osc}} = (\hbar/d)^3\,(\pi e\beta KH^2)^{-2}\,e^{-d/l}\,e^{-i\phi_L}. \quad (4.41)$$

so that the amplitude of the oscillations now falls off as H^{-4}.

Experimentally, of course, one measures not the conductivity of the film but its resistivity: for the simple FS considered in the present problem, we have by symmetry

$$\overline{\sigma}_{xx} = \overline{\sigma}_{yy}, \overline{\sigma}_{xy} = -\overline{\sigma}_{yx}, \quad \text{and} \quad \overline{\sigma}_{xz} = \overline{\sigma}_{yz} = 0,$$

so that $\overline{\rho}_{yy} = \overline{\sigma}_{xx}/\Delta$, where $\Delta = \overline{\sigma}_{xx}^2 + \overline{\sigma}_{yx}^2$. Thus for an uncompensated metal, in which $\Delta \propto H^{-2}$, $\bar{\rho}_{yy}^{N} = \bar{\sigma}_{xx}^{N}/\Delta$ is field independent, while $\bar{\rho}_{yy}^{osc} \propto H^{-2}$, $H^{-0.5}$ or H^{0} depending on whether the oscillations arise from limiting points, extremal regions or regions of constant $|\partial \mathscr{A}/\partial k_{z}|$. For a compensated metal, on the other hand, $\Delta \propto H^{-4}$, so that $\bar{\rho}_{yy}^{N} \propto H^{2}$ and $\bar{\rho}_{yy}^{osc} \propto H^{0}$, $H^{1.5}$ or H^{2} respectively. Thus the Sondheimer oscillations should be particularly marked in compensated metals, and this is confirmed by the work of Munarin & Marcus (1965, 1966) on Ga. They in fact measured $d\bar{\rho}/dH$ or $d^{2}\bar{\rho}/dH^{2}$, to reduce or remove the signal from $\bar{\rho}^{N}$, and some of their results are shown in Fig. 4.4. This

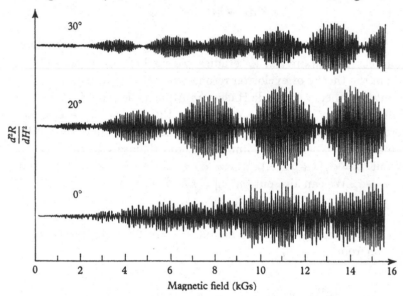

Fig. 4.4. Sondheimer oscillations in Ga (from Munarin & Marcus, 1965), for various angles of tilt between H and the normal to the sample surface.

figure shows that strong oscillations still occur when H is tilted up to $\theta = 30°$ away from the normal to the sample surface. It is not difficult to extend the theory to this case (Gurevich, 1958); as one would expect, the main effect is simply to replace d by $d/\cos\theta$ in the expression (4.31) for the period of the oscillations, and Bloomfield (1966b) has discussed the Munarin & Marcus results in some detail on this basis. The oscillations shown in Fig. 4.4 grow in amplitude with H, and are therefore presumably due to regions of extremal or constant $|\partial \mathscr{A}/\partial k_{z}|$, rather than to limiting points; the beats suggest two such regions.

So far we have supposed that only closed orbits were present in the metal. Pippard (1966a) has shown that open orbits can give rise to similar oscillations, with their period now related to the period of the open orbit in **k** space and hence to the size of the Brillouin zone. The amplitude of $\bar{\sigma}^{osc}$ is proportional to H^{-2}, as for a closed-orbit region of constant $|\partial \mathscr{A}/\partial k_z|$, and for the same reason: the width Δk_z of the effective region is independent of H. Pippard has suggested that open orbits may account for some of the strong oscillations observed by Munarin & Marcus, though as we have seen, strong oscillations are in any case to be expected from a compensated metal.

Lastly, it is worth noting the focusing effect which a uniform magnetic field H_z has on a group of electrons of common k_z. This focusing is precisely similar to that which occurs in a cathode-ray tube immersed in a uniform magnetic field, in the familiar experiment to determine e/m. Consider a group of electrons setting out from the point $\mathbf{r} = 0$ at time $t = 0$, all having the same k_z but initially at different points around the orbit, and so initially all travelling in different directions. All these electrons will pursue similar distorted helical orbits; the orbits are displaced laterally and longitudinally from one another, but all pass through $\mathbf{r} = 0$ at $t = 0$. After a time $t = n\tau_c$, all the electrons will have completed exactly n orbits in k-space, and therefore in real space they will all be refocused at the point $x = y = 0$,

$$z = n\tau_c \bar{v}_z = -(n\hbar/eH)(\partial \mathscr{A}/\partial k_z). \tag{4.42}$$

In particular, if $\partial \mathscr{A}/\partial k_z$ is constant or extremal for some value of k_z, a large number of electrons will be refocused at the same point. This effect has been elegantly demonstrated by Sharvin (1965) and Sharvin & Fisher (1965), by attaching two extremely fine probes to opposite points on the two faces of a thin single-crystal plate of Sn, and observing changes in the apparent resistance between the probes when a variable field **H** is applied parallel to the line joining the probes (see § 2.2.6).

4.5 THE ANOMALOUS SKIN EFFECT

We turn now from d.c. size effects to r.f. surface effects : the effects which arise at radio frequencies, when the skin depth becomes comparable with or smaller than the mean free path. The r.f. properties of a plane metal surface are specified completely by the surface impedance $Z = R + iX$, defined as the ratio of tangential electric field at the surface to total current (per unit width) flowing below the surface. The surface resistance R determines the rate of energy dissipation in the metal, and hence, for instance, the losses in a cavity built of the metal, and the surface reactance X determines the r.f. energy stored in the surface layers of the metal, and hence plays a part (usually a small part) in determining the resonant frequency of the cavity.

Using Maxwell's equations, Z can be written in a number of alternative forms: if the metal surface lies in the plane $z = 0$,

$$Z = R + iX = E(0) \bigg/ \int_0^\infty J(z)\,dz = 4\pi E(0)/H(0) = -4\pi i\omega E(0)/E'(0).$$

$$\text{(4.43)}$$

For a single crystal, Z will in fact be a two-dimensional tensor Z_{ij} $(i,j = x,y)$ relating the field $E_i(0)$ to the current $\int_0^\infty J_j(z)\,dz$, but we shall assume the axes x, y to be chosen to make this tensor diagonal.

From Maxwell's equations, we have $J_z(z) = 0$ and

$$E_i''(z) = 4\pi i\omega J_i(z) \quad \text{for} \quad i = x, y, \qquad \text{(4.44)}$$

and if Ohm's law holds, combination of (4.44) with $\mathbf{J} = \sigma \mathbf{E}$ at once yields the classical skin effect equations for $E_i(z)$, for the skin depth δ_c, and for Z: if $J_i = \sigma E_i$, these reduce to

$$E_i(z) = E_i(0)\,e^{-(1+i)z/\delta_c}, \quad \delta_c = (2\pi\omega\sigma)^{-\frac{1}{2}}, \quad Z = (4\pi i\omega/\sigma)^{\frac{1}{2}}. \quad \text{(4.45)}$$

Thus at a given frequency, $\Sigma = 1/R$ should be proportional to $\sigma^{\frac{1}{2}}$. It was first noticed by London (1940) that for pure Sn at 1000 Mc/s and 4 °K, Σ was much smaller than predicted by (4.45), and he suggested correctly that this was because the mean free path was then large compared with δ_c, so that the electrons no longer moved in a uniform field between collisions, and Ohm's law broke down. Pippard (1947) and Chambers (1952b) later showed that in the limit of large σ, Σ reached a limiting value independent of σ, and Pippard (1947, 1954) gave a simple 'ineffectiveness concept' argument to explain this anomalous skin effect (A.S.E.) behaviour. Though not particularly rigorous, this argument does give some physical insight into the phenomenon, and we therefore start by discussing it briefly, in slightly modified form (Chambers, 1956a).

We assume that very close to the surface, $E(z) = E(0)\,e^{-az}$, with a to be determined, and we assume that in this region $J(z) = \sigma_{\text{eff}} E(z)$. In the 'anomalous limit', when $l \gg 1/a$, an electron will stay in the skin depth for only a small fraction of a mean free path unless it is travelling almost parallel to the surface. We therefore take for σ_{eff}

the ordinary bulk d.c. conductivity which the metal *would* have, if the only effective electrons were those travelling within the range of angles $\pm \frac{1}{2}\beta/al$ to the surface. Here β is an adjustable numerical parameter, expected to be of order unity. To find σ_{eff} we can use (4.22), with $d_s = \infty$, and the integration over S_F restricted to the appropriate part of the *FS*. Over a small element δS_F of the *FS*, the direction of the velocity vector **v** will extend over a small range of solid angles $\delta\Omega$, and at any point on the *FS*, $\delta\Omega$ and δS_F are related by $\delta S_F = \delta\Omega/K$, where K is the Gaussian curvature of the *FS* at that point. If we choose polar co-ordinates so that $v_x = v\cos\phi\sin\theta$, $v_y = v\sin\phi\sin\theta$, $v_z = v\cos\theta$, $\delta\Omega = \sin\theta\,\delta\theta\,\delta\phi$, the effective electrons will be those for which θ lies within the range $\frac{1}{2}\pi \pm \frac{1}{2}\beta/al$, and over this small range we can put $K(\theta, \phi) = K(\frac{1}{2}\pi, \phi)$. Thus

$$\sigma_{ij,\,\text{eff}} = \frac{e^2}{4\pi^3\hbar}\frac{\beta}{a}\oint d\phi\, n_i n_j \Sigma K^{-1}(\tfrac{1}{2}\pi, \phi) \qquad (4.46)$$

where $n_x = \cos\phi$, $n_y = \sin\phi$. Unless the *FS* is a single closed convex surface, there may be several different points on it at which **v** has the same direction θ, ϕ; this is why we have written $\Sigma K^{-1}(\frac{1}{2}\pi, \phi)$, summed over all such points, instead of simply $K^{-1}(\frac{1}{2}\pi, \phi)$. By appropriate orientation of the xy-axes in the surface, we can always make $\sigma_{xy,\,\text{eff}}$ vanish, and we assume this done, so that the response to fields E_x and E_y can be considered separately. Note that the mean free path l has disappeared from (4.46).

For a field E_x, (4.44) then becomes $a_x^2 E_x(z) = 4\pi i\omega\sigma_{xx,\,\text{eff}} E_x(z)$, and on combining this with (4.46) we find that a_x is given by

$$a_x^3 = i\beta s_{xx}^3/\pi,$$

where

$$s_{xx}^3 = \frac{e^2\omega}{\pi\hbar}\oint d\phi\cos^2\phi\,\Sigma K^{-1}(\tfrac{1}{2}\pi, \phi), \quad s_{yy}^3 = \frac{e^2\omega}{\pi\hbar}\oint d\phi\sin^2\phi\,\Sigma K^{-1}(\tfrac{1}{2}\pi, \phi).$$

$$(4.47)$$

Since $E_x(0)/E_x'(0) = -1/a_x$, the surface impedance Z_{xx} is given at once by

$$Z_{xx} = 4\pi i\omega/a_x = 4\pi\omega(\pi/\beta)^{\frac{1}{3}}e^{\frac{1}{6}i\pi}s_{xx}^{-1} \qquad (4.48)$$

with a similar expression for Z_{yy}. Thus the surface impedance is determined by the Gaussian curvature of the *FS* at points around its 'equator'—that is, the contour defined by $\theta = \frac{1}{2}\pi$ or $v_z = 0$, on which the electrons are travelling parallel to the surface of the metal.

This result agrees completely with that given by an exact treatment, if β is chosen appropriately; for $p = 0$ (diffuse surface scattering)

$\beta = 8\pi/3 \sqrt{3}$. But the exact treatment is considerably simpler for $p = \pm 1$, in the sense defined in equations (4.18), (4.19), and we consider these cases first. We solve the problem by introducing the concept of an 'image metal', which we imagine to fill the half-space $z < 0$, so that instead of hitting the surface at $z = 0$ and being reflected back, electrons pass freely across the plane $z = 0$ into or out of the image metal. If the electric field in the image metal is arranged to be the mirror image of that in the real metal, the boundary condition (4.18) will be automatically satisfied, because of the overall central symmetry of the FS and the consequent fact that $\mathbf{v}(\mathbf{k}) = -\mathbf{v}(-\mathbf{k})$: an electron in state $-\mathbf{k}$ arriving at $z = 0$ from below will have experienced the same field as its opposite number in state $\mathbf{k}$ arriving from above, but with $\mathbf{v}$ and hence $\mathbf{E} \cdot \mathbf{v}$ reversed in sign, and hence with $\Delta \mathscr{E}$ and g reversed in sign. Conversely, if $\mathbf{E}(-z) = -\mathbf{E}(z)$, the antispecular boundary condition (4.19) will be satisfied. In either case, the required field distribution $\mathbf{E}(z)$ is set up by introducing a suitable source current sheet $\mathbf{J}_s$ in the plane $z = 0$. By using this image-metal artifice, we can forget about the surface entirely, and work in terms of the transport properties of an unbounded block of metal.

It is convenient also to work in terms of the Fourier transforms $\mathbf{E}(q)$, $\mathbf{J}_s(q)$ and $\mathbf{J}(q)$ of the field $\mathbf{E}(z)$, the source current $\mathbf{J}_s$ and the electron current $\mathbf{J}(z)$, where e.g.

$$\mathbf{E}(q) = \frac{1}{2\pi} \int_{-\infty}^{\infty} \mathbf{E}(z)\, e^{iqz} dz; \quad \mathbf{E}(z) = \int_{-\infty}^{\infty} \mathbf{E}(q)\, e^{-iqz} dq. \quad (4.49)$$

In terms of its Fourier transform, (4.44) becomes

$$-q^2 E_i(q) = 4\pi i \omega [J_{s,i}(q) + J_i(q)].$$

But $\mathbf{J}(q)$ is simply the current induced in the infinite metal by an electric field $\mathbf{E}(q)$ varying as $e^{i(\omega t - qz)}$, and these are related by a well-defined conductivity tensor $\sigma_{ij}^{\infty}(q)$, whose evaluation we return to below. (In general, σ^{∞} will depend on ω as well as q, but in the A.S.E. limit its ω-dependence vanishes, as we shall see.) For the moment, we assume the existence of $\sigma_{ij}^{\infty}(q)$, and we assume that axes have been chosen in which $\sigma_{xy}^{\infty}(q) = 0$, so that we can consider the fields E_x and E_y separately. We can then write, for instance,

$$[q^2 + 4\pi i \omega \sigma_{xx}^{\infty}(q)] E_x(q) = -4\pi i \omega J_{s,x}(q) \quad (4.50)$$

from which we can at once deduce $E_x(q)$ and hence $E_x(z)$ and Z, for given $J_s(q)$. Note that if $J_s(q) = 0$, (4.50) becomes the equation to a propagating mode in an infinite medium, and that the surface impedance Z is related to, but not equal to, the wave impedance of the propagating mode (Pippard, 1966 b). We return to this point in §4.9.

For $p = 1$, we can take as our current source a current sheet in the plane $z = 0$ of strength $2I_s$ per unit width, with Fourier transform $J_s(q) = I_s/\pi$. By symmetry, this will induce mirror-symmetric electron current distributions in the real metal and the image metal, with $J(z) = J(-z)$, and correspondingly $E(z) = E(-z)$, as required. Since the fields and currents fall to zero as $|z| \to \infty$, the total electron current must exactly screen the source current, so that

$$I_s = -\int_0^\infty J(z)\,dz = E'(0)/4\pi i\omega \tag{4.51}$$

where $E'(0) = [dE(z)/dz]_{z\to+0}$. Inserting $J_s(q) = I_s/\pi$ in (4.50) and using (4.49) and (4.51), we at once find

$$E_x(z) = -\frac{E_x'(0)}{\pi}\int_{-\infty}^\infty \frac{e^{-iqz}\,dq}{q^2 + 4\pi i\omega\sigma_{xx}^\infty(q)} \tag{4.52}$$

and in particular,

$$Z_{xx,1} = -\frac{4\pi i\omega E_x(0)}{E_x'(0)} = 4i\omega\int_{-\infty}^\infty \frac{dq}{q^2 + 4\pi i\omega\sigma_{xx}^\infty(q)} \tag{4.53}$$

with a similar expression for $Z_{yy,1}$; the subscript 1 is used to indicate $p = 1$.

For $p = -1$, we take a double current source, consisting of a current sheet of strength $2F_s/\eta$ at $z = 0$, and another of strength $-2F_s/\eta$ at $z = -\eta$, with $\eta \to 0$. This will have as its Fourier transform

$$J_s(q) = iq F_s/\pi,$$

and will clearly induce the required current and field patterns,

$$J(-z) = -J(z), \quad E(-z) = -E(z).$$

Such a double current sheet will produce a discontinuity in $E(z)$ of $-8\pi i\omega F_s$ at $z = 0$, so that

$$E(+0) = -E(-0) = -4\pi i\omega F_s. \tag{4.54}$$

Thus (4.50) now yields

$$E_x(z) = \frac{E_x(+0)}{\pi}\int_{-\infty}^\infty \frac{iq\,e^{-iqz}\,dq}{q^2 + 4\pi i\omega\sigma_{xx}^\infty(q)} \tag{4.55}$$

which is, of course, essentially just the derivative of (4.52). Differentiating again, we find

$$E_x'(z) = \frac{E_x(+0)}{\pi} \int_{-\infty}^{\infty} \frac{q^2 e^{-iqz}\, dq}{q^2 + 4\pi i \omega \sigma_{xx}^{\infty}(q)}$$

$$= \frac{E_x(+0)}{\pi} \int_{-\infty}^{\infty} \left\{ 1 - \frac{4\pi i \omega \sigma}{q^2 + 4\pi i \omega \sigma} \right\} e^{-iqz}\, dq.$$

We have to be a little careful in evaluating $E_x'(+0)$, since E' has a singularity at $z = 0$, due to the term $\int_{-\infty}^{\infty} e^{-iqz}\, dq$. But this is zero except for $z = 0$, and the second term in $E_x'(z)$ is well-behaved at $z = 0$, so that we have

$$Z_{xx,-1} = -\frac{4\pi i \omega E_x(+0)}{E_x'(+0)} = \pi \Big/\!\! \int_{-\infty}^{\infty} \frac{\sigma_{xx}^{\infty}(q)}{q^2 + 4\pi i \omega \sigma_{xx}^{\infty}(q)}\, dq. \qquad (4.56)$$

It only remains to evaluate $\sigma_{ij}^{\infty}(q)$, using (4.16) and (4.8). Since in the A.S.E. no magnetic field is applied to the metal, $\mathbf{v}$ and τ in (4.16) are independent of time, so that from (4.15), $\mathbf{r}' = \mathbf{r} - \mathbf{v}(t - t')$. If an electric field $\mathbf{E}(\mathbf{r}, t) = \mathbf{E}(q)\, e^{i(\omega t - \mathbf{q}\cdot\mathbf{r})}$ is applied to the metal, the field experienced by the electron at $\mathbf{r}', t'$ is thus

$$\mathbf{E}(\mathbf{r}', t') = \mathbf{E}(\mathbf{r}, t) \exp\{i(\omega - \mathbf{q}\cdot\mathbf{v})(t' - t)\}.$$

We can now integrate (4.16) at once, to find

$$g(\mathbf{k}, \mathbf{r}, t) = -e\frac{\partial f^0}{\partial \mathscr{E}} \frac{\mathbf{v}\cdot\mathbf{E}(\mathbf{r}, t)\tau}{1 + i(\omega - \mathbf{q}\cdot\mathbf{v})\tau}. \qquad (4.57)$$

Thus if $\omega\tau \gg 1$ or $ql \gg 1$, the denominator of (4.57) will for almost all electrons be large, and g will be small compared with the d.c. value $-e(\partial f^0/\partial \mathscr{E})\,\mathbf{E}\cdot\mathbf{v}\tau$; the electrons travel through a rapidly fluctuating field in the course of one mean free path, and pick up little net energy from it. But this is not true of electrons for which $\omega \simeq \mathbf{q}\cdot\mathbf{v}$: these electrons 'surf-ride' on the electric field (Pippard, 1960, 1965b), staying on a surface of constant phase as the wave moves through the metal, and they are as effective in producing current as they would be under d.c. conditions. If the phase velocity of the wave, ω/q, is much smaller than the electron velocity v, as it will be in the cases of interest to us, the condition $\mathbf{q}\cdot\mathbf{v} = \omega$ becomes $\mathbf{q}\cdot\mathbf{v} \simeq 0$; that is, the effective electrons are those travelling practically normal to the direction of $\mathbf{q}$.

If, as before, we take the direction of $\mathbf{q}$ as the z-axis, introduce polar co-ordinates θ, ϕ to describe the direction of $\mathbf{v}$, and write

$$d\mathbf{k} = d\mathscr{E}\, dS_F/\hbar v = d\mathscr{E}\sin\theta\, d\theta\, d\phi/\hbar v K(\theta, \phi),$$

we can at once find $\sigma_{ij}^{\infty}(q)$ from (4.8):

$$\sigma_{ij}^{\infty}(q) = \frac{e^2}{4\pi^3\hbar} \oint d\phi \int_0^{\pi} \frac{d\theta \sin\theta\, n_i n_j}{K(\theta,\phi)} \frac{v\tau}{1 + i(\omega - qv\cos\theta)\tau} \quad (4.58)$$

where $n_x = \sin\theta\cos\phi$, $n_y = \sin\theta\sin\phi$. Now if $ql \gg |1 + i\omega\tau|$, the integral over θ will be dominated by a small region close to $\theta = \tfrac{1}{2}\pi$, where $qv\cos\theta \sim 0$, and over this small region we can put

$$K(\theta,\phi) \simeq K(\tfrac{1}{2}\pi,\phi), \quad n_x = \cos\phi, \quad n_y = \sin\phi.$$

Carrying out the integration over θ, we thus find, for $i,j = x,y$,

$$\sigma_{ij}^{\infty}(q) = \frac{e^2}{4\pi^2\hbar|q|} \oint d\phi\, n_i n_j \Sigma K^{-1}(\tfrac{1}{2}\pi,\phi) \quad (4.59)$$

(where, as in (4.46), we have replaced K^{-1} by ΣK^{-1}), so that

$$\sigma_{xx}^{\infty}(q) = s_{xx}^3/4\pi\omega|q|, \quad \sigma_{yy}^{\infty}(q) = s_{yy}^3/4\pi\omega|q|, \quad (4.60)$$

with s_{xx}^3, s_{yy}^3 defined by (4.47). We see that in this limit, $\sigma_{ij}^{\infty}(q)$ becomes independent of l, like $\sigma_{ij,\mathrm{eff}}$, and for much the same reason: as l increases there is a corresponding fall in the number of electrons which can surf-ride effectively for a distance l without losing phase.

Now the main contribution to the integrals (4.53) and (4.56) comes from the region $q^2 \sim 4\pi\omega\sigma$, i.e. $|q| \sim s$, so that if $sl \gg |1 + i\omega\tau|$, we can use (4.60) over the whole range of integration, to find

$$Z_{xx,1} = 16\pi\omega e^{\frac{1}{2}i\pi}/3^{\frac{3}{2}}s_{xx}; \quad Z_{xx,-1} = 3^{\frac{1}{2}}\pi\omega e^{\frac{1}{2}i\pi}/s_{xx}, \quad (4.61)$$

with corresponding expressions for Z_{yy}. We can also evaluate $E(z)$ itself from (4.52) or (4.55), (cf. (4.116) below), and hence $J(z)$ from (4.44). The resulting curves for $p = 1$ are shown in Fig. 4.5, plotted against the dimensionless variable $w = sz$, and we see that the 'skin depth' is of order $1/s$. The curves for $p = -1$ are not very different. Thus, somewhat surprisingly, the drastic change in boundary conditions from $p = 1$ to $p = -1$ merely increases Z by a factor $\frac{27}{16}$, and for $p = 0$ we can expect Z to have some intermediate value.

The exact theory for $p = 0$ (Reuter & Sondheimer, 1948) involves appreciably more sophisticated mathematics than for $p = \pm 1$. Again, the image-metal approach is used, but now we need to arrange that the electrons crossing the plane $z = 0$ from the image metal into the real metal shall have $\Delta\mathscr{E} = 0$, in order to counterfeit the effect of diffuse reflection, and we thus need $E(z) = 0$ for $z < 0$. Thus both $E(z)$ and $E'(z)$ must fall discontinuously to zero at $z = 0$, and to achieve this we need both a current sheet of strength $I_s = E'(+0)/4\pi i\omega$ and a double current sheet of strength $F_s = -E(+0)/4\pi i\omega$ in the plane $z = 0$. But we now need more than this: electrons drifting out of the region $z > 0$ will carry r.f. currents into the region $z < 0$, and these will in turn induce electric fields in that region; to maintain $\mathbf{E}(z) = 0$, we have to

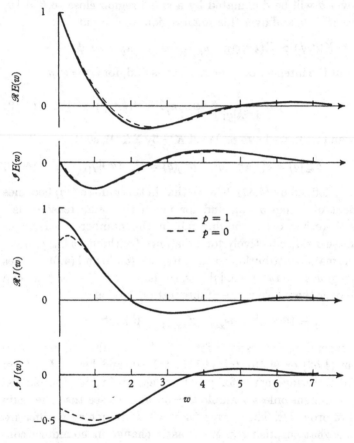

Fig. 4.5. The real and imaginary parts of $E(w)$ and $J(w)$ in the anomalous skin effect, plotted against $w = sz$. Full lines: $p = 1$. Dashed lines: $p = 0$.

introduce a continuous distribution of source currents $\mathbf{G}(z)$ in the region $z < 0$, to cancel out the electron currents there. Thus (4.50) now becomes (dropping the subscripts x or y)

$$[q^2 + 4\pi i\omega\sigma^\infty(q)]\,E(q) = -E'(+0)/2\pi + iqE(+0)/2\pi - 4\pi i\omega G(q) \quad (4.62)$$

and the presence of the term $G(q)$, which itself depends on $J(z)$ and hence on $E(z)$, appears to make this equation much more difficult to

P

solve. But the elegant Wiener–Hopf technique used by Reuter & Sondheimer to solve (4.62) in fact involves no detailed knowledge of $G(q)$: this term need not be considered explicitly at all, basically because its form is determinate if the equation is to be self-consistent. In summary, one treats q as a complex variable, expresses

$$D(q) = q^2 + 4\pi i \omega \sigma^\infty(q)$$

as a ratio ϕ_+/ϕ_-, using contour methods, and shows by an appeal to Liouville's theorem that $\phi_+(q)\,E(q)$ must be a constant. Reuter & Sondheimer's original solution apparently involved the roots of the equation $D(q) = 0$, but Dingle (1953) showed that these could be eliminated from the final expressions, and in fact if the contour used to express $D(q)$ is suitably chosen, these roots need never appear. Since these mathematical manipulations shed little light on the physics we merely quote the result,

$$Z_0 = 4\pi^2 i \omega \bigg/\!\!\int_0^\infty \ln\left[1 + 4\pi i \omega \sigma^\infty(q)/q^2\right] dq.$$

In the limit $sl \gg |1 + i\omega\tau|$, when (4.60) can be used for $\sigma^\infty(q)$, this yields

$$Z_{xx,0} = 2\sqrt{3}\,\pi\omega\, e^{\frac{1}{3}i\pi}/s_{xx}, \tag{4.63}$$

larger than $Z_{xx,1}$ by the factor $\frac{9}{8}$.

As we shall see, the image-metal approach cannot easily be extended to more complex $p = 0$ problems such as cyclotron resonance. It is therefore worth considering here an alternative approach, due to Azbel' & Kaner (1955), which does not have this limitation. Applied to the A.S.E. problem, this approach yields an equation essentially equivalent to (4.62). Azbel' & Kaner did not however solve this by the Wiener–Hopf technique; instead, they derived an expression for $G(q)$, of the form $\displaystyle\int_{-\infty}^\infty f(q,q')\,E(q')\,dq'$, and solved the resulting integral equation approximately by an iterative method.

In this approach, we still consider for convenience two semi-infinite blocks of metal, separated by the plane $z = 0$, but now this plane forms a genuine diffuse boundary between the blocks, so that electrons approaching it from either side are reflected back with $\Delta\mathscr{E} = 0$, instead of crossing it. As in the image-metal treatment of $p = 1$, we use a current sheet of strength $2I_s$ in the plane $z = 0$ to induce r.f. currents in the two blocks, and as there, $I_s = E'(z \to +0)/4\pi i\omega$, so that the Fourier transform of (4.44) is again

$$-q^2 E(q) = \pi^{-1} E'(0) + 4\pi i \omega J(q). \tag{4.64}$$

But we can no longer write $J(q) = \sigma^\infty(q) E(q)$, because the metal is no longer effectively unbounded, as it was before. For $|z|$ large, this relation will still hold, but for $|z|$ small, the presence of the diffuse boundary at $z = 0$ will reduce the current below its bulk value.

Thus the current $\mathbf{J}(z)$ induced by a periodic field $\mathbf{E}(z) = \mathbf{E}(q)\, e^{i(\omega t - qz)}$ must now be described by a position-dependent conductivity tensor $\sigma_{ij}(q, z)$:

$$J_i(z) = \sigma_{ij}(q, z)\, E_j(z).$$

For $|z|$ large, $\sigma_{ij}(q, z)$ will tend to the bulk conductivity $\sigma_{ij}^\infty(q)$. It is convenient to write $J_i(z)$ as the difference of two terms, the unperturbed bulk current and a surface correction term $J_i^s(z)$:

$$J_i(z) = J_i^\infty(z) - J_i^s(z) = [\sigma_{ij}^\infty(q) - \sigma_{ij}^s(q, z)]\, E_j(z).$$

To see the form of $J_i^s(z)$ or $\sigma_{ij}^s(q, z)$, consider the form of $\Delta\mathscr{E}(\mathbf{k}, \mathbf{r}, t)$ for an electron travelling towards $\mathbf{r}$ from the surface: from (4.16), we shall have

$$\Delta\mathscr{E} = e \int_{t - t_s}^{t} \mathbf{v}\cdot\mathbf{E}_q(\mathbf{r}', t') \exp\{-(t - t')/\tau\}\, dt'.$$

Now the corresponding contribution to $J_i^\infty(z)$ will simply have the lower limit replaced by $-\infty$, so that the contribution to $J_i^s(z)$ will be the difference between these,

$$\Delta\mathscr{E}^s = e \int_{-\infty}^{t - t_s} \mathbf{v}\cdot\mathbf{E}_q(\mathbf{r}', t') \exp\{-(t - t')/\tau\}\, dt'.$$

Thus $\mathbf{J}^s(z)$ can formally be interpreted as the current at z due to electrons which have acquired no energy from the field $\mathbf{E}_q$ since time $t - t_s$, i.e. since crossing the plane $z = 0$. Clearly, $\mathbf{J}^s(z)$ is closely related to the electron drift current $-\mathbf{G}(z)$ of the previous treatment; in fact it is simply the contribution to $-\mathbf{G}(z)$ due to the field component $\mathbf{E}_q(z) = \mathbf{E}(q)\, e^{i(\omega t - qz)}$.

For large q, $\mathbf{J}^s(z)$ will be strongly localized near the surface $z = 0$: from (4.58), $\sigma_{ij}^\infty(q)$ is dominated by electrons for which

$$qv\tau \cos\theta \gtrsim |1 + i\omega\tau|,$$

and these electrons will travel only a distance $v\tau \cos\theta \gtrsim |1 + i\omega\tau|/q$ in the z-direction before being scattered. If the surface is further away than this, they will be unaware of its presence. Thus for

$$|z| \gtrsim |1 + i\omega\tau|/q,$$

we expect $J_i(z) \simeq J_i^\infty(z)$, and hence $J_i^s(z) \simeq 0$: the surface term falls

to zero within a few wavelengths of the surface, unless $\omega\tau \gg 1$. If we
express $J_i^s(z)$ in terms of its Fourier transform

$$J_i^s(q') = (2\pi)^{-1}\int_{-\infty}^{\infty} J_i^s(z)\, e^{iq'z}\, dz, \qquad (4.65)$$

it follows that the Fourier components $J_i^s(q')$ will spread over a broad
range of q', though they are all generated by the single field component
$E_j(q)$. We can conveniently write $J_i^s(q') = \sigma_{ij}^s(q',q)\, E_j(q)$, where the
tensor $\sigma_{ij}^s(q',q)$ is given by

$$\sigma_{ij}^s(q',q) = (2\pi)^{-1}\int_{-\infty}^{\infty} \sigma_{ij}^s(q,z)\, e^{i(q'-q)z}\, dz. \qquad (4.65')$$

Just as $E_j(q)$ gives rise to currents $J_i^s(q')$ of different wave-number,
so $E_j(q')$ will give rise to a current $J_i^s(q)$, so that the *total* electron current $J_i(q)$ of wave-number q will be

$$\sigma_{ij}^\infty(q)\, E_j(q) - \int_{-\infty}^{\infty} \sigma_{ij}^s(q,q')\, E_j(q')\, dq'.$$

Thus (4.64) now becomes

$$[q^2 + 4\pi i\omega\sigma^\infty(q)]\, E(q) + E'(0)/\pi = 4\pi i\omega\int_{-\infty}^{\infty} \sigma^s(q,q')\, E(q')\, dq', \qquad (4.66)$$

where we have for simplicity assumed that $\sigma_{ij}^\infty(q)$ and $\sigma_{ij}^s(q,q')$ can
be simultaneously diagonalized (as in fact they can, in the A.S.E.
limit), and dropped the subscripts.

The apparent differences between (4.62) and (4.66) arise simply
because in (4.62), $E(q) = E_{RS}(q)$ is the transform of a field $E_{RS}(z)$ which
is zero for $z < 0$, while in (4.66), $E_{AK}(q)$ is the transform of a field
$E_{AK}(z) = E_{AK}(-z)$. For $z > 0$, $E_{RS}(z) = E_{AK}(z)$, and it follows that
$E_{AK}(q) = E_{AK}(-q) = E_{RS}(q) + E_{RS}(-q)$. Now from the central
symmetry of the FS, it is easily shown that $\sigma_{ij}^\infty(q) = \sigma_{ij}^\infty(-q)$, so that if
we add to (4.62) the corresponding equation with q replaced by $-q$, we
at once obtain (4.66), with the right-hand side written in the form

$$-4\pi i\omega[G_{RS}(q) + G_{RS}(-q)] = -4\pi i\omega G_{AK}(q)$$

precisely as expected.

In the limit $ql \gg |1 + i\omega\tau|$, where $\sigma_{ij}^\infty(q)$ is given by (4.60), it is
not difficult using (4.16) to evaluate $\sigma_{ij}^s(q,q')$, or more conveniently
$\sigma_{ij}^s(q,q') + \sigma_{ij}^s(q,-q')$. Equation (4.66) then becomes

$$[q^2 + is^3/|q|]\, E(q) + \frac{1}{\pi}E'(0) = \frac{2}{\pi}is^3\int_0^\infty \frac{\ln|q/q'|}{q^2 - q'^2}\, E(q')\, dq' \qquad (4.67)$$

where, as usual, we have assumed s_{ij}^3 to be diagonalized, and dropped the subscripts on s^3 and $E(q)$. To find Z_0, Azbel' & Kaner solve this equation by iteration. Neglecting the right-hand side, (4.67) becomes identical with the equation for $p = 1$, so that

$$E(q) = -E'(0)/\pi[q^2 + is^3/|q|], \quad \text{and} \quad Z = Z_1,$$

as given by (4.61). Inserting this first approximation for $E(q)$ on the right-hand side, we obtain as a second approximation to Z_0

$$Z_0^{(2)} = Z_1\left[1 + \frac{2}{\pi^2}\int_0^\infty \frac{u\,du}{\sinh u + \sinh 2u}\right] = 1\cdot0926Z_1. \quad (4.68)$$

The convergence of the iterative process is therefore fairly rapid: this second approximation is already close to the correct value,

$$Z_0 = 1\cdot125Z_1$$

(cf. (4.63)). We can also use (4.67) to find a second approximation to the form of the field $E_0(z)$ in terms of the $p = 1$ field $E_1(z)$. (The Wiener–Hopf solution, of course, yields an exact expression for $E_0(z)$, but in an exceedingly intractable form for computation.) Using the dimensionless variable $w = sz$, we find

$$E_0^{(2)}(w) = E_1(w) - \frac{1}{1\cdot0926\pi^2}\int_{-\infty}^\infty \frac{u\,du}{\sinh u + \sinh 2u}\frac{E_1(w) - E_1(we^{-u})}{1 - e^{-u}}$$
$$(4.69)$$

from which $J_0^{(2)}(w)$ follows (probably not very reliably) by differentiating twice. The resultant curves are shown in Fig. 4.5. As one would expect, $J_0(w)$ reaches its peak a little way in from the surface, rather than at the surface itself, where the emergent electrons contribute nothing.

4.6 CYCLOTRON RESONANCE

In A.S.E. measurements, no magnetic field is applied to the metal. If we apply a steady field H, accurately parallel to the surface of the sample, and measure the variation of Z with H, we are doing quite a different experiment: we are observing Azbel'–Kaner cyclotron resonance (A.K.C.R.), first discussed by Azbel' & Kaner (1957, 1958). This yields quite different information about the FS, and in practice involves somewhat different experimental techniques, but the theoretical treatment is practically identical: again, we are concerned with evaluating Z, and again this depends on the conductivity tensors

$\sigma_{ij}^{\infty}(q)$ and $\sigma_{ij}^{s}(q, q')$, but now these tensors are of course affected by the field H.

Qualitatively, the origin of a.k.c.r. is clear enough. The electrons no longer travel in straight lines between collisions, but in the helical orbits defined by (4.28) and (4.29). Electrons in suitable orbits will thus return periodically into the r.f. skin depth with frequency ω_c (Fig. 4.6), and if $\omega = n\omega_c$, they will see the r.f. field in the same phase each time, and gain energy coherently from it, whereas if $\omega \neq n\omega_c$, this

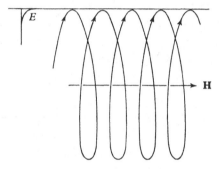

Fig. 4.6. Showing the origin of a.k.c.r. (from Chambers, 1956a).

resonant effect will be lost. We thus expect anomalies in $Z(H)$ whenever $\omega_c = \omega/n$, that is, whenever $1/H = ne/m_c^* \omega$. In practice the electron will pass through the surface layer only a limited number of times before being scattered, and this will limit the sharpness of the resonance.

Again, it is helpful to start by looking at the problem from the ineffectiveness concept viewpoint (Heine, 1957; Chambers, 1965). From this viewpoint, the argument leading to (4.48) is still applicable, but we now need to revise our estimate of $\sigma_{ij, \text{eff}}$ (eq. (4.46)) and s_{ij}^{s} (eq. (4.47)). Figure 4.7a shows a possible electron orbit in real space, projected onto the yz-plane normal to the field H_x. The position of this orbit in the yz-plane is such that it just comes up into the surface layer at the extremal point 1, without hitting the surface, and it can thus contribute resonantly to Z, in the way shown in Fig. 4.6, if $\omega = n\omega_c$. Other electrons having the same value of k_x will be pursuing similar orbits, but displaced in the yz-plane, and for some of these electrons the extremal points 2, 3 or 4 will lie in the skin depth. According to the ineffectiveness concept, these electrons will also contribute to σ_{eff} and to Z, since they are moving parallel to the surface at these points, but they will clearly contribute in a non-resonant manner, since their orbits intersect the surface and therefore cannot be completed without scattering if $p = 0$. In the A.S.E. problem, we assumed that the angular range of effective electrons was $\Delta\theta = \beta/al$, and that each electron travelled a full free path l in the skin depth $1/a$, to give a product $l\Delta\theta = \beta/a$ independent of l. But now, if the local radius of curvature of the orbit at point 1 is r_1, the electron will only travel for a distance $\sim (r_1/a)^{\frac{1}{2}}$ within the skin depth before passing through

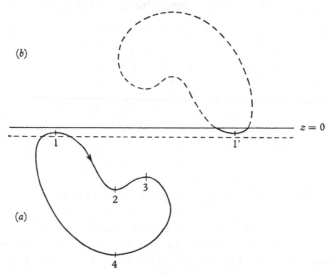

Fig. 4.7. Projections of a real-space electron orbit on a plane normal to H. (a) The orbit at $k_z = +k_1$; (b) the conjugate orbit at $k_z = -k_1$, showing the non-resonant extremal point $1'$; conjugate to 1 (from Chambers, 1965).

point 1, and if $1/a \ll r_1 < l$ this distance will be much less than l; correspondingly the energy $\Delta\mathscr{E}$ picked up from the r.f. field will only be $\sim e\mathbf{E}.\mathbf{v}(r_1/a)^{\frac{1}{2}}$ instead of $e\mathbf{E}.\mathbf{v}l$. Against this, the angular range of effective electrons will now be much greater: $\Delta\theta \sim \beta(r_1/a)^{\frac{1}{2}}/r_1 = \beta/(ar_1)^{\frac{1}{2}}$ instead of β/al. Thus the product of angular range and effective path length becomes $(r_1/a)^{\frac{1}{2}}.\beta/(ar_1)^{\frac{1}{2}} = \beta/a$—precisely the same as in zero field. This argument is perhaps not very convincing, but the conclusion is in fact correct: the electrons travelling parallel to the surface at extremal points such as 2, 3, 4 contribute just as much to σ_{eff} and to Z as they do when $H = 0$. The only difference arises for the electrons at the resonant extremal point, 1. These electrons will have acquired energy $\Delta\mathscr{E}$ in traversing the distance $\sim (r_1/a)^{\frac{1}{2}}$, just before passing through point 1, but we must now add to this the energy they have acquired in previous traverses through the surface layer. In each of these previous traverses, the electron has travelled the full distance $(\sim 2(r_1/a)^{\frac{1}{2}})$ through the surface layer, and between each traverse and the next, the phase of the r.f. field has changed by $2\pi\omega/\omega_c$; finally, the chance of the electron surviving without collision from one traverse to the next is $\exp(-2\pi/\omega_c\tau)$, so that the *total* excess energy with which the average electron arrives at point 1 will be

$$\Delta\mathscr{E}(1 + 2e^{-2\pi\lambda} + 2e^{-4\pi\lambda}...) = \Delta\mathscr{E}\frac{1 + e^{-2\pi\lambda}}{1 - e^{-2\pi\lambda}} = \Delta\mathscr{E}\coth\pi\lambda \quad (4.70)$$

where $\lambda = i\omega/\omega_c + 1/\omega_c\tau$. In other words, the contribution of these electrons to σ_{eff} has been enhanced by the factor $\coth\pi\lambda$.

This argument, then, leads us to expect that for $p = 0$ the surface impedance $Z_0(H)$ will still be given by (4.48) or (4.63), but with (4.47) for s^3 now replaced by

$$s_{ij}^3 = \frac{e^2\omega}{\pi\hbar}\oint d\phi\, n_i n_j\left\{\sum_R \coth\pi\lambda\frac{1}{K_R} + \sum_N\frac{1}{K_N}\right\}. \quad (4.71)$$

As in (4.47), $K = K(\frac{1}{2}\pi, \phi)$ is the Gaussian curvature of the FS at the point ϕ on the equatorial contour $\theta = \frac{1}{2}\pi$, and as there, we sum over all points on the FS having a common azimuth angle ϕ, including the resonance factor for all points where the real-space orbit can resonate (such as point 1 in Fig. 4.7) and omitting it otherwise.

Now the k-space orbit corresponding to the real-space orbit of Fig. 4.7a will have the same shape, but rotated through $\frac{1}{2}\pi$, and from (4.28) the sense of rotation depends on the sign of H. Thus this k-space orbit may correspond in real space either to orbit (a) or to orbit (b) of Fig. 4.7, depending on the sign of H, so that if H is reversed, point 1 ceases to be resonant, and point 4 becomes resonant instead (not for orbit (b) itself, of course, but for a similar orbit displaced downwards). Nevertheless, the value of (4.71) is independent of the sign of H, because of the central symmetry of the FS. Suppose that in a field $+H_x$ the real-space orbit (a) of Fig. 4.7 corresponds to the k-space orbit at $k_x = +k_1$. Then the k-space orbit at $k_x = -k_1$ will be of precisely the same shape, but rotated through π, and in the field $+H_x$ this orbit will give rise to the real-space orbit of Fig. 4.7b. We can thus always combine the contributions from the resonant point 1 and the 'conjugate' non-resonant point $1'$: at both these points the FS will have the same Gaussian curvature, and since $\phi(-k_1) = \phi(k_1) + \pi$, the factor

$n_i n_j$ ($= \cos^2 \phi, \sin^2 \phi$ or $\sin \phi \cos \phi$) will be the same for both. Equation (4.71) then takes on the symmetrized form

$$s_{ij}^3 = \frac{e^2 \omega}{\pi \hbar} \oint d\phi \, n_i n_j \left\{ \sum_B \frac{1}{1 - e^{-2\pi\lambda}} \frac{1}{K_B} + \sum_I \frac{1}{K_I} \right\}$$

$$= s_{ij}^3(0) + \frac{e^2 \omega}{\pi \hbar} \oint d\phi \, n_i n_j \sum_B \frac{1}{e^{2\pi\lambda} - 1} \frac{1}{K_B}, \qquad (4.72)$$

where the points B are the bounding extremal points, such as 1 and 4, and the points I are the internal extrema, such as 2 and 3, which are unable to resonate in either $+H$ or $-H$ if $p = 0$. Since the definition of bounding and internal extrema is independent of the sign of H, this is a more convenient form for s_{ij}^3 than (4.71). In the second form of (4.72), we have simply written $(1 - e^{-2\pi\lambda})^{-1} = 1 + (e^{2\pi\lambda} - 1)^{-1}$, and thus separated out the zero-field term $s_{ij}^3(0)$, given by (4.47).

As in A.S.E. theory, the ineffectiveness-concept approach leads to essentially the same results as the more exact treatment, if the constant β in (4.48) is chosen appropriately. To develop the more exact theory, we need first to consider the form of $\sigma_{ij}^\infty(q)$ in a magnetic field, for $i, j = x, y$. It is not difficult to write down a formal expression for $\sigma_{ij}^\infty(q, H)$, if we describe the position of an electron in its k-space orbit at any time t by the angle variable

$$\alpha = \int^t \omega_c dt = (\hbar/eH) \int dk_l/v_n,$$

where v_n is the velocity component normal to H, and dk_l is measured around the orbit. We take $\mathbf{q}$ to be along the z-axis, as before, but for generality we shall not assume $\mathbf{H}$ to be along the x-axis, normal to z; instead, we assume $\mathbf{H}$ to lie in the xz-plane at an angle ψ to x. Each electron will then pursue a helical orbit with the axis of the helix inclined at an angle ψ to the plane $z = 0$, so that its velocity $v_z(t)$ includes a steady drift component $\bar{v}_z = \bar{v}_H \sin \psi$, with $\bar{v}_H$ given by (4.30). After a time τ_c, the electron has thus moved a distance $\bar{v}_z \tau_c$ in the z-direction.

Consider now the evaluation of $g_\mathbf{k}(\mathbf{r}, t)$ from (4.16) for an electron whose position at time t is given by the orbit angle α, and which is subjected to a field $E_j(q) e^{i(\omega t - qz)}$. For simplicity, we assume τ to be constant around the orbit. Writing

$$t - t' = (\alpha - \alpha')/\omega_c \quad \text{and} \quad z - z' = \int_{\alpha'}^{\alpha} d\alpha'' v_z(\alpha'')/\omega_c,$$

we then have

$$g_\alpha(z,t) = -e\frac{\partial f^0}{\partial \mathscr{E}}E_j(q)\,e^{i(\omega t - qz)}\,\omega_c^{-1}\int_{-\infty}^\alpha d\alpha'\,v_j(\alpha')\,e^{-i\omega(\alpha-\alpha')/\omega_c}\,e^{-(\alpha-\alpha')/\omega_c\tau}$$

$$\times \exp\left\{iq\int_{\alpha'}^\alpha d\alpha''v_z(\alpha'')/\omega_c\right\}. \quad (4.73)$$

Since the metal is unbounded, the integration runs from $-\infty$ to α, but v_j and v_z vary periodically with α, with period 2π, so that we can conveniently split the integration up into the ranges $\alpha - 2\pi$ to α, $\alpha - 4\pi$ to $\alpha - 2\pi$, etc. Between each range and the next, the integrand changes by a factor $e^{-2\pi\lambda'}$, where

$$\lambda' = i\omega/\omega_c - iq\bar{v}_z/\omega_c + 1/\omega_c\tau = \lambda - iq\bar{v}_z/\omega_c;$$

apart from this, the integrand varies in exactly the same way within each range. Summing the contributions from all these ranges, we get

$$g_\alpha(z,t) = -e\frac{\partial f^0}{\partial \mathscr{E}}E_j(q)\,e^{i(\omega t - qz)}\frac{1}{\omega_c}\frac{1}{1-e^{-2\pi\lambda'}}\int_{\alpha-2\pi}^\alpha d\alpha'\,v_j(\alpha')\,e^{-\lambda(\alpha-\alpha')}$$

$$\exp\left\{iq\int_{\alpha'}^\alpha d\alpha''v_z(\alpha'')/\omega_c\right\}. \quad (4.74)$$

Thus, using $d\mathbf{k} = |m_c^*|\hbar^{-2}dk_H\,d\mathscr{E}\,d\alpha$ (cf. (4.34)) to evaluate (4.8), we find

$$\sigma_{ij}^\infty(q,H) = \frac{e^2}{4\pi^3\hbar^2}\int dk_H\frac{|m_c^*|}{\omega_c}\frac{1}{1-e^{-2\pi\lambda'}}\int_0^{2\pi}d\alpha\,v_i(\alpha)$$

$$\times \int_{\alpha-2\pi}^\alpha d\alpha'\,v_j(\alpha')\,e^{-\lambda(\alpha-\alpha')}\exp\left\{iq\int_{\alpha'}^\alpha d\alpha''v_z(\alpha'')/\omega_c\right\} \quad (4.75)$$

$$= \frac{e^2}{4\pi^3\hbar^2}\int dk_H\frac{|m_c^*|}{\omega_c}\frac{1}{1-e^{-2\pi\lambda'}}F_{ij}(k_H), \quad (4.76)$$

where

$$F_{ij}(k_H) = \int_0^{2\pi}d\alpha\,v_i(\alpha)\,e^{iqz(\alpha)-\lambda\alpha}\int_{\alpha-2\pi}^\alpha d\alpha'\,v_j(\alpha')\,e^{\lambda\alpha'-iqz(\alpha')} \quad (4.77)$$

and

$$z(\alpha) - z(\alpha') = \int_{\alpha'}^\alpha d\alpha''v_z(\alpha'')/\omega_c.$$

Thus the contribution to σ_{ij}^∞ from the orbit in the plane k_H is determined by $F_{ij}(k_H)$. In the Azbel'–Kaner limit, $ql \gg |1+i\omega\tau|$ and $q\Delta z \gg 1$, where $\Delta z = z_{\max} - z_{\min}$ is the caliper diameter of one turn of the orbit, F_{ij} is dominated by the regions where the phases $qz(\alpha)$ and and $qz(\alpha')$ are varying only slowly; in other words, the regions where $v_z(\alpha)$, $v_z(\alpha') \sim 0$—the extremal points of Fig. 4.7 or Fig. 4.8. Around the rest of the orbit, the phase fluctuates rapidly with α or α', and the integrand of (4.77) averages to zero. It is clear from Fig. 4.8 why the regions $v_z(\alpha') \sim 0$ dominate the integration over α': it is only here that the electron, on its way to the point α, sees the electric field in stationary phase, and therefore picks up a significant amount of energy $\Delta\mathscr{E}$ from it. It is less clear why the regions $v_z(\alpha) \sim 0$ dominate the integral over α, until we recall that electrons passing through a given point in the metal at time t with different values of α will lie on orbits which are

Fig. 4.8. An electron moving through a spatially varying transverse field will interact with it most strongly at the extremal points, when it is travelling parallel to the wave-fronts.

similar in shape, but displaced from each other in real space. This displacement of the whole orbit relative to the field pattern will lead to a rapid variation in phase of $\Delta\mathscr{E}$ with α unless the displacement happens to be parallel to the wave-fronts—that is, unless α is at or near an extremal point. We note that since the effective parts of the orbit are those with $v_z \sim 0$, the conductivity components σ_{iz}, σ_{zz} will be very small, both here and in A.S.E. theory, and can be neglected.

Near the extremal point α_l, we can write

$$z(\alpha) = z(\alpha_l) + (y - y_l)^2/2r_{x,l} \simeq z(\alpha_l) + (\alpha - \alpha_l)^2 v_{y,l}^2/2\omega_c^2 r_{x,l},$$

where $r_{x,l}$ is the local radius of curvature of the real-space orbit, projected on the yz-plane. Thus

$$\lambda\alpha - iqz(\alpha) \simeq \lambda\alpha_l - iqz(\alpha_l) - i\beta_l(\alpha - \alpha_l)^2 \qquad (4.78)$$

where $\beta_l = qv_{y,l}^2/2\omega_c^2 r_{x,l}$. (More precisely, the phase term $i\alpha\omega/\omega_c$ in $\lambda\alpha$ displaces the stationary-phase point from α_l, where $v_z = 0$, to some adjacent point α_L, where $v_z = \omega/q$, and the effect of this is to replace α_l by α_L in (4.78). But in the AK limit, α_L is very close to α_l, and we can neglect this refinement.) Each of the extremal points around the orbit will contribute separately to the integrals in (4.77), and we can thus write $F_{ij} = \sum_l \sum_m F_{ij,lm}$, where $F_{ij,lm}$ is the contribution from electrons passing through the point α_l, due to the energy which they have gained from the field near α_m. If $l \neq m$, the two integrals in (4.77) become essentially independent, and if we assume that

$$v_i, v_j \ (i,j = x,y)$$

are approximately constant over the small effective region near the extremal point, the integrals have the form

$$v_{i,j}(\alpha_l)\, e^{\pm\{\lambda\alpha_l - iqz(\alpha_l)\}} \int d\alpha\, e^{\mp\{i\beta_l(\alpha-\alpha_l)^2\}}.$$

If $\beta_l \gg 1$, we may formally extend the limits of integration to $\pm\infty$, and use (just as on p. 196) the fact that $\displaystyle\int_{-\infty}^{\infty} e^{-x^2}\, dx = \pi^{\frac{1}{2}}$. We thus find

$$F_{ij,lm} = e^{-\lambda(\alpha_l-\alpha_m)}\, e^{iq\{z(\alpha_l)-z(\alpha_m)\}}\, v_i(\alpha_l)\, v_j(\alpha_m)\, \pi(\beta_l\beta_m)^{-\frac{1}{2}}. \quad (4.79)$$

Thus the contribution to F_{ij}, and so to σ_{ij}^{∞}, depends on the phase difference $q\Delta z_{lm}$ between the field at $z(\alpha_l)$ and $z(\alpha_m)$. Now Δz_{lm} will depend on the size of the real-space orbit and hence on k_H, so that if $q\Delta z_{lm} \gg 1$, as our treatment assumes, we can plausibly argue that on integrating over k_H to find σ_{ij}^{∞}, these rapidly fluctuating terms will average to zero. This is not quite true, and we shall look at these terms again in §4.8, but for the moment we neglect them.

We are thus left with the contributions $F_{ij,ll}$. Here the two integrals in (4.77) both run over the same part of the orbit, and they can no longer be treated as independent. But if we write $\alpha - \alpha_l = \gamma$, and

$$F_{ij,ll} \simeq v_i(\alpha_l)\, v_j(\alpha_l) \int_{-\infty}^{\infty} d\gamma\, e^{i\beta_l\gamma^2}\left\{\int_{-\infty}^{\gamma} + e^{-2\pi\lambda'}\int_{\gamma}^{\infty}\right\} d\gamma'\, e^{-i\beta_l\gamma'^2} \quad (4.80)$$

where we have made use of the fact that

$$\int_{\gamma-2\pi}^{\gamma} d\gamma' \ldots = e^{-2\pi\lambda'} \int_{\gamma}^{\gamma+2\pi} d\gamma' \ldots,$$

we can integrate at once (combining the contributions from $\pm\gamma$) to find

$$F_{ij,ll} = v_i(\alpha_l)\, v_j(\alpha_l)\, (1 + e^{-2\pi\lambda'})\, \pi/2|\beta_l|. \quad (4.79')$$

Now if the tilt angle ψ between H and x is zero (so that $\lambda' = \lambda$), it follows from (4.28) that $r_x = \hbar\rho_x/eH$, where ρ_x is the radius of curvature of the k-space orbit at the corresponding extremal point, so that

$|\beta_l| = |qm_c^*/\omega_c\rho_{x,l}|v_{y,l}^2/2\hbar$, and this will remain a good approximation if ψ is not too large. Thus from (4.76) we have

$$\sigma_{ij}^\infty(q,H) = \frac{e^2}{4\pi^2\hbar|q|}\int dk_H \coth\pi\lambda' \sum_l \frac{v_{i,l}v_{j,l}}{v_{y,l}^2}|\rho_{x,l}| \qquad (4.81)$$

where the summation extends over all extremal points on each orbit. Finally, we can rewrite this expression in a form similar to (4.71). The extremal-point electrons, with $v_z = 0$, are just the equatorial electrons, with $\theta = \frac{1}{2}\pi$, of our previous discussion, and for these electrons a little geometry shows that $dk_H|\rho_{x,l}|/v_{y,l}^2 \equiv d\phi/v^2\,K(\frac{1}{2}\pi,\phi)$, so that

$$\sigma_{ij}^\infty(q,H) = \frac{e^2}{4\pi^2\hbar|q|}\oint d\phi\, n_i n_j \sum \coth\pi\lambda' \frac{1}{K(\frac{1}{2}\pi,\phi)}, \qquad (4.82)$$

and if $s_{ij}^3 = 4\pi\omega|q|\sigma_{ij}^\infty(q,H)$,

$$s_{ij}^3 = \frac{e^2\omega}{\pi\hbar}\oint d\phi\, n_i n_j \sum \coth\pi\lambda' \frac{1}{K(\frac{1}{2}\pi,\phi)}. \qquad (4.83)$$

(Note that different extremal points on the same orbit k_H will in general occur at different values of the azimuth angle ϕ.)

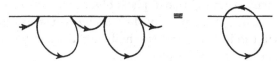

Fig. 4.9. 'Specular reflection', in the k-reversed sense of (4.18), can be treated by the image-metal technique, even for $H \neq 0$.

The results (4.82), (4.83) differ from our previous expression (4.71) in just the way we should expect. We are now dealing with bulk metal, so that the question of diffuse surface scattering does not arise, and all extremal points can contribute to σ_{ij}^∞ in a resonant fashion. Now for $p = \pm 1$, the image-metal treatment of the surface impedance problem, discussed in §4.5, remains completely applicable when $H \neq 0$. If an electron approaching the surface in state $\mathbf{k}$ is reflected into state $-\mathbf{k}$, with or without reversal of $\Delta\mathscr{E}$, it will simply bounce back and forth between the orbits at $+k_H$ and $-k_H$, and its behaviour will be completely equivalent to that of an electron which passes freely to and fro across the plane $z = 0$ between the real metal and the image metal, with $E(-z) = \pm E(z)$ (Fig. 4.9). Thus the solutions (4.61) remain completely valid for $p = \pm 1$, with s_{ij}^3 now given by (4.83). There is one slight complication: (4.61) is only valid if σ_{ij}^{∞} and s_{ij} are diagonal tensors, and to make them diagonal we may need to choose axes x', y' rotated relative to the field direction H_x. Moreover, it may not be possible to diagonalize the real and imaginary parts of σ_{ij}^{∞} simultaneously, if m_c^* and hence λ vary over the FS. To get an exact solution, it is then necessary to evaluate Z when σ_{ij}^{∞} is non-diagonal, starting from a coupled pair of equations of the form (4.50). The resulting expressions (Azbel' & Kaner, 1958) are cumbersome, and in practice this complication is unimportant; we shall not consider it further.

In the A.S.E. problem, the expression Z_0 for the surface impedance with diffuse surface reflection was of precisely the same form as Z_1 and Z_{-1}, and intermediate in magnitude between them. In A.K.C.R., this is no longer true: for a free-electron metal, for instance, in which λ is constant over the FS, the non-resonant background in (4.71), from orbits which hit the surface, appreciably alters the form of the $Z(H)$ curve from that predicted by (4.83). The exact solution of the A.K.C.R. problem for $p = 0$ is a very lengthy business (Bloomfield, 1966a), and we shall only sketch the argument here. The image-metal concept is no longer helpful, because even if we arrange $E(z) = 0$ for $z < 0$, an electron which crosses from $z > 0$ to $z < 0$ is no longer lost to the real metal, as it is when $H = 0$, and as it should be to represent $p = 0$; instead, it will cross back after a short time from the image metal into the real metal. We therefore need to adopt the Azbel'–Kaner technique, in which the perturbation due to the surface is represented by a surface correction current $J^s(z)$ and an associated conductivity $\sigma(q, q')$, as in (4.65). Whether we choose to make the problem symmetrical, with metal filling all space and $E(-z) = E(z)$, as AK do, or leave the region $z < 0$ empty and put $E(z) = 0$ there, as Bloomfield does, is largely a matter of taste. In either case, the problem is to derive an expression for $\sigma_{ij}^s(q, q')$, and then to solve the resultant integral equation, which for the symmetric case is still (4.66).

To derive $\sigma_{ij}^s(q, q')$, we first have to derive the current $J_i(z)$ due to a field $E_j(q) \, e^{i(\omega t - qz)}$, starting again from (4.73), where the lower limit of integration is now α_s; $\alpha - \alpha_s$ is the total orbit angle traversed by the electron after its last collision with the surface before reaching the point z. Thus α_s is, formally, the largest angle less than α which satisfies the equation $z = \int_{\alpha_s}^{\alpha} v_z(\alpha') \, d\alpha' / \omega_c$. For given k_H, α_s is clearly a rather complicated function of α and z, particularly if the orbit has more than two extremal points, or if the field is tilted at an angle to the surface.

These complications are fully discussed by Bloomfield (1966a). Using this modified form of (4.73), it is simple enough to show that the current flowing at depth z can again be written in the form

$$J_i(z) = [\sigma^{\infty}_{ij}(q) - \sigma^{s}_{ij}(q,z)] E_j(q) \, e^{i(\omega t - qz)},$$

where σ^{∞}_{ij} differs from σ^{∞}_{ij} (4.75) only in the replacement of α by α_s in the limits of integration over α'. To find $\sigma^{s}_{ij}(q,q')$, we then have to take the Fourier transform of $J^{s}_i(z)$; that is, we have to evaluate

$$\sigma^{s}_{ij}(q',q) = \frac{1}{2\pi} \int_{-\infty}^{\infty} \sigma^{s}_{ij}(q,z) \, e^{i(q'-q)z} \, dz, \qquad (4.84)$$

where z can be expressed in terms of α and α_s through

$$z = \int_{\alpha_s}^{\alpha} d\alpha' v_z(\alpha')/\omega_c, \quad dz = |v_z(\alpha_s)|/\omega_c|d\alpha_s.$$

The resulting somewhat intimidating expression for $\sigma^{s}_{ij}(q',q)$ is essentially identical with the function $Q_{\mu\nu}(k',k)$ of Azbel' & Kaner (1957).

We need not discuss the form of $\sigma^{s}_{ij}(q',q)$ in any detail, but this function has one important property, not explicitly mentioned by AK, which needs comment. Whereas in $H = 0$, $\sigma^{s}_{ij}(q',q)$ is a smooth function of q', showing no particular structure at $q' = q$ (cf. (4.67)), a very strong maximum—essentially a delta-function—develops at $q' = q$ for $H \neq 0$, so that for practical purposes we can subtract out this delta-function and incorporate it in $\sigma^{\infty}_{ij}(q)$:

$$\sigma'_{ij}(q',q) = \sigma^{s}_{ij}(q',q) - \sigma^{\delta}_{ij}(q)\,\delta(q'-q); \quad \sigma'_{ij}(q) = \sigma^{\infty}_{ij}(q) - \sigma^{\delta}_{ij}(q). \quad (4.85)$$

The remaining term, $\sigma'_{ij}(q',q)$, behaves smoothly at $q' = q$ and is relatively small; like $\sigma^{s}_{ij}(q',q)$ in the $H = 0$ case, it modifies the surface impedance only slightly. It is the delta-function term $\sigma^{\delta}_{ij}(q)$ which essentially 'corrects' $\sigma^{\infty}_{ij}(q)$ and hence $s^{\delta}_{ij}(q)$ from the form (4.83) to the form (4.71); that is, it is this term which cancels out the spurious resonance factor $\coth \pi\lambda$ from the non-resonant extrema.

It is easy enough to see physically how this delta-function arises. Consider the contribution ΔJ to $J_i(z)$, or the contribution $\Delta\sigma$ to $\sigma_{ij}(q,z) = \sigma^{\infty}_{ij}(q) - \sigma^{s}_{ij}(q,z)$, from electrons with a particular value of k_H. We assume for simplicity that H is along the x-axis, so that $\bar{v}_z = 0$, and that the orbit has only two extrema, 1 and 2. It follows from (4.80) and (4.81) that $\Delta\sigma(z) = \Delta\sigma_1(z) + \Delta\sigma_2(z)$, where $\Delta\sigma_1$ comes from real-space orbits which just rise to the depth z at extremum 1, and

$\Delta\sigma_2$ from orbits which just fall to z at extremum 2 (Fig. 4.10). Clearly, if $|z| \gg \Delta z$, the caliper diameter of the orbit, neither of these orbits will intersect the surface, and $\Delta\sigma_1$ and $\Delta\sigma_2$ will both contain the resonant factor $\coth \pi\lambda$. But if $|z| \lesssim \Delta z$, either $\Delta\sigma_1$ or $\Delta\sigma_2$ will become non-resonant, because one or other of the orbits will intersect the surface. Finally, when $|z| \ll 1/q$, fewer and fewer of the resonant orbits will survive—the range of integration over γ in (4.80) will become more and more restricted for resonant orbits—until at $z = 0$ the resonant

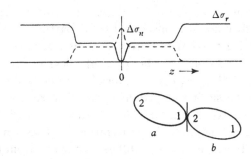

Fig. 4.10. The resonant ($\Delta\sigma_r$) and non-resonant ($\Delta\sigma_n$) contributions to σ, from the orbits a and b. For $z > 0$, $\Delta\sigma_r$ comes from extremal point 2 on orbit b, and $\Delta\sigma_n$ from extremal point 1 on orbit a; for $z > \Delta z$, the orbit diameter, this point also contributes to $\Delta\sigma_r$, and $\Delta\sigma_n$ falls to zero. For $z < 0$, orbits a and b interchange roles.

term vanishes. Thus if we now divide $\Delta\sigma$ into resonant and non-resonant contributions $\Delta\sigma_r$ and $\Delta\sigma_n$, these will vary with z as shown in Fig. 4.10.

The difference between $\Delta\sigma_r + \Delta\sigma_n$ and the bulk value $\Delta\sigma^\infty$ represents the contribution from these electrons to the surface correction term $\sigma_{ij}^s(q, z)$. Clearly, from Fig. 4.10, this difference $\Delta\sigma^s$ is virtually constant and independent of z over the wide range $2\Delta z$, and correspondingly its contribution to the transform (4.84) will extend only over the narrow range $q' \sim q \pm 1/\Delta z$. If $E(q')$ in (4.66) varies only slowly with q' in this range, we can indeed replace $\Delta\sigma^s(q', q)$ by a delta-function contribution at $q' = q$, as in (4.85). The strength of this delta-function is $\Delta\sigma^\infty - (\Delta\sigma_r' + \Delta\sigma_n')$, where $\Delta\sigma_r'$ and $\Delta\sigma_n'$ are the values in the plateau region $1/q \lesssim |z| \lesssim \Delta z$, and when this is subtracted from $\Delta\sigma^\infty$, as in (4.85), the net effect is to leave a contribution $\Delta\sigma_r' + \Delta\sigma_n'$ to the 'effective' bulk conductivity $\sigma_{ij}'(q)$. Physically, this is just what we might expect: since $E(z)$ only extends a short way from the surface compared with Δz, the true bulk conductivity $\sigma_{ij}^\infty(q)$ is irrelevant; what matters is clearly the modified value $\sigma_{ij}'(q)$ in the plateau region.

And we can write down $\sigma'_{ij}(q)$ immediately, clearly, simply by removing the resonant factor coth $\pi\lambda$ from the appropriate terms in (4.82), to give

$$\sigma'_{ij}(q) = \frac{e^2}{4\pi^2\hbar|q|} \oint d\phi \, n_i n_j \left\{ \sum_R \frac{\coth \pi\lambda}{K_R} + \sum_N \frac{1}{K_N} \right\}$$

$$= \frac{e^2}{4\pi^2\hbar|q|} \oint d\phi \, n_i n_j \left\{ \sum_B \frac{1}{1-e^{-2\pi\lambda}} \frac{1}{K_B} + \sum_I \frac{1}{K_I} \right\} \quad (4.86)$$

in complete agreement with (4.71), (4.72).

Within the narrow region $|z| \ll 1/q$, $\Delta\sigma_r$ falls to zero and $\Delta\sigma_n$ correspondingly increases; these localized departures from the plateau values $\Delta\sigma'_r$ and $\Delta\sigma'_n$ contribute to the residual surface term $\sigma'_{ij}(q, z)$, whose Fourier transform will be $\sigma'_{ij}(q', q)$. Clearly, $\sigma'_{ij}(q', q)$ will now vary smoothly with q and q', just as for $H = 0$, and lead to only a small correction to the surface impedance. Azbel' & Kaner (1957, (5.11)) have derived an expression for $\sigma'_{ij}(q', q)$, and have used it to estimate the size of the correction to Z, by iteration of (4.66) (their equations (5.12), (6.7)). The equation now differs somewhat from (4.67) on the right-hand side, because $\sigma'_{ij}(q', q)$ is somewhat different from its zero-field form, but for free electrons AK (1958, (3.19)) find that $Z_0^{(2)}$ is again about 15 % larger than Z_1; that is, Z_0 should be given to good approximation by (4.63). More recently, Hartmann & Luttinger (1966) have solved the free-electron equation (AK 1957, (6.7)) exactly, by an elegant Laplace-transform method. They find

$$Z_0 = 2\sqrt{3} A^2 \pi \omega e^{i\pi/3}/s \quad (4.87)$$

where

$$A(\lambda) = \frac{4\sin(\pi z_1/3)\sin[\pi(1-z_1)/3]}{\sin(\pi z_1)}; \quad \cos \pi z_1 = \tfrac{1}{2}(1+e^{-2\pi\lambda}). \quad (4.88)$$

This result should apply to any metal in which m_c^*, τ, and therefore $\lambda(= i\omega/\omega_c + 1/\omega_c\tau)$ are constant over the FS, and there are no internal extrema on the orbits, so that (4.72) reduces to

$$s^3 = s^3(H = 0)/(1 - e^{-2\pi\lambda}).$$

Thus the exact value of Z_0 differs from the approximation (4.63) by the factor A^2. As λ varies, this factor oscillates about a value ~ 1 with the same period as the main resonance factor $(1 - e^{-2\pi\lambda})^{\frac{1}{3}}$, but in antiphase and with considerably smaller amplitude (Hartmann, 1966), so that the net effect is to leave the oscillations of R and X diminished slightly in amplitude, but otherwise virtually unaffected.

Q

For $H = 0\,(\lambda \to \infty)$, $A^2 = 1\cdot031$, so that the 'zero-field' impedance predicted by this equation is slightly greater than that predicted by (4.63), which is exact for $H = 0$. But this is to be expected: as remarked above, the A.K.C.R. form of $\sigma'_{ij}(q',q)$ is somewhat different from the A.S.E. form and (4.87), (4.88) are not exact in the limit $H = 0$. The change-over in the form of $\sigma'_{ij}(q',q)$ occurs at fields so low that $\omega_c \tau \ll 1$, and the A.K.C.R. oscillations have therefore vanished.

In this low-field region, a series of small peaks occurs in Z, first studied by Khaikin (1960) and later by Koch & Kuo (1966), who proposed a qualitative explanation for them in terms of the electron trajectories near the surface. In terms of our present treatment, they suggested that the peaks arose when the extremal point on the orbit α_l and the stationary-phase point α_L (cf. the remarks below (4.78)) could no longer be regarded as coincident, but became separated by a distance $\delta z \sim 1/s$. This argument predicts that the peaks will occur at fields $H \sim 10-50$ Gs for $\omega/2\pi \sim 50$ Gc/s, in reasonable agreement with experiment, but predicts $H \propto \omega^{\frac{2}{3}}$, whereas the experiments of Koch & Kuo indicate $H \propto \omega^{\frac{3}{4}}$. Khaikin (1966) has proposed a somewhat similar explanation, which also predicts $H \propto \omega^{\frac{3}{4}}$.

Prange & Nee (1968) have suggested an interesting alternative explanation of these low-field peaks in quantum-mechanical terms. We recall from §3.2 that the wave-function of a free electron in a field H_x can be written in the form $\psi = e^{ik_x x} e^{ik_y y} F(z-z_0)$, where F is a simple harmonic oscillator wave-function, oscillatory in the region $|z-z_0| < r_c$ and exponentially decaying outside this region. Here $z_0 = \hbar k_y/eH_x$, and r_c is the classical orbit radius. This is the solution appropriate to an unbounded metal. If the metal occupies only the region $z > 0$, and we impose the boundary condition $F = 0$ at $z = 0$ (instead of $F \to 0$ at $z \to -\infty$), the allowed energy levels are considerably altered, particularly if we choose k_y so that $z_0 \sim -r_c$. Then only the exponential tail and perhaps the last few nodes of the oscillator-like wave-function lie within the metal, and an electron in such a state is confined to a region very close to the surface. Prange & Nee show that for given k_x, k_y the energy levels of these surface states, near $\mathscr{E}_F$, are spaced (nonuniformly) at intervals of order $(k_F r_c)^{\frac{1}{3}} \hbar\omega_c$, very much larger than the spacing $\hbar\omega_c$ for states in bulk metal. (For a simple semi-classical derivation of the energy levels of these surface states, see Koch, 1968.) They show that the various low-field peaks of dR/dH observed by Koch & Kuo can be identified with transitions between various pairs of surface levels $\mathscr{E}_m$ and $\mathscr{E}_n$, occurring when $\mathscr{E}_n - \mathscr{E}_m = \hbar\omega$, an moreover the theory predicts that for any given transition $H_{\text{peak}} \propto \omega^{\frac{2}{3}}$, as observed experimentally. Prange & Nee in fact consider a general FS, rather than a free-electron metal, and show that the level spacing for electrons of given k_x is then determined by $v_{y0}/\rho_{z0}^{\frac{1}{3}}$, where v_{y0} is the electron velocity at the point on the orbit where $v_z = 0$, so that the electron is travelling parallel to the surface, and ρ_{z0} is the radius of curvature of the orbit at the corresponding point in k-space.

This explanation of the low-field peaks has interesting implications: in putting $F = 0$ at $z = 0$, we are implicitly assuming that the electron is specularly reflected at the surface (in some sense), so that the outgoing wave interferes coherently with the incoming wave to produce a well-defined standing wave with a node at $z = 0$. If reflection were completely diffuse, no such standing waves would be expected, and if this is indeed the explanation of the low-field peaks, reflection must be at least partially specular. The main evidence for diffuse reflection, perhaps, comes from Pippard's (1957) A.S.E. measurements on Cu, from which he deduced the shape and size of the FS: assuming $p = 0$, he found that the FS was of just the right size to fit within the Brillouin zone, whereas the assumption $p = 1$ would have yielded a FS of quite the wrong size. But it may be that the surfaces now being used in A.K.C.R. studies are sufficiently perfect to show appreciable specular reflection; if so, the line-widths of the low-field peaks should enable the value of p to be estimated, as pointed out by Prange & Nee.

4.7 A.K.C.R. LINE-SHAPES, POLARIZATION EFFECTS, TIPPING EFFECTS

Assuming that $p \sim 0$, and accepting (4.63) and (4.72) as an adequate approximation for Z when the field H_x is parallel to the sample surface, it is a straightforward matter to work out the form of $Z(H)$ in the limiting cases $\omega_c \tau < 2\pi$, $\omega_c \tau > 50$. If $\omega_c \tau < 2\pi$, so that $|e^{-2\pi\lambda}| < e^{-1}$, we can put $(1 - e^{-2\pi\lambda})^{-1} \sim 1 + e^{-2\pi\lambda}$, and if m_c^* is constant over the FS and there are no internal extrema on the orbits, we have simply $Z \propto e^{\frac{1}{3}i\pi}(1 - \frac{1}{3}e^{-2\pi\lambda})$. Thus R and X vary sinusoidally with ω/ω_c, and the oscillation amplitude falls as $e^{-2\pi/\omega_c \tau}$. The quantities usually measured experimentally are dR/dH or dX/dH, and these will vary as $\omega_c^{-2} e^{-2\pi\lambda}$. Typical experimental results obtained under these conditions are shown in the upper curve of Fig. 4.11.

If m_c^* is not constant, but passes through extremal values m_c^* at certain azimuth angles ϕ_e, the observed oscillations will occur at the field intervals $\Delta(1/H) = e/m_e^* \omega$, corresponding to the extremal masses m_e^*. Near ϕ_e, we can write $m_c^* \simeq m_e^*[1 - \alpha(\phi - \phi_e)^2]$, so that

$$\omega/\omega_c \simeq (\omega/\omega_{c,e})\,[1 - \alpha(\phi - \phi_e)^2], \qquad (4.89)$$

and if $\omega_c \tau < 2\pi$ we can write (4.72) in the form

$$s_{xx}(0)/s_{xx}(H) = Z_{xx}(H)/Z_{xx}(0) \simeq 1 - \frac{1}{3s_{xx}^3(0)}\,\frac{e^2\omega}{\pi\hbar}\oint d\phi\, n_x^2 K_B^{-1} e^{-2\pi\lambda}. \qquad (4.90)$$

Near ϕ_e, the integral has the form

$$\int d\phi\, n_x^2 K_B^{-1} e^{-2\pi/\omega_c \tau} e^{-2\pi i\omega/\omega_{c,e}} \exp\left[\frac{2\pi i\omega\alpha}{\omega_{c,e}}(\phi - \phi_e)^2\right] \qquad (4.91)$$

and if $\omega\alpha/\omega_{c,e} > 1$ we can in the usual way approximate this by

$$[n_x^2 K_B^{-1}]_{\phi_e}\, e^{-2\pi/\omega_{c,e}\tau} e^{-2\pi i\omega/\omega_{c,e}} \int_{-\infty}^{\infty} d\phi\, \exp\left[\frac{2\pi i\omega\alpha}{\omega_{c,e}}(\phi - \phi_e)^2\right]$$

$$= [n_x^2 K_B^{-1}]_{\phi_e}\left(\frac{\omega_{c,e}}{2i\omega\alpha}\right)^{\frac{1}{2}} e^{-2\pi/\omega_{c,e}\tau} e^{-2\pi i\omega/\omega_{c,e}}. \qquad (4.92)$$

That is, we argue that the distorted Cornu spiral represented by the integrand of (4.91) can be replaced by the undistorted spiral (4.92) without great error, provided that the distortion of the first turn of the spiral is not too severe (cf. Pippard 1960, p. 228). Each extremal mass m_e^* will contribute a similar term to (4.90), so that $[Z_{xx}(H)/Z_{xx}(0)] - 1$ will consist of a set of superposed damped oscillatory terms, each having a different period in $1/H$, and in general it will be difficult to resolve them clearly (Häussler & Welles, 1966).

In the opposite limit, $\omega_c \tau > 50$, it becomes much simpler to resolve the contributions from different extremal masses. From the identity

$$\coth \pi\lambda = \pi^{-1} \sum_{n=-\infty}^{\infty} (\lambda - in)^{-1} = \pi^{-1} \sum_{-\infty}^{\infty} \left\{\frac{1}{\omega_c \tau} + i\left(\frac{\omega}{\omega_c} - n\right)\right\}^{-1},$$

we see that the factor $\coth \pi\lambda$ is equivalent to a sum of resonance curves, centred on the points $\omega = n\omega_c$, and if $\omega_c \tau \gg 1$ these resonances become very sharp, so that near any given resonance we can neglect the remaining terms in the series, and write

$$\coth \pi\lambda = \frac{\omega_c \tau}{\pi}\,\frac{1}{1 + i(\omega - n\omega_c)\tau}. \qquad (4.93)$$

If m_c^* varies with ϕ, ω_c will likewise vary, and there may be a number of orbits at different ϕ which are resonant simultaneously in a given field H, with different values of n. But if these resonances occur in regions where $dm_c^*/d\phi \neq 0$, it is easy to see that they do not lead to any corresponding resonances in s_{ij}^3 (4.71); on integrating over ϕ, the resonant effect vanishes, and for these non-extremal mass regions we can effectively replace $\coth \pi\lambda$ by unity. Physically, this is because a slight change of H does not remove the resonant contribution to s_{ij}^3; it merely displaces it to an adjacent orbit of slightly higher

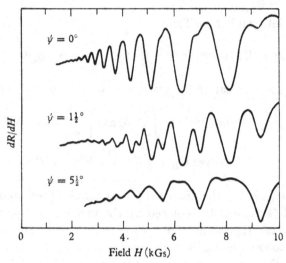

Fig. 4.11. The effect of tip angle ψ on the A.K.C.R. signal
(from Grimes & Kip, 1963).

or lower m_c^*. As usual, the observable effects come from the regions on the FS where m_c^* becomes extremal, so that a large number of orbits come into resonance simultaneously. By substituting (4.89) into (4.93) and integrating over ϕ, we find that whenever an extremal-mass orbit is near resonance, s_{ij}^3 is enhanced by an amount

$$\Delta s_{ij}^3 = \frac{e^2\omega}{\pi\hbar}\left[\frac{n_i n_j}{K_R|\alpha|^{\frac{1}{2}}}\right]_{\phi_e}\frac{1}{n}\left(\frac{\pm i\omega\tau}{1+i\mu}\right)^{\frac{1}{2}} \qquad (4.94)$$

where $\mu = (\omega - n\omega_{c,e})\tau$. The positive sign is to be taken if $\alpha > 0$ in (4.89), corresponding to a maximum of m_c^*, and the negative sign if $\alpha < 0$. Clearly, the form of Δs_{ij}^3 does not change discontinuously at $\alpha = 0$. (4.94) is valid for $|\alpha| \gtrsim 1/n$; if $|\alpha| \ll 1/n$, so that m_c^* remains effectively equal to m_c^* over some finite range of azimuth angles $\Delta\phi$, we have simply

$$\Delta s_{ij}^3 \simeq \frac{e^2\omega}{\pi\hbar}\left[\frac{n_i n_j}{K_R}\right]_{\phi_e}\frac{\Delta\phi}{\pi n}\frac{\omega\tau}{1+i\mu}. \qquad (4.94a)$$

If $\omega = n\omega_{c,e}$ when $H = H_n$, we can write $\mu = -\omega\tau(H - H_n)/H_n$, so that if $\omega\tau$ is large the line-width of the resonance is correspondingly narrow, and the resonances due to different extremal-mass orbits are readily resolved. Away from resonance, s_{ij}^3 reverts approximately to its zero-field value $s_{ij}^3(0)$, given by (4.47), and if this is asumed to be somewhat larger than Δs_{ij}^3, the line-shapes calculated using (4.94) or (4.94a) agree quite well with those observed experimentally in high-purity metals (Chambers, 1965; Moore, 1966).

The relaxation time τ can be determined from the line-width if $\omega_c\tau$ is large, or from the decay in oscillation amplitude if $\omega_c\tau$ is small ((4.90) and (4.92)). If τ varies around the orbit, the measurements will clearly yield the mean scattering rate $\overline{\tau^{-1}}$ averaged over the cyclotron period τ_c:

$$\overline{\tau^{-1}} = \tau_c^{-1} \int_0^{\tau_c} \tau^{-1} dt.$$

Experimentally it is found that $\overline{\tau^{-1}}$ varies with temperature as $a + bT^3$ (Moore, 1966; Häussler & Welles, 1966). The first term is due to scattering by static imperfections, and the second to phonon scattering. This can be compared with the effective free paths for d.c. thermal and electrical transport, L_T and L_E ((4.12) and (4.5)): at low temperatures, $L_T^{-1} \simeq a + b_T T^3$ and $L_E^{-1} \simeq a + b_E T^3(T/\theta_D)^2$, where θ_D is the Debye temperature. The extra factor $(T/\theta_D)^2$ appears in L_E^{-1} because collisions with low-energy phonons, though they rapidly randomize the energy of the electrons, only deflect their velocity vectors $\mathbf{v}$ through small angles. But in A.K.C.R., even a small deflection of $\mathbf{v}$ will be enough to displace the electron into an orbit which no longer passes through the skin depth, or one which hits the surface. Thus even small-angle collisions are effective in reducing the A.K.C.R. signal amplitude, just as they are in reducing the d.c. heat flow, and τ^{-1} is a measure of the total scattering rate, including all collisions. It therefore varies like L_T^{-1} rather than L_E^{-1}. This sensitivity of A.K.C.R. signals to small-angle scattering was pointed out by Azbel' & Kaner (1958).

Three types of extremal-mass orbit can be distinguished: central orbits, limiting-point orbits and intermediate orbits. If the symmetry of the FS is such that the orbit area $\mathscr{A}$ is an even function of k_H, then the orbit in the plane $k_H = 0$ will necessarily be of extremal area, with $\partial\mathscr{A}/\partial k_H = 0$, and consequently the orbit mass m_c^* (4.27) will also be extremal. From (4·30), it follows that $\bar{v}_H = 0$ for central orbits. At an elliptic limiting point, where $\mathbf{v}$ is parallel to $\mathbf{H}$, m_c^* is again extremal (and is given by (4.32)): although $\partial m_c^*/\partial k_H$ does not in general vanish at these points, $\partial k_H/\partial \phi$ clearly does, so that $\partial m_c^*/\partial \phi = 0$. An electron exactly at the limiting point, travelling exactly parallel to H, will of course travel in a straight line rather than a helix, so that it will not undergo cyclotron resonance (cf. Fig. 4.6); the resonance from the limiting-point regions will be produced by electrons which are near but not at the limiting points, whose real-space radii are somewhat greater than the skin depth. Lastly, there may be other intermediate orbits, in addition to central and limiting-point orbits, for which m_c^* happens to be extremal.

One can achieve some discrimination between these different types of extrema by looking at the dependence of the resonance on the r.f. field polarization. If the steady field H is in the x-direction, limiting-points electrons will have $v_y/v = n_y \simeq 0$, so that they contribute nothing to σ_{yy} or to s_{yy}^3: an r.f. field E_y does not interact appreciably with these electrons. Limiting-point resonances should therefore vanish if the r.f. field is perpendicular to H. Conversely, central orbits have $\bar{v}_x = 0$, so that they are likely to have $v_x < v_y$ at most points around the orbit, including the point where $v_z = 0$, where they are travelling parallel to the surface. Thus, we expect that for these orbits $n_x^2 \ll n_y^2$, so that they contribute less to s_{xx}^3 than to s_{yy}^3: the resonances from these orbits should be much less marked if the r.f. field is parallel to H than if it is perpendicular, and should vanish entirely if the r.f. field is perpendicular to the direction in which the electrons move while they are passing through the skin depth. For intermediate orbits, we might in general expect $v_x \sim v_y$, so that the strength of resonance will be comparable for both polarizations. Moreover, these orbits always occur in conjugate pairs at $\pm k_H$ (cf. Fig. 4.7), and the electrons on these two orbits will in general be moving in different directions in the xy-plane as they pass through the skin depth, so that the resonance should now appear for any direction of polarization. But these rules no longer apply if the tensor $s_{ij}^3(0)$ is highly anisotropic, with its principal axes at an angle to the xy-axes (whose orientation is now fixed by the direction of H); polarization studies then need to be interpreted with some caution.

The different types of extremal-mass orbit should also behave differently when H is tipped at a small angle ψ to the sample surface. The axes of the helical orbits are then likewise tilted, so that after one revolution an electron has moved a distance $2\pi \bar{v}_z/\omega_c = 2\pi \bar{v}_H \sin \psi/\omega_c$ in the z-direction, and may no longer come up into the skin depth. If $\bar{v}_H$ is large, as for limiting-point orbits, we expect the effect of tipping to be more severe than for central orbits, with $\bar{v}_H \simeq 0$. Even for central orbits, we expect some effect; though the extremal-mass orbit itself has $\bar{v}_H = 0$, so that the real-space orbit is a closed loop rather than a helix, contributions to the observed signal come from a small range of orbits around the central orbit, and these will be affected by tipping.

Azbel' & Kaner originally predicted that A.K.C.R. signals should therefore vanish for tip angles greater than a fraction of a degree. Experimentally, signals are often observable out to $\psi > 5°$, and it was Khaikin (1962 b) who first realized the importance of accurate field alignment: unless the field is parallel to the surface to within a few minutes of arc (and the sample surface is correspondingly flat), the resonance pattern is significantly distorted, and can no longer be interpreted in terms of the unmodified AK theory. A variety of tipping effects have been observed experimentally, and we shall not discuss them all. One of the most interesting is the 'mass doubling' observed for example in Al and K (Grimes et al. 1963; Spong & Kip, 1965; Grimes & Kip, 1963). As Fig. 4.11 shows, the A.K.C.R. peaks tend to split and double in number for $\psi \sim 1°$, and for still larger tip angles the peaks become inverted.

In principle, the tipped-field A.K.C.R. problem can be solved in precisely the same way as the parallel-field problem: we merely have to substitute in (4.66) the forms of $\sigma^\infty(q)$ and $\sigma^s(q',q)$ appropriate to a tipped field, and solve the resulting equation. In practice, the form of $\sigma^s(q',q)$, or of the modified quantity $\sigma'(q',q)$ of (4.85), becomes very difficult to handle in a tipped field, but if we assume as before that $\sigma'(q',q)$ is small and can be neglected as a first approximation, the problem becomes somewhat simpler. We then merely have to modify $\sigma'(q)$ to take account of the field tipping, and this simply involves replacing λ by $\lambda' = \lambda - iq\bar{v}_z/\omega_c$. Again, it is convenient to combine the contributions from the conjugate orbits at $\pm k_H$ (Fig. 4.7), which have equal and opposite values of $\bar{v}_H$ and hence of $\bar{v}_z = \bar{v}_H \sin \psi$. The second form of (4.86) can be obtained from the first, as before, by combining the term from the non-resonant extremal point at $-k_H$ with that from the resonant extremal point at k_H, but now it is more convenient to combine the terms from the resonant points at $\pm k_H$. We assume for simplicity that the FS has mirror symmetry about the plane containing x and $\mathbf{H}$, so that the two resonant points both have the same value of K_R and of n_x^2, n_y^2. Writing $\pi\bar{v}_H \sin \psi/\omega_c = a$, so that $\pi\lambda' = \pi\lambda - iqa$, we can thus replace $\coth(\pi\lambda - iqa)$ by

$$\tfrac{1}{2}[\coth(\pi\lambda - iqa) + \coth(\pi\lambda + iqa)]$$
$$= \coth \pi\lambda / [1 + \sin^2 qa \, (\coth^2 \pi\lambda - 1)].$$

We thus find

$$\sigma'_{ii}(q) = \frac{e^2}{4\pi^2\hbar|q|} \oint d\phi\, n_i^2 \left\{ \sum_R \frac{\coth\pi\lambda}{1 + \sin^2 qa\,(\coth^2\pi\lambda - 1)}\frac{1}{K_R} + \sum_N \frac{1}{K_N} \right\} \quad (4.95)$$

which is a generalization of the expression obtained by Pincus (cf. Grimes *et al.* 1963).

If we neglect the right-hand side of (4.66), Z is given as usual by (4.53), but evaluation of this equation is now complicated by the q-dependence of σ', which no longer varies simply as $|q|^{-1}$. If the tip angle is large, so that $|qa| \gg 1$ for the dominant wave-numbers $|q| \sim 1/s$, $\sin^2 qa$ will oscillate rapidly over the range of integration of (4.53), and it is tempting as a very crude approximation to replace $\sin^2 qa$ by $\frac{1}{2}$. This yields

$$\langle \sigma'_{ii}(q) \rangle \sim \frac{e^2}{4\pi^2\hbar|q|} \oint d\phi\, n_i^2 \left\{ \sum_R \frac{\tanh 2\pi\lambda}{K_R} + \sum_N \frac{1}{K_N} \right\}. \quad (4.96)$$

This expression is of precisely the same form as the expression (4.86) for $\psi = 0$, except that $\coth \pi\lambda$ has become $\tanh 2\pi\lambda$. It follows at once that if (4.96) has any validity, $Z(H)$ will fluctuate twice as fast with field, and that the impedance maxima will be replaced by minima, just as found experimentally (Fig. 4.11) for $\psi \sim 1°$. But in fact this agreement is somewhat spurious: what we need is not the average of $\sin^2 qa$ over the range $\delta q = \pi/a$, nor the average of $\sigma'(q)$ over that range (which in fact yields the zero-field expression (4.59)), but if anything the average of $[q^2 + 4\pi i\omega\sigma'(q)]^{-1}$. Nevertheless, the form of $Z(H)$ computed by Pincus using the correct expression (4.95) turns out to be very similar to that predicted by the dubious (4.96), and shows good qualitative agreement with the experimental results of Fig. 4.11. Pincus assumed a 'lens-shaped' FS, with the axis of the lens parallel to the untipped field, so that the values of $\bar{v}_H$ were concentrated closely around the two limiting-point values $\pm v_L$, and the values of a were correspondingly concentrated around $\pm \pi v_L \sin\psi/\omega_c$. Grimes *et al.* (1963) and Spong & Kip (1965) have given a physical picture of the origin of the mass-doubling effect on this model: electrons on one face of the lens, spiralling down into the metal away from the surface, are refocused periodically at depths $2a, 4a, 6a, \ldots$ (cf. (4.42)), to produce 'current sheets' flowing parallel to the surface at these depths. Electrons on the conjugate orbits on the other face of the lens, spiralling upwards through these current sheets, will pick up energy from them (or from the fields induced by them) on their way to the surface, and it is not difficult to show that the net effect of this two-stage process is to produce the doubling of frequency exhibited by (4.96) and observed experimentally. Analytically, one can see from (4.52) that if $\sigma(q)$ includes a component which fluctuates as $\cos 2qa$, anomalies may occur in $E(z)$ at depths $z = 2na$, where this fluctuation and its harmonics synchronize with the e^{-iqz} factor in the Fourier transform. We discuss these effects more fully in the next section.

Smith (1967) has made an extensive survey of the behaviour of Z in tipped fields, using (4.95) or an equivalent expression in the approximate evaluation of (4.53), and shown that the mass-doubling effect is not confined to the lens-shaped FS. It can also occur if the FS is spherical (as in K): the operative current sheets are then produced by electrons near the limiting points on the FS, where $\bar{v}_H$ is extremal. For more general forms of FS, he finds that the theoretical predictions rival in their variety the effects observed experimentally, and in some cases show reasonably good agreement with them. We should not expect close agreement at large tip angles, because the surface term $\sigma'(q',q)$ then becomes increasingly important. Consideration of the tipped-field equivalent of Fig. 4.10 shows that the surface term $\sigma'(q,z)$ will extend increasingly far from the surface as the tip angle increases, because electrons spiralling away from the surface will have to travel a distance $|z| \sim 2\bar{v}_H \tau \sin \psi$ before their resonance factor builds up to its full value $\coth \pi\lambda'$. (At smaller distances, when the electron has only completed n revolutions of its orbit since leaving the surface, the summation involved in passing from (4.73) to (4.74) will extend over only n terms, and this will have the effect of replacing $\coth \pi\lambda'$ by

$$\coth \pi\lambda'[1 - e^{-2\pi(n+1)\lambda'}].)$$

If the surface term $\sigma'(q,z)$ extends out to a depth $|z| \gg 1/q$, its Fourier transform $\sigma'(q,q')$ will show a sharp peak at $q = q'$. Just as the sharp peak we previously encountered in $\sigma^s(q,q')$ had the effect of subtracting out the 'unwanted' resonances from $\sigma^\infty(q)$, to yield $\sigma'(q)$, so this peak will further modify $\sigma'(q)$ by removing the resonance factor for electrons spiralling outwards from the surface, and leaving this factor only for those spiralling towards the surface. Thus we can no longer combine the resonance factors from $\pm k_H$, as we did in (4.95), since one of them is absent, and the two-stage process involved in the mass-doubling effect can no longer occur. It would be tedious to work out a detailed theory for the transition region, but for large ψ we can expect that the only surviving A.K.C.R. signals will come from regions where $\bar{v}_H \sim 0$ over an appreciable range of k_H, so that a significant number of electrons pursue essentially closed rather than helical orbits in real space. In Cu, for example, Koch *et al.* (1964) find that some parts of the *FS* appear to have this property, since A.K.C.R. signals from these regions persist out to large tip angles.

4.8 R.F. SIZE EFFECTS

In § 4.5 to 4.7, we have assumed that the metal sample was effectively a semi-infinite block. If the thickness is large compared with l, this is a valid assumption. But it is clear from Fig. 4.6 that if A.K.C.R. is observed in a thin parallel-sided plate of thickness d, the resonance signals will abruptly vanish when the caliper diameter of the orbit $z_{max} - z_{min} = \Delta z$ becomes equal to d. At lower fields, the electrons can complete only a fraction of an orbit before they collide with one or other of the sample surfaces. The cut-off field H_c is thus a direct measure of the caliper diameter Δk_y of the k-space orbit, since

$$\Delta k_y = eH_x\Delta z/\hbar \qquad (4.97)$$

from (4.28). This A.K.C.R. cut-off was first observed by Khaikin (1962a). Shortly afterwards Gantmakher (1963) showed that surface impedance anomalies could be observed at the cut-off field at much lower frequencies, of order 1 Mc/s, and since then he has shown that these r.f. size effects provide a powerful new tool for *FS* studies (Gantmakher, 1967). Once again, it is easiest to see the physical origin of these effects by considering the electric field distribution E(z) in real space and applying ineffectiveness concept arguments, though once again a detailed solution has to be phrased in terms of the Fourier components of the field E(q).

In a pure enough metal at low temperatures, A.S.E. conditions ($sl \gg 1$) may persist down to $\omega/2\pi < 10^3$ c/s. At a frequency of say 1 Mc/s, the skin depth $1/s$ will typically be about 3μ, much smaller than the typical orbit radius in a field of say 1 kGs, so that the electrons will still see the r.f. field only briefly when they rise into the skin depth, as shown in Fig. 4.6. Since $\omega \ll \omega_c$, the electron sees the field in essentially the same phase each time, and thus picks up energy coherently from it. Effectively the cyclotron resonance condition $\omega = n\omega_c$ is now satisfied with $n = 0$, and the resonance factor $\coth \pi\lambda$ in (4.71) becomes $\coth(\pi/\omega_c\tau)$. The value of s_{ij}^2 is correspondingly enhanced above its zero-field value, and Z is likewise reduced. In a thin plate, the enhancement factor will only operate for orbits which are 'viable' in the sense that their caliper diameter is less than d. If the caliper diameter Δk_y passes through an extremal value $\Delta k_{y,e}$ for some value of k_H, there will be an abrupt change in the number of viable orbits at the field

$$H_c = \hbar \Delta k_{y,e}/ed, \qquad (4.98)$$

and a corresponding abrupt change in Z (Kaner, 1958). In fact the effect is a little more subtle than this: impedance anomalies can also occur when $H \simeq nH_c$, with n integral. These arise because all the electrons of extremal diameter Δz_e, which pass through the skin depth with $v_z = 0$, will again have $v_z = 0$ at a depth Δz_e below the surface, and will give rise to a 'current sheet' at that depth. This r.f. current sheet generates an electric field 'spike' at that depth, which in turn excites a current sheet at depth $2\Delta z_e$, and so on (Azbel', 1960). We thus have a chain of extremal-diameter orbits extending down into the metal, together with an associated chain of field spikes, as shown schematically in Fig. 4.12. Whenever one of these field spikes passes through the far side of the plate, an anomaly will occur in Z, and this will happen whenever $H = nH_c$.

The electrons on these extremal-diameter orbits will usually have $v_y > v_x$ as they pass through the skin depth, so that an r.f. field E_y will produce stronger size-effect signals than E_x. A field E_x will couple most strongly to the limiting-point electrons, and these can also produce size-effect signals, of a rather different type, if H is tipped at an angle ψ to the sample surface. As discussed at the end of the last section, the limiting-point electrons will then be refocused at intervals

$$\Delta z = 2\pi v_H \sin \psi / \omega_c = 2\pi \hbar \sin \psi / eHK^{\frac{1}{2}} \qquad (4.99)$$

(using (4.32)) to produce a sequence of current sheets and field spikes with this spacing. As before, size-effect signals will be seen whenever refocusing occurs at the far side of the plate (Fig. 4.13), i.e. whenever $n\Delta z = d$ or $H = nH_c$, where now

$$H_c = 2\pi \hbar \sin \psi / edK^{\frac{1}{2}}. \qquad (4.100)$$

If the limiting-point mass $|m_c^*| = \hbar/v_L K^{\frac{1}{2}}$ is also known from A.K.C.R. measurements, the velocity v_L of the limiting-point electrons can thus be determined (Gantmakher & Kaner, 1963).

Clearly, both types of size-effect signal yield very direct information on the FS geometry; in addition, the amplitude of the limiting-point signals gives direct information on the mean free path of the electrons near the limiting point. These electrons have to travel a distance

$d/\sin\psi$ to cross from one side of the plate to the other, and their chance of doing so without collision is $\exp(-d/l\sin\psi)$. If we assume, plausibly, that for small tip angles this is the major factor determining the variation of signal amplitude A, a plot of $\ln A$ vs $1/\sin\psi$ at once yields d/l (Gantmakher & Kaner, 1963). Here again, as in A.K.C.R. measurements of $\overline{\tau^{-1}}$ (§4.7), we expect $l^{-1} \sim a+bT^3$, because even

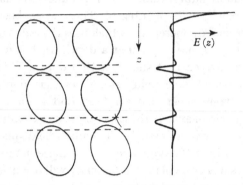

Fig. 4.12. Illustrating (very schematically) the field spikes set up below the surface of the metal by chains of interacting orbits.

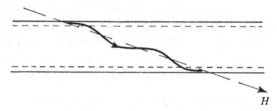

Fig. 4.13. The r.f. size effect in a tilted field.

small-angle scattering by phonons will remove an electron from the narrowly defined group which contributes to the signal, and this temperature-dependence is indeed observed experimentally (Gantmakher & Sharvin, 1965).

To treat the r.f. size-effect problem theoretically, we first need a suitable definition of the surface impedance Z of a parallel-sided plate occupying the region $0 \leqslant z \leqslant d$. If the fields $\mathbf{E}$ and $\mathbf{H}$ in the plate vary only with z, Poynting's theorem shows that energy is absorbed by the plate from the applied fields at a rate (per unit area) given by the instantaneous value of

$$W = \mathbf{E}(0).\mathbf{I}(0) + \mathbf{E}(d).\mathbf{I}(d). \qquad (4.101)$$

Here we have written

$$H(0) \times n/4\pi = I(0) \quad \text{and} \quad H(d) \times n/4\pi = -I(d),$$

where n is the unit vector in the z-direction, so that

$$I(0) + I(d) = \int_0^d J(z)\, dz.$$

Equation (4.101) clearly separates the contributions to W from the two sides of the plate. If we write $E_i(0) = Z_{ij}(0)\, I_j(0)$, etc., we have

$$W = Z_{ij}(0)\, I_i(0)\, I_j(0) + Z_{ij}(d)\, I_i(d)\, I_j(d).$$

Since $\partial E/\partial z = \mp 4\pi i\omega I$ at $z = 0, d$, we can again define Z_{ij}, as in (4.43), through

$$-4\pi i\omega E_i(0) = Z_{ij}(0)\, E_j'(0), \quad 4\pi i\omega E_i(d) = Z_{ij}(d)\, E_j'(d). \quad (4.102)$$

But if the plate is thin, the surface impedance so defined will depend on the mode of excitation. If, for example, we apply r.f. only to the face $z = 0$, some energy may pass through the plate and be radiated from the face $z = d$, so that the energy $E(d) \cdot I(d)$ absorbed by this face is negative, and correspondingly the 'output impedance' $Z^0(d)$ is negative; in fact it follows from (4.102) that $-Z^0(d) = 4\pi c$, the impedance of free space (in e.m.u.). The energy drawn from the applied field at $z = 0$ will be correctly given by the input impedance $Z^I(0)$, but only part of this energy will be taken up by the plate; the rest will be transmitted through it.

It is difficult to calculate this input impedance $Z^I(0)$ directly, because $I(0)$ depends on $E(d)$ as well as on the applied field $E(0)$, and $E(d)$ is not initially specified; it must be determined in the course of the problem. It is simpler (in principle, at least) to calculate the impedances Z^S and Z^A, corresponding to symmetric excitation of the plate $(E(d) = E(0))$ or antisymmetric excitation $(E(d) = -E(0))$. In the latter case,

$$I(0) + I(d) = \int_0^d J\, dz = 0 \quad \text{and} \quad H(0) = H(d),$$

and this corresponds to the usual experimental arrangement, in which both faces of the plate are exposed to a uniform r.f. magnetic field. Clearly $Z^A(0) = Z^A(d)$ and $Z^S(0) = Z^S(d)$. Since a suitable combination of symmetric and antisymmetric excitation is equivalent to one-sided excitation, the value of Z^I and $I(d)/I(0)$ for the one-sided case can be found, if necessary, in terms of Z^S and Z^A. Assuming for simplicity that Z^A and Z^S are both diagonal, we find

$$I(d)/I(0) = (Z^A - Z^S)/(8\pi c + Z^A + Z^S),$$

which can be approximated by

$$I(d)/I(0) = (Z^A - Z^S)/8\pi c \qquad (4.103)$$

since typically Z^A, $Z^S \sim 10^{-4} \times 4\pi c$, and to the same accuracy

$$Z^I = \tfrac{1}{2}(Z^A + Z^S). \qquad (4.104)$$

We note that Gantmakher & Kaner (1963) have used the approximation $Z^I \simeq -4\pi i\omega[E(0) - E(d)]/E'(0)$, and that other approximations were used earlier by Azbel' (1960) and Kaner (1963). In these approximations, $E(z)$ is assumed to have the same form as in the semi-infinite metal. It is not clear that this expression will be a good approximation to Z^I, though it may be a somewhat better approximation to Z^A (Kaner & Fal'ko, 1966), if the antisymmetric field distribution is thought of as the superposition of two unperturbed fields, one starting out from each face of the plate. Even then, the assumption that the form of $E(z)$ is unperturbed by the second face is clearly a dubious one.

The detailed computation of Z^A or Z^S is a laborious process, which has not yet been attempted, but we can sketch the steps involved. We again use the AK technique described in §4.5, but to generate the required antisymmetric or symmetric field distribution we now need a source-current sheet of strength $\mp 2I_s$ at $z = d$, as well as a sheet $2I_s$ at $z = 0$. In fact to preserve the symmetrical disposition of $E(z)$ and $J(z)$ about each source sheet, we now need an infinite array of sheets, of strength $2I_s$ at $z = 2nd$ and of strength $\mp 2I_s$ at $z = (2n+1)d$. The Fourier transform of this array will be

$$J_s(q) = \pi^{-1} I_s \sum_{n=-\infty}^{\infty} (e^{2niqd} \mp e^{(2n+1)iqd}).$$

In place of (4.51), we now have

$$I_s = -\int_0^{\frac{1}{2}d} J(z)\,dz = [E'(0) - E'(\tfrac{1}{2}d)]/4\pi i\omega. \qquad (4.105)$$

For symmetric excitation, $E'(\tfrac{1}{2}d) = 0$, and even for antisymmetric excitation, we shall usually have $E'(\tfrac{1}{2}d) \ll E'(0)$. Thus in place of (4.66), we now have

$$[q^2 + 4\pi i\omega\sigma^\infty(q)]E(q) + \frac{1}{\pi}[E'(0) - E'(\tfrac{1}{2}d)]\sum_n (e^{2niqd} \mp e^{(2n+1)iqd})$$

$$= 4\pi i\omega \int_{-\infty}^{\infty} \sigma^s(q,q')E(q')\,dq'. \qquad (4.106)$$

To find the value of Z^A or Z^S it only remains to insert appropriate expressions for $\sigma^\infty(q)$ and $\sigma^s(q,q')$, solve (4.106), and hence evaluate (4.102). As before, we in fact replace $\sigma^\infty(q)$ and $\sigma^s(q,q')$ by modified functions $\sigma'(q)$ and $\sigma'(q,q')$, and as before we deduce the form of these from the form of $\sigma(q,z) = \sigma^\infty(q) - \sigma^s(q,z)$.

Consider first the parallel (untipped) field problem. We now have to take into account the terms $F_{ij,lm}$ of (4.79), which we have so far neglected. These terms give the contribution to the current at one extremal point on the orbit, $z(\alpha_l)$, due to the energy acquired from the field at another such point, $z(\alpha_m)$. We ignored these terms in §4.6, because they fluctuate rapidly with Δz_{lm}, and indeed they only have a slight effect on the A.K.C.R. signals in bulk metal (Azbel', 1960). But clearly it is these terms which give rise to the parallel-field r.f. size effect, by producing current sheets at the depths where Δz_{lm} takes on extremal values Δz_e. If in the usual way we write

$$\Delta z_{lm} = \Delta z_e[1 - \gamma(k_H - k_{H,e})^2]$$

and assume that the other factors in the integrand of (4.76) remain substantially constant over the first turn or so of the Cornu spiral, we find that each extremum Δz_e contributes to $\sigma^\infty(q)$ an amount of the form

$$\Delta\sigma_{ij}^\infty(q) = \frac{e^2}{4\pi^2\hbar|q|}\frac{2e^{-\lambda\Delta\alpha}}{1-e^{-2\pi\lambda}}\frac{v_{il}v_{jm}}{|v_{yl}v_{ym}|}\left(\frac{\pi\rho_{xl}\,\rho_{xm}}{iq\gamma\,\Delta z_e}\right)^{\frac{1}{2}}e^{iq\Delta z_e}, \quad (4.107)$$

where $\Delta\alpha = \alpha_l - \alpha_m$ is the difference in orbit angle between the two extremal points $(v_z = 0)$ on the extremal orbit $(\Delta z_{lm} = \Delta z_e)$. Adding these contributions to the main term (4.81) or (4.82), we can write

$$\sigma_{ij}^\infty(q) = \frac{A_{ij}}{|q|}\left\{\frac{1+e^{-2\pi\lambda}}{1-e^{-2\pi\lambda}} + \sum_n \frac{a_n}{\sqrt{(qD_n)}}\frac{e^{-\lambda\Delta\alpha_n}}{1-e^{-2\pi\lambda}}e^{i(qD_n+\eta_n)}\right\}, \quad (4.108)$$

where $D_n = \Delta z_{e,n}$ and the coefficients a_n are of order unity. For simplicity we have assumed that $\lambda = 1/\omega_c\tau$ is the same for all orbits; since $e^{-2\pi\lambda}$ is now monotonic rather than oscillatory, this will not materially affect the form of $\sigma(q)$. The phase factors η_n (which depend on the sign of $\rho_{x,l}, \rho_{x,m}$) are also unimportant and will be neglected from now on.

We note in passing that (4.108) is relevant to the theory of magnetoacoustic oscillations, though the condition $qD \gg 1$, assumed in deriving (4.108), is seldom attainable experimentally in ultrasonic work. The theory of ultrasonic attenuation falls outside the scope of this chapter; it has been well reviewed by Pippard (1960, 1965b), Stolz (1963) and Mackintosh (1964).

If $p = 1$, so that all electrons suffer specular reflection at the surface of the plate in the sense of (4.18), the conductivity will as usual be unaffected by the presence of the surfaces, and we can at once deduce $E(q)$ by inserting (4.108) in (4.106), with $\sigma^s(q, q') = 0$. But the surface impedance $Z(H)$ will then show additional size-effect lines, for instance when $D = md$, with m integral, which will certainly not be observed if the surface reflection is diffuse. Even if reflection is partly specular for electrons arriving at the surface at glancing angles, with $v_z \sim 0$, it is likely to be effectively diffuse for electrons arriving at large angles, and the lines at $D = md$ will only be observed if these large-angle electrons are specularly reflected. To treat the more realistic $p = 0$ problem, we first consider the form of $\sigma(q, z)$. For a plate of thickness d occupying the region $0 \leqslant z \leqslant d$, this can be deduced at once from (4.108), simply by noting that for orbits which intersect either surface of the plate, $e^{-2\pi\lambda}$ must be replaced by zero. Formally, we can put $\tau = 0$ in the term $e^{-2\pi\lambda}$ for these electrons, because they can never complete an orbit without collision. Similarly, if the part of the orbit between α_l and α_m intersects the surface, we must also replace $e^{-\lambda\Delta\alpha}$ by zero. Note that the second term in (4.108) may still survive even if the overall orbit diameter is more than the plate thickness, if the orbit contains internal extrema. Thus in Fig. 4.7, the term in D_{12} may survive even when $d < D_{14}$, and will give rise to a size-effect signal when $d = D_{12}$. Signals of this sort have indeed been observed by Gantmakher & Krylov (1965), but for simplicity we consider here only overall extremal diameters such as D_{14}.

We consider, in fact, the simplest possible situation: a closed convex FS on which the orbit diameter D varies from zero at the limiting points to a maximum value of D_e on the central section. We can then write

$$\sigma_{ij}(q, z) = \frac{A_{ij}}{|q|} \left\{ \frac{1 + e^{-2\pi\lambda}}{1 - e^{-2\pi\lambda}} f(z) + [1 - f(z)] \right.$$

$$\left. + \frac{a}{\sqrt{(qD_e)}} \frac{e^{-\lambda\Delta\alpha}}{1 - e^{-2\pi\lambda}} [g_L(z) e^{iqD_e} + g_R(z) e^{-iqD_e}] \right\}, \quad (4.109)$$

where $g_L(z)$ and $g_R(z)$ represent the contributions from orbits lying to the left and to the right of z, for which $\Delta z_e = D_e, -D_e$ respectively. The factor $f(z)$ can likewise be written $f_L(z) + f_R(z)$. In bulk metal, $g_L = g_R = f = 1$, and $f_L = f_R = 0.5$. A little thought shows that $f_L(z)$ and $g_L(z)$ will vary with z and with field as shown in Fig. 4.14, and that $f_R(d - z) = f_L(z)$, $g_R(d - z) = g_L(z)$. At high enough fields,

R

where $D_e < d$, all orbits whose right-hand extremal points fall in the region $d - D_e < z < d$ will be viable— that is, they will lie completely within the metal—and f_L attains its full value of 0·5. For $z < d - D_e$, the extremal-diameter orbits will intersect the boundary at $z = 0$, so that at this point f_L abruptly falls; it then continues to fall smoothly to zero at $z = 0$, where essentially all orbits intersect the surface. The fall at $z = d - D_e$ will only be pronounced if the extremal-diameter orbits contribute significantly to σ_{ij}^∞; thus we expect the fall to be more pronounced for σ_{yy} than for σ_{xx}, since we expect $v_y^2 \gg v_x^2$ at the extremal points on the extremal orbits.

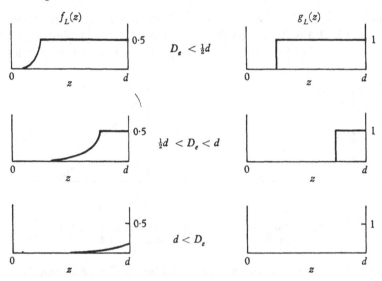

Fig. 4.14. The variation of the functions $f_L(z)$ and $g_L(z)$ across a thin plate (cf. (4.109)).

Since only the extremal-diameter orbits contribute to g_L, this function falls to zero for $z < d - D_e$; smaller diameter orbits do not contribute a tail at smaller z, as they do to f_L. In both f_L and g_L, the transition at $z = d - D_e$ will not be completely abrupt, but will extend over a distance $\sim 1/q$, and similarly f_L and g_L will fall to zero within a distance $\sim 1/q$ of $z = d$, for the reasons discussed in connection with Fig. 4.10. As H tends to zero, the functions f and g also tend to zero, and $\sigma_{ij}(q, z)$ tends to $A_{ij}(q)/|q|$ in (4.109). Thus $A_{ij}(q)/|q|$ is simply the zero-field conductivity of (4.59).

To find $\sigma'(q)$ and $\sigma'(q', q)$, we now need merely to take the Fourier transform of $\sigma(q, z)$: $\sigma'(q)$ is simply the average value of $\sigma(q, z)$ across the plate, and $\sigma'(q',q)$ is given by an expression of the form (4.84), with $\sigma^s(q,z)$ replaced by $\sigma'(q) - \sigma(q,z)$. We note that in (4.106) we are implicitly considering an infinite block of metal, with scattering boundaries at intervals d, so that $\sigma(q,z)$ is periodic with period d, and $E(z)$, $J(z)$ are likewise periodic with period d or $2d$ for Z^S or Z^A. Thus their Fourier transforms will in fact be Fourier series, though we shall continue to use the Fourier transform notation.

For $\sigma'(q)$, we thus have

$$\sigma'_{ij}(q) = \frac{A_{ij}}{|q|}\left\{1 + \frac{1}{1 - e^{-2\pi\lambda}}\left[2e^{-2\pi\lambda}\bar{f} + \frac{2a}{\sqrt{(qD_e)}}\,e^{-\lambda\Delta\alpha}\bar{g}\cos qD_e\right]\right\}, \quad (4.110)$$

where $\bar{f} = \bar{f}_L + \bar{f}_R$, averaged over the plate, and $\bar{g} = \bar{g}_L = \bar{g}_R$; qualitatively, $\bar{f}$ and $\bar{g}$ will vary as shown in Fig. 4.15. Replacing $\sigma^\infty(q)$ by $\sigma'(q)$ and $\sigma^s(q,q')$ by $\sigma'(q,q')$ in (4.106), and neglecting the right-hand

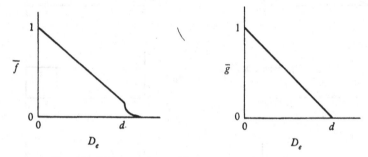

Fig. 4.15. The variations of $\bar{f}$ and $\bar{g}$, the mean values of $f_L + f_R$ and of g_L, with extremal orbit diameter D_e.

side, we can thus obtain a first approximation to $E(q)$, $E(z)$ and Z in the usual way. But in fact $\sigma'(q,q')$, as we have defined it here, will become large near $q = q'$, just as $\sigma^s(q,q')$ did in the A.K.C.R. problem, and for similar reasons; and this time, we cannot legitimately absorb this peak into $\sigma'(q)$, as we did there. Consequently, the zero-order solution of (4.106), neglecting the right-hand side, yields certain spurious r.f. size-effect 'signals', as we shall see, which would be removed in higher approximation.

Using (4.49) to invert from $E(q)$ to $E(z)$, the zero-order solution becomes

$$E(z) = -\frac{1}{\pi}[E'(0) - E'(\tfrac{1}{2}d)]\int_{-\infty}^{\infty}\frac{\sum_n[e^{2niqd} \mp e^{(2n+1)iqd}]\,e^{-iqz}}{q^2 + i|q|^{-1}s^3(I + J\cos qD_e)}\,dq, \quad (4.111)$$

where $s^3 = 4\pi\omega A$, $I = [1 + 2e^{-2\pi\lambda}\bar{f}/(1 - e^{-2\pi\lambda})]$ and

$$J = 2ae^{-\lambda\Delta\alpha}\bar{g}/\sqrt{(qD_e)}\,(1 - e^{-2\pi\lambda}).$$

240 R. G. CHAMBERS

As usual, the main contribution to the integral comes from the regions $|q| \sim s$. If D_e is large compared with the skin depth $1/s$, we therefore have $qD_e \gg 1$ and $I \gg J$, so that we can expand the denominator of (4.111) and write

$$E(z) = -\frac{1}{\pi}[E'(0) - E'(\tfrac{1}{2}d)] \int_{-\infty}^{\infty} F^{-1}\Sigma[e^{2niqd} \mp e^{(2n+1)\,iqd}]\, e^{-iqz}$$
$$[1 - G\cos qD_e + G^2\cos^2 qD_e - \dots]\,dq, \quad (4.112)$$

where $F = q^2 + i|q|^{-1}s^3I$ and $G = i|q|^{-1}s^3J/F$. Now just as (4.52) yields a field $E(z)$ which is localized near $z = 0$, so here the terms in $\exp(-iq)(z - n'd \pm mD_e)$ (where $n' = 2n$ or $2n+1$) will produce fields localized near $z = n'd \pm mD_e$. In this approximation there will therefore be an appreciable perturbation of the surface field $E(0)$, and hence of Z, whenever $n'd = mD_e$, with $n' \leqslant m$. (For $n' > m$, $d < D_e$, so that $\bar{g} = 0$ and $G = 0$.) But it is clear that only the $n' = 1$ term has physical significance: the other terms correspond to chains of orbits extending through the scattering surfaces, and they would be eliminated in an exact solution, taking into account the $\sigma'(q, q')$ terms.

It would not be prohibitively difficult to iterate (4.106) once, to find the first-order correction due to $\sigma'(q, q')$; alternatively, we can apply a crude 'correction' to the zero-order treatment, which eliminates these spurious terms very simply, merely by restricting the summation in (4.111) or (4.112) to the two $n = 0$ terms. If we also neglect $E'(\tfrac{1}{2}d)$, we then arrive at essentially the approximation used by Kaner & Fal'ko (1966):

$$Z(H) = -4\pi i\omega E(0)/E'(0) \simeq -4\pi i\omega[E_0(0) \mp E_d(0)]/E'(0), \quad (4.113)$$

where $E_0(z)$ is the field in a semi-infinite metal block excited by a current sheet at $z = 0$, if the conductivity of the block has the modified form (4.110), and $E_d(z)$ is the field excited by a current sheet at $z = d$. In this approximation, the form of $Z(H)$ directly reflects the form of $E(z)$ at the depth $z = d$ below the surface: as H is increased and D_e shrinks, the field splashes at $z = d - mD_e$ due to excitation of the face $z = d$ are 'pulled in' through the face $z = 0$, to produce anomalies in $E(0)$ and in $Z(H)$. Clearly, these anomalies will be localized around the fields $H = mH_c$, with H_c given by (4.98). Using this approximation, Kaner & Fal'ko have calculated the shape of the differentiated line dZ/dH, but only for a simplified form of $\sigma'(q)$, and a precise treatment of the line-shape would be of considerable interest.

Experimentally, each line is found to show an oscillatory structure, as would be expected if the lines reflect the form of the field splashes at $z \sim mD_e$ or $d - mD_e$. Figure 4.5 shows that near $z = 0$, $E(z)$ oscillates with a 'period' Δz of order $4/s$, so that if the field splashes near $z = mD_e$ show a similar period, the oscillations in Z would be expected to have a period $\Delta H \sim 4H/sD_e$. This agrees reasonably well with the period of the observed oscillations. In order to measure Δk_y accurately, using (4.98), one needs to know precisely what point in the line corresponds to the critical field H_c. In the absence of a detailed theory of the line-shape, Koch & Wagner (1966) have studied this question experimentally, and shown that in K, where the FS is practically spherical, H_c occurs at the low-field edge of the line: as the frequency of the r.f. is varied, the line width changes, but this point remains unshifted.

We need not discuss the theory of the tipped-field r.f. size effect in any detail; it follows precisely the same course as the parallel-field treatment, and the only change is in the form of $\sigma(q, z)$ and hence of $\sigma'(q)$, $\sigma'(q, q')$. The extremal contribution to $\sigma_{yy}(q, z)$ may now come from orbits which are displaced somewhat from the extremal-diameter orbits, because of the effect of the drift term $\bar{v}_z$ (Koch & Wagner, 1966), and the extremal contribution to $\sigma_{xx}(q, z)$ will come from electrons near the limiting points, which produce field splashes at the intervals Δz given by (4.99). As in the parallel-field case, an exact treatment of the problem has not yet been attempted, though Gantmakher & Kaner (1963) have given an approximate treatment based on (4.113); see also Kaner (1967).

4.9 SOME PROPAGATING MODES

In the phenomena we have discussed so far, the r.f. electric field falls off rapidly below the surface of the metal, though isolated field splashes may also exist at greater depths. But the high-frequency properties of a metal in a magnetic field show a remarkable variety, and in some circumstances propagating modes can exist which penetrate to much larger distances without appreciable attenuation. The nature of the surface scattering is then relatively unimportant, and it is sufficient to consider only the relatively simple $p = 1$ problem. In fact it is sometimes possible to neglect surface effects entirely, and consider simply the condition for wave propagation in bulk metal. To illustrate this, consider the form of $E(z)$ in the A.S.E. problem, given by (4.52)

$$E(z) = -\frac{E'(0)}{\pi} \int_{-\infty}^{\infty} \frac{e^{-iqz} dq}{q^2 + 4\pi i \omega \sigma^{\infty}(q)}. \qquad (4.52)$$

(As we have seen, this equation also represents an approximate solution to $p = 0$ problems, if we replace $\sigma^\infty(q)$ by $\sigma'(q)$ where necessary.) For $z = 0$, (4.52) can be evaluated directly, but for $z \neq 0$ the oscillatory integrand makes it simpler to use contour methods. If we treat q as a complex variable $q' - iq''$ and evaluate (4.52) by completing the contour in the lower half-plane, it is clear that $E(z)$ will be determined (partly at least) by the poles of the integrand; that is, by the roots of the equation

$$D(q) \equiv q^2 + 4\pi i \omega \sigma^\infty(q) = 0. \qquad (4.114)$$

As remarked below (4.50), this is simply the equation satisfied by a propagating mode $E \sim \exp i(\omega t - qz)$ in the unbounded metal. Such a mode will satisfy (4.44) everywhere except near the surface. Near the surface, the relationship between E and J is perturbed, and for $p = 1$ the perturbation is represented by the discontinuity in E' at $z = 0$ and the associated source current $J_s(q)$ in (4.50), which leads to the more complicated field distribution (4.52).

Each of the roots of (4.114) in the lower half-plane of q will contribute to $E(z)$ an amount $iE'(0)q_{r,n}^{-1}e^{-iq_n z}$. Here $(2q_{r,n})^{-1}$ is the residue of $D(q)^{-1}$ at q_n, so that $D(q) \to 2q_{r,n}(q - q_n)$ for $q \to q_n$, or $2q_{r,n} = (dD/dq)_{q=q_n}$. If $\sigma^\infty(q)$ varies only slowly with q near q_n, we therefore have $q_{r,n} = q_n$, but not otherwise.

In the A.S.E. problem, the form of $\sigma^\infty(q)$ is such that there is only one root of (4.114) in the lower half-plane. But $\sigma^\infty(q)$ also contains a branch-point in the lower half-plane, and this will also contribute to $E(z)$. The origin of this branch-point is clear enough. If $q'' > 1/l$, integration of (4.16) will no longer yield (4.57) for electrons travelling in the direction of $\mathbf{q}$, because the integral diverges as $t \to -\infty$: the amplitude of the electric field varies as $e^{-q''z}$, and rises rapidly for negative z, so that the small fraction of electrons ($\sim e^{z/l}$) arriving from large negative distances contributes an unlimited amount to the current. Thus a bulk conductivity $\sigma^\infty(q)$ does not exist for $q'' > 1/l$; such a field distribution can only exist near the surface (cf. Pippard, 1966b). Nevertheless, the formal expression (4.58) remains well-behaved for $q'' > 1/l$, except when $q' = \omega\tau q''$. As q' passes through this value, the value of the integrand at the point $\cos\theta = 1/q''l$ changes abruptly from $+i\infty$ to $-i\infty$. To render $\sigma^\infty(q)$ analytic, we must therefore introduce a cut from the branch-point at $q = q_b = -i(1 + i\omega\tau)/l$ to $-i(1 + i\omega\tau)\infty$ (and a similar cut in the upper half-plane, which does not concern us here).

For a free-electron metal, (4.58) can be integrated explicitly to give

$$\sigma^\infty(q) = \frac{\sigma_0}{1+i\omega\tau}\frac{3}{2a^3}[(1+a^2)\tan^{-1}a - a] \qquad (4.115)$$

where $a = ql/(1+i\omega\tau)$. Since $\tan^{-1}a = \frac{1}{2}i\ln[(1-ia)/(1+ia)]$, this function does indeed have branch-points at $a = \pm i$, and for more complex metals, the behaviour of (4.58) will be qualitatively similar. In the extreme anomalous limit $sl \to \infty$, it is enough to replace (4.60) by $4\pi\omega\sigma^\infty(q) = s^3/(q^2 + 1/l^2)^{\frac{3}{2}}$, which reduces to (4.60) for all relevant q on the real axis but remains analytic at $q = 0$, and which has branch-points at $q'' = \pm 1/l$. (For simplicity, we assume $\omega\tau \ll 1$.) For $l \to \infty$, evaluation of (4.52) then yields

$$\frac{E(w)}{(dE/dw)_{w=0}} = \frac{2i}{3\beta}e^{-i\beta w} - \frac{1}{3}e^{-w} - \frac{2i}{\pi}\int_0^\infty \frac{ue^{-uw}}{1-u^6}du, \qquad (4.116)$$

where $w = sz$ and $\beta = q_1/s = e^{-i\pi/6}$. The first term arises from the pole at q_1, and represents the contribution of the propagating mode; the remaining terms arise from the integration along the sides of the cut from $q = -i/l(\to 0)$ to $q = -i\infty$, and represent the effect of the surface perturbation. The last term in (4.116) can be represented approximately by $-2i/(\pi w^2 + 6\sqrt{3})$, so that for large w this term becomes dominant, and $E(w) \sim w^{-2}$.

In this approximation, the main effect of finite l is to replace the lower limit of integration in (4.116) by $1/sl$, so that at large distances we have $E(w) \sim w^{-2}e^{-w/sl} \sim z^{-2}e^{-z/l}$. This is confirmed by a more careful analysis (Reuter & Sondheimer, 1948, equation (A 42)). Thus if $sl \gg 1$ the term in $E(w)$ due to the surface perturbation is comparable with or greater than the propagating term at all depths, and this surface term dies away at a depth $\sim l$, as we might expect on physical grounds.

The situation is quite different in the classical skin-effect limit, where ωl^2 is so small that the root q_1 satisfies $|q_1| \ll 1/l$. For $\omega\tau \ll 1$, we can then put $\sigma^\infty(q) \simeq \sigma_0$, independent of q, and the propagating-mode contribution to (4.52) becomes $iE'(0)q_1^{-1}e^{-iq_1z}$, falling off much more slowly than $e^{-z/l}$. At the same time, the branch-point at $q = -i/l$ recedes towards $-i\infty$, and the surface term becomes vanishingly small —as we might expect, since the propagating mode by itself has the correct slope $E'(0)$ at $z = 0$. Thus the propagating mode is now dominant at all depths, and the surface impedance becomes equal simply to the wave impedance $Z_W = 4\pi\omega/q_1$ of the propagating mode.

Generalizing, we conclude that whenever (4.114) has a root q_n

with $q_n'' < 1/l$, the corresponding weakly attenuated propagating mode will become dominant at large distances from the surface (since on physical grounds the surface term is always expected to die out in a distance l or less), and that if $\sigma^\infty(q)$ varies only slowly near $q = q_n$ this mode is likely to dominate the surface impedance too.

Equation (4.114) for the propagating mode and (4.52) for the form of $E(z)$ are valid only if $\sigma_{ij}^\infty(q)$ is diagonal. It is simple enough to find the appropriate generalization of these equations when $\sigma_{ij}^\infty(q)$ is not diagonal (see e.g. Kaner & Skobov, 1966, equations (2.4) and (9.12)), and hence also to find the generalization of (4.53) for Z_{ij}. Even in the A.S.E. limit, where $4\pi\omega\sigma_{ij}^\infty$ takes the simple form $s_{ij}^3/|q|$, the form of Z_{ij} becomes very much more complicated than (4.61) (Azbel' & Kaner 1958, equation (3.17)), and the form of $E(z)$ is likewise much more complicated than (4.116). And when a magnetic field is applied, the dispersion equation may have a bewildering variety of roots. Even on a free-electron model, the positions of these will depend on ω, ω_c, τ, l, the electron density n_e and the angle between $\mathbf{q}$ and $\mathbf{H}$. In a two-band model, containing n_e and n_h free-electron-like electrons and holes, the roots will also depend on the corresponding hole parameters. Many of these roots depend on the average properties of the whole electron assembly, and give little information on the detailed form of the FS, so that it is usually sufficient to consider a free-electron model or a simple two-band model. The form of $\sigma^\infty(q, \omega_c)$ for a free-electron model, for arbitrary $\mathbf{q} \wedge \mathbf{H}$, has been given by Cohen, Harrison & Harrison (1960) in the form of an infinite series of integrals involving Bessel functions, and also by Kaner & Skobov (1964) for $\mathbf{q} \perp \mathbf{H}$. The apparent differences between these two formulations can largely be removed by suitable application of Bessel-function sum rules. Kaner & Skobov (1966) have given an extensive review of the propagating modes which can arise under various conditions; here we consider just two situations of physical interest: $\mathbf{H} \| \mathbf{q}$ and $\mathbf{H} \perp \mathbf{q}$.

For $\mathbf{H} \perp \mathbf{q}$, we have the familiar A.K.C.R. geometry of §§4.6–8. If $\omega_c\tau$ is large, the value of $\sigma'(q)$ varies rapidly near resonance (cf. (4.86), (4.94)), and the position of the root q_1 of (4.114) will correspondingly vary rapidly. Kaner & Skobov (1964) point out that if the whole FS resonates simultaneously and $\omega_c\tau$ is large enough, q'' may become small near resonance so that the mode is only weakly attenuated, but this is unlikely to occur often in practice. They also point out that if account is taken of the extremal-diameter contribution (4.107) to $\sigma'(q)$, so that $4\pi\omega\sigma'(q) \sim s^3[1 + b\cos qD]/|q|$ with $b < 1$ (cf. (4.110)), the oscillatory term in $\sigma'(q)$ will give rise to many additional roots of (4.114), closely spaced in q. (In their equation (1.7), for free electrons, they neglect the oscillatory term, but it will clearly exist here too, with $b \sim (qD)^{-\frac{1}{2}}$.) From this viewpoint the field splashes and therefore the r.f. size-effect signals arise from interference between these propagating modes.

In an earlier paper (1963), Kaner & Skobov concluded that propagating modes of low damping could not exist for $E = E_x$, unless the field H was tipped at an angle to the x-axis. However, this conclusion was based on a truncated form of $\sigma^\infty(q, \omega_c)$ for free electrons, retaining only the lowest-order terms in the Bessel function expansion. Walsh & Platzman (1965) found experimentally that at microwave frequencies a weakly attenuated mode was strongly excited in K under these conditions, with H parallel to x, and showed that such a mode would indeed be expected to exist, if the full expression for $\sigma^\infty(q, \omega_c)$ is used. This mode has very small q, of order $1/R$ where $R = v_F/\omega$, so that (4.114) reduces essentially to $\sigma^\infty(q, \omega_c) \simeq 0$. Walsh & Platzman show that for free electrons σ^∞_{xx} does indeed become zero, if collisions are neglected, when ω_c is somewhat larger than ω. The precise value of ω_c at which σ^∞_{xx} vanishes depends on q, and the resultant dispersion relation is approximately
$$q^2 R^2 \simeq 5(\omega_c^2/\omega^2 - 1), \tag{4.117}$$
in good agreement with experiment. Experimentally, q was determined from the fluctuations in $Z(H)$ of a thin plate (of thickness d) on the high-field side of the fundamental A.K.C.R. peak at $\omega_c = \omega$; these fluctuations were assumed to occur when q satisfied $q = 2\pi n/d$, so that a standing-wave pattern was set up across the thickness of the plate.

Clearly, these fluctuations in $Z(H)$ are quite distinct from those occurring in the r.f. size-effect: those arose from localized field spikes, themselves generated by the interference of propagating modes of high q; here the field E_x is sinusoidal rather than localized, and of very much lower q. Physically, the condition for σ^∞_{xx} to vanish (neglecting collisions) is that the average electron—averaging over all orbits on the FS—shall pick up zero net energy from the spatially varying wave E_x: the metal then becomes effectively transparent to this wave. The precise form of the dispersion relation will therefore depend on the properties of the whole FS, and (4.117) applies only to a free-electron metal.

More recently, Platzman & Walsh (1967) have shown that (4.117), and the corresponding expression for the other r.f. polarization

($E = E_y$), are substantially modified if many-body effects are taken into account. The Landau coefficients which describe these many-body effects can thus be measured by studying the dispersion relation of these modes. This result is of great interest: electron–electron and electron–phonon interactions have remarkably little effect on most of the phenomena discussed in this chapter, and for most purposes the one-electron approximation, which we have used throughout, is perfectly adequate. At this one point, however, the one-electron approximation becomes inadequate, so that many-body effects have observable consequences. (See also Mermin & Cheng, 1968.)

We turn now to the other magnetic-field orientation, $\mathbf{H}\|\mathbf{q}$. For this field orientation, two quite distinct types of propagating mode exist— the 'helicon mode', if $n_e \neq n_h$, and the 'Alfvén wave' if $n_e = n_h$. The conditions for exciting Alfvén waves are difficult to satisfy in metals, and we refer the reader elsewhere (Buchsbaum, 1964; Kaner & Skobov, 1966) for a discussion of them. The helicon mode—a very weakly attenuated, circularly polarized mode—is readily excited in uncompensated metals. Similar modes have been familiar in magneto-ionic and plasma physics for many years, but in metals this mode was first explicitly discussed by Konstantinov & Perel' (1960) and by Aigrain (1960), who named it the helicon, and first observed by Bowers, Legendy & Rose (1961). This observation stimulated considerable interest among plasma physicists as well as solid-state physicists (see e.g. Buchsbaum, 1964, and other papers in the same volume), and also stimulated interest in propagating modes generally.

To discuss helicon propagation, we shall consider a FS having rotational symmetry about the z-axis (where $\mathbf{H}\|\mathbf{q}\|z$), as in our discussion of Sondheimer oscillations (§4.4); as there, this simplifies the theory without unduly restricting its applicability. We consider a circularly polarized transverse electric field

$$\mathbf{E} = E_x + iE_y = E_0 e^{i(\omega t - qz)}$$

in an unbounded block of metal, and calculate the resultant current $\mathbf{J} = J_x + iJ_y = J_0 e^{i(\omega t - qz)}$. Note that the sign of ω now determines the sense of rotation of the field. As in discussing Sondheimer oscillations, we write $\mathbf{V} = v_x + iv_y = v_r e^{-i(\alpha + \omega_c t)}$, where $v_r = v_r(k_z)$, and a straightforward combination of the methods used to derive (4.35) and (4.57) then shows that $\mathbf{J} = \sigma_c(q)\,\mathbf{E}$, with

$$\sigma_c(q) = \frac{e^2}{4\pi^2\hbar^2} \int dk_z \frac{|m_c^*|\,v_r^2 \tau}{1 + i(\omega + \omega_c - qv_z)\tau}. \tag{4.118}$$

It is easy to show that (4.114) now gives the condition for propagation of circularly polarized waves, if σ is replaced by σ_c, and that (4.52) gives $\mathbf{E}(z)$ (and hence Z) when the surface $z = 0$ is excited with circularly polarized r.f. The response to a linearly polarized excitation can of course be found by decomposing it into two circular waves of opposite sense of rotation.

Consider first the behaviour of (4.118) in the classical limit, $ql \ll 1$. As usual, the propagating modes we shall be concerned with have phase velocities ω/q much less than the electron velocities v, so that $|\omega\tau| \ll ql$, and in the classical limit we automatically have $|\omega\tau| \ll 1$. If $\omega_c = 0$, (4.118) then yields the d.c. conductivity σ_0, and for finite $\omega_c\tau$ we therefore have

$$\sigma_c = \sigma_0/(1 + i\omega_c\tau) \qquad (4.119)$$

where we have assumed for simplicity that $\omega_c\tau$ is independent of k_z. Inserting (4.119) into (4.114), we find that for $\omega = \pm|\omega|$,

$$q_1 = 2^{\frac{1}{2}}/\delta_c(\mp\omega_c\tau \pm i)^{\frac{1}{2}}, \qquad (4.120)$$

where $\delta_c = (2\pi|\omega|\sigma_0)^{-\frac{1}{2}}$ is the classical zero-field skin depth. Thus if $\omega_c\tau$ is large and positive and ω is positive, q_1 is almost purely imaginary, and the wave decays exponentially without phase change in a distance $(\frac{1}{2}\omega_c\tau)^{\frac{1}{2}}\delta_c$. But if either ω or ω_c is reversed in sign, q_1 becomes almost purely real. The wave then propagates with a wavelength $2\pi\delta_c|\frac{1}{2}\omega_c\tau|^{\frac{1}{2}}$ and very little attenuation: the attenuation length $1/q'' \simeq 2^{\frac{1}{2}}\delta_c|\omega_c\tau|^{\frac{1}{2}}$ will be many times the classical skin depth. This is the helicon mode. It arises essentially as a consequence of the Hall effect, because of which the current $\mathbf{J}$ in an uncompensated metal flows almost at right angles to the applied field $\mathbf{E}$ if $|\omega_c\tau|$ is large. In fact the behaviour in the classical limit can be discussed entirely in terms of the resistivity and Hall coefficient, without reference to any microscopic model (Chambers & Jones, 1962), and the possible existence of a helicon mode could have been predicted at any time since 1879, when the Hall effect was discovered. For any metal in which the d.c. resistivity ρ and the Hall coefficient A_H are isotropic in the xy-plane, we have $\sigma_c = 1/(\rho + iA_H H)$ in the classical limit, and (4.114) yields

$$q_1^2 = 4\pi\omega/(i\rho - A_H H) \qquad (4.121)$$

so that $A_H(H)$ and $\rho(H)$ can be determined from measurements of $q_1(H)$. $q_1(H)$ can be determined from the behaviour of a thin plate, in which standing-wave helicon modes are set up when $q_1'd = n\pi$, though care has to be taken to avoid errors due to edge effects in a plate of finite lateral dimensions (Amundsen, 1966).

As ω increases or H decreases, q_1 will grow, until the condition $|q_1|l \ll 1$ is no longer satisfied. But as long as $|q_1|l \ll \omega_c \tau$, the behaviour of (4.118) is not materially affected, and σ_c will continue to have essentially the value (4.119), independent of q_1. Contrast this with the behaviour when $\omega_c = 0$: σ then becomes appreciably q-dependent for $ql \sim 1$, and for $ql > 1$ is increasingly dominated by the surf-riding electrons, corresponding to the transition from classical to A.S.E. conditions. But for $|\omega_c \tau| \gg 1$, no electrons can surf-ride until

$$q_1'l \simeq (\omega_c + \omega)\tau \simeq \omega_c \tau,$$

and until that point is reached the 'classical' expressions (4.120), (4.121) remain approximately valid.

As soon as $q_1'l = \omega_c \tau$, or more precisely as soon as $q_1'v_z = \omega_c$ for some group of electrons, these electrons will be able to surf-ride (or to undergo 'Doppler-shifted cyclotron resonance') and will contribute strongly to the real part of σ_c; consequently the imaginary part of the root q_1'' becomes large, and the propagating mode becomes very much more strongly attenuated. In effect, then, the helicon mode abruptly vanishes at a critical value of ω or H, and the position of this 'helicon edge' in principle determines the value of ω_c/v_z and hence of $m_c^* v_z$ for the electrons which first become able to surf-ride. If these are limiting-point electrons, we can thus find the Gaussian curvature K at the limiting point, from (4.32) (Stern, 1963). But to determine $m_c^* v_z$, we need to know the value of q_1' at the edge, and the difficulty is of course that the simple dispersion equation (4.120) or (4.121) become inexact as the edge is approached, so that q_1' cannot be computed precisely. Neither can it be measured precisely, because for finite τ the attenuation remains too large to observe the helicon mode experimentally for some distance beyond the edge. And in this region, we can no longer equate Z with $Z_W = 4\pi\omega/q_1$; instead we must use (4.52) to find $E(0)$ and hence Z. For free electrons, (4.118) can be integrated explicitly, to yield again an expression of the form (4.115), but with ω replaced by $\omega + \omega_c$ and with $a = ql/[1 + i(\omega + \omega_c)\tau]$. For Na and K, Taylor (1965) has shown that the Gaussian curvature can be measured reasonably accurately by using this expression to estimate q_1' at the edge. For more complex Fermi surfaces, the form of $Z(H)$ near the edge has been investigated in some detail by Overhauser & Rodriguez (1966), McGroddy et al. (1966) and Stanford & Stern (1966), and the last authors conclude that the helicon edge is of limited value for FS determination, because of the uncertainty in the value of q_1' at the edge, and in the precise position of the edge itself.

More interesting in this respect are the surface impedance oscillations observed in thin plates with H normal to the surface at much lower fields, $H \ll H_{\text{edge}}$, by Gantmakher & Kaner (1965). These authors show that these oscillations are essentially just r.f. Sondheimer oscillations, and that their period in H is determined by (4.31) or (4.32): that is, by the extremal values of $m_c^* v_z$ on the FS. At first sight this conclusion is surprising: for $H \ll H_{\text{edge}}$ (or $\omega \gg \omega_{\text{edge}}$) we are essentially back in the A.S.E. region, where Z is dominated by the surf-riding electrons, travelling almost parallel to the surface, and one would expect no appreciable contribution to Z from the limiting-point electrons, travelling normal to the surface. And indeed the surface impedance of a bulk sample in this region varies quite smoothly with H and, for instance, shows no appreciable structure in the region $\omega \sim \omega_c$ (Chambers, 1956b). Nevertheless the (very small) r.f. field $E(z)$ at large distances from the surface does show Sondheimer oscillations, and as Gantmakher and Kaner show, these arise analytically from the branch-point contributions to (4.52). If $\omega \ll \omega_c$, the denominator of (4.118) will vanish when $q = (1 + i\omega_c \tau)/iv_z \tau$, and this equation determines the path of the cut in the complex q-plane needed to make $\sigma_c(q)$ analytic. The cut extends from $-i(1 + i\omega_c \tau)\infty$ for $v_z = 0$ to the branch-point $q_b = (1 + i\omega_e \tau_e)/iv_{ze}\tau_e$, where ω_e/v_{ze} is the smallest value of ω_c/v_z at the limiting point. Now as we saw earlier, the expression (4.116) for $E(z)$ is valid only in the limit $q_b = 0$, and as a good approximation to the behaviour for $q_b \neq 0$ we can simply replace the lower limit of integration by iq_b/s. If $q_b = (1 + i\omega_e \tau_e)/iv_{ze}\tau_e$, this at once yields

$$E(z) \sim z^{-2} \exp\left(-z/v_{ze}\tau_e\right) \exp\left(-i\omega_e z/v_{ze}\right) \qquad (4.122)$$

for $z \gg 1/s$. If we use the approximation (4.113) for $Z(H)$, we see that sinusoidal oscillations in $Z(H)$ will indeed occur, as observed by Gantmakher, with a period in H determined by $\omega_e d/v_{ze} = 2\pi n$, i.e. by (4.31) or (4.32). The limiting-point oscillations will be rather weak, just as in the d.c. case, because the transverse velocity v_r in (4.118) tends to zero at the limiting points; just as before, much stronger oscillations may arise from intermediate extrema of ω_c/v_z, if any, for which $v_r \neq 0$. These may give rise to additional branch points in $\sigma_c(q)$, and generate oscillations in $Z(H)$ in essentially the same way as the limiting points.

CHAPTER 5

THE ORDINARY TRANSPORT PROPERTIES OF METALS

by J. M. ZIMAN†

Despite our excellent understanding of the electronic structure of simple metals, it is not very easy to make exact quantitative calculations of such familiar properties as the electrical resistivity. It is difficult to sustain precision in an argument passing through several stages, from the construction of wave-functions through the calculation of transition probabilities to the solution of the Boltzmann equation. Each step may be well understood in principle, but one does not always have sufficient detailed information or computing power to avoid the errors inherent in the use of simplified models, such as the Debye lattice spectrum or the point charge impurity.

To be quite rigorous, the whole theory ought to be based upon general quantum-statistical foundations, such as the density matrix and temperature Green function formulations, in which all transport coefficients are defined as expectation values of abstract operators, independent of the state functions used to represent them. This subject is now fairly well understood (see, e.g. Abrikosov, Gorkov & Dzyaloshinski, 1963; Kubo, 1959; Kohn & Luttinger, 1957; Greenwood, 1958, etc.) but it is too difficult for the present work. For a system such as a liquid metal or a concentrated disordered alloy these more powerful techniques are indispensable, but in the cases discussed here, where the mean free path of an electron is long compared with its wavelength, the Boltzmann method is perfectly satisfactory.

In this chapter we first discuss the mechanisms by which electrons are scattered—by phonons, by lattice imperfections, and by impurities—and then go on to consider the way in which different mechanisms contribute to the macroscopic transport properties, such as the electrical and thermal resistivities, thermo-electric power, and Hall coefficient. There is not space here to consider all aspects of this large subject (see Ziman, 1960); topics have been selected where significant progress has been made in recent years.

---

† Dr Ziman is Professor of Theoretical Physics at the H. H. Wills Physics Laboratory, University of Bristol, Tyndall Avenue, Bristol 2.

5.1 ELECTRON-PHONON INTERACTION: BLOCH FORMULATION

The field seen by an electron in a solid must depend on the configuration of the ions. It is convenient to suppose that the periodic potential $\mathscr{V}(\mathbf{r})$ of the perfect lattice is changed by an amount

$$\Delta\mathscr{V}(\mathbf{r}) = \mathscr{V}(\mathbf{r}; l+\mathbf{u}_l) - \mathscr{V}(\mathbf{r})$$
$$= \sum_l \mathbf{u}_l . \nabla_l \mathscr{V}(\mathbf{r}) \qquad (5.1)$$

by the motion of the ions. This merely says that the total change is the sum of independent effects, each linear in the displacement $\mathbf{u}_l$ of an ion from its proper lattice site.

This formulation does not tell us how $\Delta\mathscr{V}(\mathbf{r})$ is to be calculated, but has important general implications because of the translational symmetry of the crystal. Thus, we must have, for example

$$\nabla_l \mathscr{V}(\mathbf{r}) = \nabla_{l+l'} \mathscr{V}(\mathbf{r}+l') \qquad (5.2)$$

when we move to another equivalent site by a lattice vector l'.

We treat $\Delta\mathscr{V}$ as a perturbation on the one-electron Bloch states of the perfect crystal, satisfying

$$\psi_\mathbf{k}(\mathbf{r}+l) = e^{i\mathbf{k}.l}\psi_\mathbf{k}(\mathbf{r}). \qquad (5.3)$$

The lattice displacement $\mathbf{u}_l$ may be expressed in terms of creation and annihilation operators acting on the phonon states $|n_\mathbf{q}\rangle$ of the solid (Ziman, 1964a, §§2.1, 2.11)

$$\mathbf{u}_l = -i\sum_\mathbf{q} \left(\frac{\hbar}{2NMv_\mathbf{q}}\right)^{\frac{1}{2}} e^{-i\mathbf{q}.l}\mathbf{e}_\mathbf{q}(a_\mathbf{q}^* - a_{-\mathbf{q}}). \qquad (5.4)$$

For simplicity we assume a Bravais lattice. The polarization type of the phonon is supposed to be included as an extra index carried by the symbol for the wave-vector $\mathbf{q}$.

From these expressions we may define a standard matrix element for the interaction between electrons and phonons

$$\mathscr{M}_{\mathbf{k}'\mathbf{k}} \equiv \sum_\mathbf{q} \langle n_\mathbf{q}+1| \int \psi_{\mathbf{k}'}^*(\mathbf{r})\Delta\mathscr{V}(\mathbf{r})\psi_\mathbf{k}(\mathbf{r})\,d\mathbf{r} \,|n_\mathbf{q}\rangle$$

$$= -i\sum_\mathbf{q}\sum_l \left(\frac{\hbar(n_\mathbf{q}+1)}{2NMv_\mathbf{q}}\right)^{\frac{1}{2}} e^{-i\mathbf{q}.l} \int \psi_{\mathbf{k}'}^*(\mathbf{r})\{\mathbf{e}_\mathbf{q}.\nabla_l\mathscr{V}(\mathbf{r})\}\psi_\mathbf{k}(\mathbf{r})\,d\mathbf{r}. \quad (5.5)$$

Using the Bloch theorem (5.3) we can shift the origin of $\mathbf{r}$ to the site l of the ion that is being moved. We then get

$$\mathscr{M}_{\mathbf{k}'\mathbf{k}} = -i\sum_\mathbf{q} \left(\frac{\hbar(n_\mathbf{q}+1)}{2NMv_\mathbf{q}}\right)^{\frac{1}{2}} \mathscr{I}_\mathbf{q}(\mathbf{k}',\mathbf{k})\frac{1}{N}\sum_l e^{i(\mathbf{k}-\mathbf{k}'-\mathbf{q}).l}, \qquad (5.6)$$

where the translational symmetry of the lattice makes

$$\mathscr{I}_q(\mathbf{k}', \mathbf{k}) = N \int \psi_{\mathbf{k}'}^*(\mathbf{r} - \boldsymbol{l}) \{ \mathbf{e}_q \cdot \nabla_l \mathscr{V}(\mathbf{r}) \} \psi_{\mathbf{k}}(\mathbf{r} - \boldsymbol{l}) \, d\mathbf{r} \qquad (5.7)$$

independent of $\boldsymbol{l}$.

But the sum of wavy terms over all lattice sites automatically vanishes unless the argument adds up to a reciprocal lattice vector $\mathbf{g}$. The matrix element thus reduces to

$$\mathscr{M}_{\mathbf{k}'\mathbf{k}} = -i \sum_{\text{polarizations}} \left(\frac{\hbar(n_q + 1)}{2NMv_q} \right)^{\frac{1}{2}} \mathscr{I}_q(\mathbf{k}', \mathbf{k}), \qquad (5.8)$$

where the wave-vector of the phonon modes involved is given by the condition

$$\mathbf{q} = \mathbf{k} - \mathbf{k}' - \mathbf{g}. \qquad (5.9)$$

We at once recognize the basic selection rule for the process: the crystal momentum of the phonon produced in the transition is equal to the change in electron wave-vector, reduced (if necessary) to the first Brillouin zone. This rule, which allows for electron–phonon Umklapp processes as well as for 'normal' processes in which wave-vector is conserved, is a general consequence of the translational symmetry of the lattice, and is not special, to, say, the elementary diffraction model (Ziman, 1964 a, § 2.7).

In the transition defined by (5.5) a phonon is created; this is only possible if energy is conserved, so that

$$\mathscr{E}(\mathbf{k}) = \mathscr{E}(\mathbf{k}') + \hbar v_q. \qquad (5.10)$$

It is obvious from (5.4) that there is a conjugate process in which a phonon of wave-vector $-\mathbf{q}$ may be destroyed, with the same matrix element but with n_q instead of $n_q + 1$, and

$$\mathscr{E}(\mathbf{k}) = \mathscr{E}(\mathbf{k}') - \hbar v_q. \qquad (5.11)$$

This formulation of the electron–phonon interaction is due to Bloch (1928), and is still the best we have. Attempts to improve upon it, by, for example, allowing the electron wave-functions themselves to 'follow' the motion of the ions turn out to give the same result, at least to first order in perturbation theory (Sham & Ziman, 1963). The objection that the perturbing potential seen by a given electron must allow for the motion of the 'other' electrons can be met by showing that it is quite adequate to calculate such effects in the adiabatic approximation, as if $\Delta \mathscr{V}(\mathbf{r})$ were the change in the potential produced by some *frozen* set of displacements $\mathbf{u}_l$, including self-consistent contributions from all electron states (Ziman, 1955: see Chester, 1961; Sham & Ziman, 1963).

A more serious weakness may be the assumption that only terms linear in the lattice displacements are important in the scattering of electrons. Here we are encroaching on the field of problems that are best discussed by more advanced mathematical techniques. Terms of higher order in $\mathbf{u}_l$ obviously correspond to multiphonon processes, with the possibilities of intermediate states in which energy is not conserved, etc. Some of these effects can be taken into account by putting a Debye–Waller factor into the elementary diffraction formula (Bailyn, 1961), but these may well be largely cancelled by other terms (Sham & Ziman, 1963). The only general advice offered on this subject is Baym's theorem (1964), which says that one should calculate the scattering of electrons as if they saw exactly the same structure factor as would be measured directly by neutrons or X-rays; but this probably contains hidden assumptions, such as that the electrons are nearly free, and that ionic potentials move rigidly, which cannot be justified except in special cases.

5.2 RIGID IONS AND PORTABLE PSEUDO-POTENTIALS

The simplest assumption one can make in the calculation of the factor $\mathscr{I}_q(\mathbf{k}, \mathbf{k}')$ in the electron–phonon matrix element is that the total field in the crystal is a sum of potentials carried rigidly by the ions. Thus, if

$$\mathscr{V}(\mathbf{r}) = \sum_l \mathscr{V}_a(\mathbf{r} - l), \qquad (5.12)$$

we have at once, from (5.7),

$$\mathscr{I}_q(\mathbf{k}, \mathbf{k}') = -N \int \psi_\mathbf{k}^*(\mathbf{r}) \{\mathbf{e}_q . \nabla \mathscr{V}_a(\mathbf{r})\} \psi_{\mathbf{k}'}(\mathbf{r}) \, d\mathbf{r}, \qquad (5.13)$$

where we simply differentiate the potential $\mathscr{V}_a$ carried by each ion.

For plane waves, trivial integration by parts produces the elementary diffraction result (Ziman, 1964 a, §§ 2.7, 6.12)

$$\mathscr{I}_q(\mathbf{k}, \mathbf{k}') = \mathbf{e}_q . \int \nabla\{\psi_\mathbf{k}^*(\mathbf{r}) \psi_\mathbf{k}(\mathbf{r})\} \mathscr{V}_a(\mathbf{r}) \, d\mathbf{r}$$

$$\approx \mathbf{e}_q . (\mathbf{k} - \mathbf{k}') \mathscr{V}_a(\mathbf{k}' - \mathbf{k}), \qquad (5.14)$$

where

$$\mathscr{V}_a(\mathbf{K}) \equiv N \int \mathscr{V}_a(\mathbf{r}) e^{-i\mathbf{K} . \mathbf{r}} d\mathbf{r} \qquad (5.15)$$

s

is just the Fourier transform of the rigid ion potential, normalized to the volume $1/N$ of a unit cell.

As we shall see, the best theories of the electron–phonon interaction can often be expressed in the form (5.14)—but only after several intuitive or deductive steps. All that we can say at this point is that the polarization factor $\mathbf{e_q} \cdot (\mathbf{k} - \mathbf{k'})$ is important in ensuring that for N-processes, where $\mathbf{q} = \mathbf{k} - \mathbf{k'}$, only longitudinally polarized phonons contribute to the scattering of electrons.

To improve on (5.14) we may suppose that our Bloch functions are expanded in plane waves, i.e.

$$\psi_{\mathbf{k}}(\mathbf{r}) = \sum_{\mathbf{g}} \alpha_{\mathbf{g}} e^{i(\mathbf{k}+\mathbf{g}) \cdot \mathbf{r}}. \tag{5.16}$$

We may then use (5.14) to get

$$\mathscr{I}_{\mathbf{q}}(\mathbf{k}, \mathbf{k'}) = \sum_{\mathbf{g}, \mathbf{g'}} \alpha_{\mathbf{g}} \alpha_{\mathbf{g'}} \{ \mathbf{e_q} \cdot (\mathbf{k} + \mathbf{g} - \mathbf{k'} - \mathbf{g'}) \mathscr{V}_a(\mathbf{k'} + \mathbf{g'} - \mathbf{k} - \mathbf{g}) \}. \tag{5.17}$$

But this sort of expression is really much too complicated for practical computation, unless there are only a few important terms in the series (5.16). In the N.F.E. model of electronic band structure (Ziman, 1964 a, §3.2), Bloch states with k-vectors near the faces of the Brillouin zone are of this form. The calculation of electron–phonon processes involving such states can go wrong if this mixing of waves is not respected (Sham & Ziman, 1963). An excellent exercise in the theory of Bloch functions and zones is to derive the matrix element for transitions between, say two 'neck' states on the Fermi surface of copper, showing that the result is the same whether one regards it as an N-process or a U-process.

But (5.14) cannot be used as it stands. The true potential carried by an ion is very strong, and would cause rapid transitions in processes corresponding to large Fourier components of $\mathscr{V}_a$. These components are important because they would give rise to large higher-order terms in the full perturbation series of which our matrix element $\mathscr{M}_{\mathbf{kk'}}$ is the leading term. In the diffraction model, to which (5.14) is equivalent, we have to allow for multiple scattering of the electron from the same ion, which would profoundly modify our estimate of its total scattering power. In other words, by using (5.14) we should be treating each individual ion as if its potential were so weak that the Born approximation would be quite satisfactory; this assumption is quite unjustifiable for electrons at the Fermi level in real metals.

The usual trick now is to replace the true ion potential by the corresponding Fourier component $\mathscr{V}_{ps}(\mathbf{K})$ of a pseudo-, model, or effective potential of the type used in the calculation of the band structure (see Chapter 1). We merely assert that the pseudo-potential not only gives the band gaps, but moves in space along with the true potential that it replaces.

For a formal justification of this procedure we may go back to (5.13) and allow for the fact that the Bloch functions that occur in the integral are by no means true plane waves, even in the N.F.E. approximation. For example, in the o.p.w. formulation (Ziman, 1964a, §3.8)

$$\psi_{\mathbf{k}}(\mathbf{r}) = e^{i\mathbf{k}\cdot\mathbf{r}} - \sum_t \langle e^{i\mathbf{k}\cdot\mathbf{r}}, b_t \rangle b_t(\mathbf{r}),$$
(5.18)

where $b_t(\mathbf{r})$ is a core function, whose 'wiggles' are important precisely in the core region where $\nabla \mathscr{V}_a(\mathbf{r})$ is large. Putting (5.18) into (5.13) produces three types of terms, which can be systematically transformed (Sham, 1961; Sham & Ziman, 1963) into various contributions to what turns out to be a Fourier component of the pseudo-potential,

$$\mathscr{V}_{ps}(\mathbf{k}', \mathbf{k}) = \mathscr{V}_a(\mathbf{k}' - \mathbf{k}) + \sum_t (\mathscr{E} - \mathscr{E}_t) \langle e^{i\mathbf{k}\cdot\mathbf{r}}, b_t \rangle \langle b_t, e^{i\mathbf{k}'\cdot\mathbf{r}} \rangle.$$
(5.19)

Nevertheless, one must be cautious in identifying plane waves with pseudo-plane waves and true potentials with pseudo-potentials. There is a great arbitrariness in the definition of the pseudo-potential, which must not be allowed to infect the calculation of the electron–phonon interaction. If, for example, we had decided to use a model potential for the band structure different from (5.19) then we should be very careful to put the correct wave-functions, with all mixings as in (5.16), into the integral (5.13).

Perhaps the proper thing to do is to put in place of $\mathscr{V}_a(K)$ the corresponding element $t_a(k)$ of the t-matrix of the ion potential (Ziman, 1964b; Greene & Kohn, 1965), which can be expressed in terms of phase shifts by well-known procedures. Such a step is intuitively obvious from the diffraction point of view, and has been justified for N.F.E. systems by appeal to the work of Langer (1960) and Baym (1964). But when the potentials are supposed to overlap, or when one wants to take account of screening one runs into formal difficulties that are veiled in the less precise notion of a pseudo-potential.

The t-matrix is equivalent to a 'quasi-potential' (Ziman, 1964b)

$$\mathscr{V}_{qu}(K) = -\frac{2\pi\hbar^2 N}{mk} \sum_{l=0}^{\infty} (2l+1) \sin \eta_l \, e^{i\eta_l} P_l(\cos\theta)$$
(5.20)

taking the place of $\mathscr{V}_a(K)$ for scattering through the angle θ. But this matrix element is not Hermitian, contradicting a necessary property of the Bloch expression (5.7). It must be remembered that the scattering of an electron in a Bloch state—already orthogonalized, etc., to the potential that ought to have been at the site from which the ion is displaced—is not equivalent to the scattering of an external beam of electrons. It may be that one should replace the t-matrix by the 'reactance' or 'K'-matrix of the ion potential; this is Hermitian, and is also deeply implicated in the general theory of Bloch states (Ziman, 1965). For small phase shifts this makes little difference; we simply replace $\sin \eta_l \exp (i\eta_l)$ by $\tan \eta_l$ in (5.20).

5.3 SCREENING THE ELECTRON–PHONON INTERACTION

One of the most difficult problems in the theory of metals has seemed to be to allow for the effect of the charge shift of the electron gas when it moves in sympathy with the ions. This problem is tamed, but by no means entirely domesticated.

The general procedure is essentially that used by Bardeen (1937); the field produced by charge shift is treated self-consistently as part of the very field to which the charge shift is due. In a linearized Hartree approximation, this is equivalent to the use of a dielectric function $\epsilon(K)$ which divides the Fourier component of the bare potential to give the corresponding component of the screened potential,

$$\mathscr{V}_{\text{scr}}(K) = \frac{\mathscr{V}_{\text{bare}}(K)}{\epsilon(K)}. \tag{5.21}$$

The formal complexities of a general theory along these lines have been explored by Sham & Ziman (1963); there have not yet been any real calculations going further than assuming that the dielectric function is that of a simple free-electron gas

$$\epsilon(K) = 1 + \frac{4\pi e^2}{K^2} \frac{ZN}{\tfrac{2}{3}\mathscr{E}_F} \left\{ \frac{1}{2} + \frac{4k_F^2 - K^2}{8k_F K} \ln \left| \frac{2k_F + K}{2k_F - K} \right| \right\}. \tag{5.22}$$

This type of expression can be improved to allow for exchange effects (Rice, 1965), but does not really take account of the fact that the screening electrons are in Bloch states which might not be described by simple plane waves. It would thus be quite inappropriate in problems involving narrow, tightly bound bands, as in transition metals.

The usual argument (e.g. Harrison, 1966) is to start from a lattice of bare rigid ions, as in (5.12), (5.13) and then to replace the matrix element of the bare potential $\mathscr{V}_a(K)$ of each ion by its pseudo-potential $\mathscr{V}_{ps}(K)$. We then allow for screening by dividing by the dielectric function as in (5.21)

$$\mathscr{I}_q(\mathbf{k}, \mathbf{k}') = \mathbf{e}_q \cdot (\mathbf{k} - \mathbf{k}') \frac{\mathscr{V}_{ps}(\mathbf{k}' - \mathbf{k})}{\epsilon(\mathbf{k}' - \mathbf{k})}$$

$$= (\mathbf{e}_q \cdot \mathbf{K}) \mathscr{U}_{\text{N}\Psi\text{A}}(K). \tag{5.23}$$

We thus arrive at the simple theory already discussed in Chapter 1. Each ion is given the screened pseudo-potential or atomic form factor $\mathscr{U}_{\text{N}\Psi\text{A}}$ which plays the role of an atomic potential in calculations of band structure, the electron–phonon interaction, and many other phenomena.

The general behaviour of this function is now well understood in principle. The main effect of screening is to eliminate the long-range Coulomb field of the ion. Thus if

$$\mathscr{V}_a(r) \sim -\frac{Ze^2}{r} \tag{5.24}$$

for large r, then (Ziman, 1964a, §§5.3, 6.12)

$$\mathscr{U}_{\text{N}\Psi\text{A}}(K) \to -\tfrac{2}{3}\mathscr{E}_F \tag{5.25}$$

for small K. Putting this into (5.8) and (5.23) gives the standard 'deformation potential' associated with dilatation of the electron gas (Ziman, 1960, §§5.6, 5.7; Ziman, 1964a, §6.13). As K increases, however, $\mathscr{U}_{\text{N}\Psi\text{A}}(K)$ becomes smaller in magnitude, changes sign, oscillates, etc., in a way that depends in detail on the actual atomic potential.

This behaviour is very similar to the formula for the electron–phonon interaction obtained by Bardeen (1937), after allowing for screening corrections to a formula derived by Mott & Jones (1936). It is interesting to notice (Sham & Ziman, 1963) that the Bardeen–Mott–Jones theory, although too crude for present-day calculations, was based upon the pseudo-potential concept, long before that was given a name. The details of the potential inside the core are eliminated by comparing the properties of the wave-function at the boundary of the Wigner–Seitz cell with those of a shallow square-well potential whose depth is adjusted to give the correct gradient at the bottom of the band.

The screened pseudo-potential $\mathscr{U}_{\text{N}\Psi\text{A}}(K)$ can be given a simple physical significance (Ziman, 1964b). Let us transform it back into real space, by the inversion of an integral like (5.15) to a function

$\mathscr{U}_{\text{NΨA}}(\mathbf{r})$. The whole theory up to (5.23) may then be obtained by the direct application of first-order perturbation theory to a hypothetical total potential.

$$\mathscr{V}(\mathbf{r}) = \sum_{l} \mathscr{U}_{\text{NΨA}}(\mathbf{r} - l), \qquad (5.26)$$

as if each ion carried, rigidly, this simple potential function. For many purposes this is a very convenient fiction. The linearity of the screening theory implies that the potentials of neighbouring ions may be over-lapped without interference. All charge shifts of the electron gas have been taken into account, so that we may treat each 'neutral pseudo-atom' as if it were, say, an atom of argon. It can be even shown that the forces between two ions are the same as we should calculate for the electrostatic forces between the two charge clouds corresponding to the pseudo-potential, immersed in a medium of dielectric response $\epsilon(K)$.

The difficulty with the whole of this theory is that it is limited to the circumstances of 'weak' pseudo-potentials and linearized screening. We have no general mandate to write down (5.23) as if a weak pseudo-potential were equivalent to a weak real potential. What should we do, for example, for a metal such as Cu with its d-band near the Fermi level and the fierce distortion of its Fermi surface?

An alternative route may present fewer formal barriers. From the concept of neutral pseudo-atoms we have learnt that the external Coulomb fields of ions, being weak real potentials in which the linear screening theory is reasonably reliable, are superposible after screening. Suppose, therefore, that we tackle the Hartree (or Hartree–Fock) problem for an isolated ion, immersed in the electron gas, before we calculate pseudo-potentials or bring the ions together into a crystal. From the spherically symmetrical bare ion potential $\mathscr{V}_a(\mathbf{r})$ we should construct a self-consistent potential $\mathscr{U}_{\text{s.c.}}(\mathbf{r})$, such as would be seen by any one electron after allowing for the electron-electron interaction. We could then construct a pseudo-potential equivalent to $\mathscr{U}_{\text{s.c.}}$, to be used in place of $\mathscr{U}_{\text{NΨA}}$ in the above theory.

This route has not yet been explored, although the techniques developed by March and his pupils (March & Murray, 1960) are steps on the way. But even in principle we can learn something from it. In the first place, we know that the screening cloud around a charge in a degenerate electron gas does not fall off exponentially, but has Friedel oscillations of density (Ziman, 1964a, §5.5) at relatively large distances: each neutral pseudo-atom, even in a pure metal, must be supposed to carry those halos of change around with it—a most important point in the calculation of interatomic forces and other cohesive phenomena. We also know that in order to solve the Hartree problem we should have to find solutions to the radial Schrödinger equation in the self-consistent potential. These solutions would, by their behaviour at the boundary of a sphere surrounding the ion, define the phase shifts in a 'quasi-potential' such as (5.20) which could be used as $\mathcal{U}_{\mathrm{N\Psi A}}$ in the electron–phonon interaction. But the condition that the screening cloud must have the same total charge as the ion imposes the Friedel sum rule (Ziman, 1964a, §5.5) on these phase shifts

$$Z = \frac{2}{\pi} \sum_l (2l+1)\, \eta_l(k_F). \tag{5.27}$$

We note with satisfaction that if all the phase shifts are small this condition makes (5.20) consistent with (5.25), i.e. $\mathcal{V}_{qu}(0) \to -\tfrac{2}{3}\mathcal{E}_F$ (Ziman, 1964b; Greene & Kohn 1965).

5.4 VACANCY SCATTERING

At first sight, the calculation of the resistance due to lattice vacancies in a metal crystal seems simple; we just calculate the scattering cross-section of a region of empty space at each lattice site. But this implies a thorough understanding of the theory of Bloch states in the perfect lattice, to provide a standard of comparison.

The simplest argument relies upon the fact that the removal of an ion breaches the local average neutrality of the perfect crystal. The vacancy is thus equivalent to a charge $-Z|e|$ in an otherwise empty continuum. But this charge in its turn acquires a screening cloud (actually a 'positive' cloud of 'anti-electrons') in the electron gas. Thus, we ought to treat it as a localized perturbing potential of the form

$$\mathcal{U}_v(r) = -\frac{Z|e|}{r}\exp(-\lambda r), \tag{5.28}$$

where λ is the usual screening parameter (Ziman, 1964a, §5.3).

Unfortunately, this potential gives rise to phase shifts which do not satisfy the Friedel sum rule for its assumed charge, and is therefore not fully self-consistent. This difficulty may be avoided by solving the screening and scattering problem precisely (Lee & March, 1960) or by varying the parameter λ in (5.28) until the sum rule is satisfied (see Seeger, 1962).

But a more serious error remains. There is no guarantee that the hypothetical charge assigned to the vacancy is closely localized to a single point. The way in which this charge is distributed will have important effects on the scattering cross-section that would be assigned to the vacancy. The whole model is too specific, but cannot be improved without making arbitrary, unjustifiable assumptions.

The general theory of neutral pseudo-atoms discussed in §5.3 provides a way out (Harrison, 1963; Ziman, 1964 b). A vacancy is simply the absence of one of these hypothetical objects in the otherwise perfect crystal. The perfect crystal does not scatter its Bloch electrons. Therefore, by an elementary application of Babinet's principle, the diffraction pattern of the vacancy is the same as the diffraction pattern of a single neutral pseudo-atom in empty space. In other words, instead of (5.28) we write

$$\mathcal{U}_v(r) = -\mathcal{U}_{\text{NΨA}}(r); \tag{5.29}$$

the hypothetical distribution of charge imputed to the vacancy is exactly the negative of the screening and 'pseudo-charge' carried by the ion that was removed. Because of (5.27) our definition of $\mathcal{U}_{\text{NΨA}}$ ensures that the Friedel sum rule is satisfied for $\mathcal{U}_v(r)$, when measured relative to the perfect lattice.

There are other complications; for example, the Bloch states may not be simple pseudo-plane-waves, but might have to be constructed out of linear combinations of such functions, as in (5.16). The matrix element for scattering between two such states would then be of the form (5.17), i.e.

$$\int \psi_{\mathbf{k}'}^* \, \mathcal{U}_v(r) \, \psi_{\mathbf{k}} \, d\mathbf{r} = -\sum_{\mathbf{g}, \mathbf{g}'} \alpha_{\mathbf{g}} \alpha_{\mathbf{g}'} \, \mathcal{U}_{\text{NΨA}}(\mathbf{k} + \mathbf{g} - \mathbf{k}' - \mathbf{g}'). \tag{5.30}$$

Another familiar phenomenon is the dilatation $(\delta V / V)$ of the lattice around a vacancy. To allow correctly for this† we should calculate the diffraction pattern in the displaced, imperfect lattice, as compared with the perfect crystal. Thus, the effective matrix element for scattering through the vector $\mathbf{K}$ would be (cf. Ziman, 1964 a, §2.7)

$$\mathcal{U}_v(\mathbf{K}) = -\mathcal{U}_{\text{NΨA}}(\mathbf{K}) + \mathcal{U}_{\text{NΨA}}(\mathbf{K}) \frac{1}{N} \sum_{l \neq 0} (e^{i\mathbf{K} \cdot \mathbf{R}_l} - e^{i\mathbf{K} \cdot l}), \tag{5.31}$$

assuming that the lth ion has moved to $\mathbf{R}_l$ relative to the position of the vacant site. To evaluate the second term in general requires a

† The argument given by Ziman (1964 b) is not valid.

model of the pattern of displacements in the neighbourhood of the vacancy. But in the limit of small K, we must have (Faber & Ziman, 1965)

$$\frac{1}{N}\sum_{l \neq 0}(e^{i\mathbf{K}\cdot\mathbf{R}_l} - e^{i\mathbf{K}\cdot\mathbf{l}}) \to \frac{1}{N}\left(\underset{\text{displaced lattice}}{\sum 1} - \underset{\text{undisplaced lattice}}{\sum 1}\right)$$

$$= -\frac{\delta N}{N} = -\frac{\delta V}{V}. \tag{5.32}$$

The point is that the calculation implies a fixed standard volume (which we have taken to be unity in many formulae) within which the electron states are normalized. The dilatation δV decreases by δN the number of terms in the summation. For small K, therefore, the vacancy behaves as if it had the effective potential

$$\mathcal{U}_v = -\mathcal{U}_{\text{NΨA}}\left(1 + \frac{\delta V}{V}\right). \tag{5.33}$$

This is a formal proof of the heuristic principle, suggested by Blatt (1957) and Harrison (1958), of multiplying the effective charge of a vacancy in the continuum model by this same factor. It is easy to verify that the Friedel sum of the charge shifts produced by (5.33) equals the effective valency

$$-Z' = -Z\left(1 + \frac{\delta V}{V}\right). \tag{5.34}$$

Faber & Ziman (1965) and Harrison (1966) show by model calculations that this simple correction factor is only valid for small K, and that the dilatation effect becomes smaller—even changes sign—as K increases.

The theory of scattering by interstitial ions of the parent metal is obviously equivalent to the theory of vacancy scattering except for a change of sign, and careful watch (especially in the case of mixed Bloch functions, such as (5.30)) must be kept for any effects due to the interstitial not being at a true lattice site of the perfect crystal.

Despite the fairly specific formulae of the theory of vacancy and interstitial scattering, it has not been well-tested experimentally. The trouble is that the best experiments (e.g. Simmons & Balluffi, 1962) are on noble metals, where the pseudo-potential model is not in principle, very satisfactory, and where there are, as yet, no good calculations beyond the continuum theory.

In any case, a vacancy does slightly stretch the bounds of the model. At the centre of the vacant cell the electron density cannot really be near the average density in the metal, so the linearized screening theory would be inaccurate. This is a problem that must eventually require a complete self-consistent solution of the electron distribution around the actual arrangement of ions.

5.5 IMPURITIES

Scattering by dilute admixtures of impurities can easily be discussed in the same language as for vacancies. It is intuitively obvious that the effective potential for an ion of metal B substituted for an ion of the parent metal A must be of the form

$$\mathcal{U}(K) = \mathcal{U}^B(K) - \mathcal{U}^A(K). \tag{5.35}$$

But we cannot immediately substitute for $\mathcal{U}^B(K)$ the standard potential of a neutral pseudo-atom of B, for this depends for its definition on the electron density in the bulk pure metal B, which is by no means necessarily the same as in the matrix where the ion now sits. To allow for this, we should first calculate the unscreened pseudopotential $\mathcal{V}_{ps}(K)$ of B, and then divide by a dielectric function (5.22) calculated at the electron density $Z_A N$, and Fermi level $\mathscr{E}_F^A$, of the bulk metal A. If element B is of valency Z_B, then, by the argument leading to (5.25),

$$\mathcal{U}(K) \to -\tfrac{2}{3}\mathscr{E}_F^A \left(\frac{Z_B}{Z_A} - 1\right) \tag{5.36}$$

as $K \to 0$, which is just what we should expect for the effective potential of a localized charge $(Z_B - Z_A)|e|$ in a continuum model. For phase shifts, this result also ensures that the Friedel sum rule is satisfied for scattering by this potential. The Blatt–Harrison principle for the effects of dilatation can easily be applied, as in (5.34), so that we should give the impurity the effective relative valency

$$Z' = Z_B - Z_A(1 + \delta V/V). \tag{5.37}$$

We thus recover some of the properties of the standard valence-charge model of impurities (Mott, 1936, see Mott & Jones, 1936, § 12; Ziman, 1960, §§ 6.3, 9.2; Ziman, 1964a, §§ 6.9, 7.4) leading to Linde's rule, etc. But the limitations of that model are now apparent. The impurity could only be treated as a point charge in a continuum if $\mathcal{U}(K)$ happened to be at just the right functional form. It would only be true in general if all metallic ions were themselves just point charges. We know now that $\mathcal{U}_{N\Psi A}$, the form factor of a neutral pseudo-atom, falls off rather more rapidly at large K than the Fourier transform of a screened Coulomb potential, and this has important consequences in the calculation of the resistivity. The detailed differences between pseudo-potentials of metals of the same valency must therefore be seen as differences in the resistivity they will produce when dissolved in the same parent metal.

This would explain the 'period effect', found when elements of the same valency are alloyed—especially Au in Ag, where there is no change of atomic volume to produce an effective charge. According to our general theory, the scattering would be seen only for relatively large values of K, approaching $2k_F$, where the pseudo-potentials for the two elements would have moved apart. The observed effects are consistent with such an explanation (Ziman, 1964b). Indeed, this was proved long ago by Mott (1936) using the solution of the Wigner–Seitz calculation as a pseudo-potential just as in the theory of the electron–phonon interaction.

We can also understand (Faber & Ziman, 1965) how it is that scattering by impurities dissolved in a polyvalent metal such as Pb is not described very well by a valence charge model leading to something like Linde's rule. At high electron densities the Fermi radius $2k_F$ is so large that most of the resistance comes from values of K far away from the region where (5.36) is valid. In such cases we ought also to make allowances for the mixing of wave-functions at zone boundaries as in (5.30) a very complicated problem in general.

Unfortunately the best systematic information we have on the resistivity of dilute alloys is again for the alloys of the noble metals, Cu, Ag, Au, for which we do not have proper pseudo-potential models. It has not yet proved possible to test the ideas of this section with such precision as to make good quantitative comparisons with experiment (see e.g. Harrison, 1966). The special properties of transition metal impurities are discussed in Chapter 8 of the present work. We ought also to consider concentrated alloys where the arrangement of solute atoms becomes important. It is not difficult to set up a diffraction model (Faber & Ziman, 1965; Harrison, 1966) by which many interesting phenomena may be understood qualitatively such as the dependence of the resistivity upon the concentration and degree of order of the components, but it could not be said that there is much to show yet of a quantitative kind. The difficulty of defining the Bloch functions in such a system have still to be overcome; even the validity of such a concept as the rigid band model (see e.g. Friedel, 1954) is still in doubt.

Strictly speaking, one ought to allow for the fact that a proportion of the events in which an electron is scattered by an impurity also involve the emission or absorption of a phonon (Koshino, 1960) but this turns out to have a negligible effect on the transport properties, of the metal (Taylor, 1964).

5.6 EXTENDED IMPERFECTIONS: STACKING FAULTS AND DISLOCATIONS

The theory of the scattering of conduction electrons by typical crystal imperfections of the kind introduced by plastic deformation is obviously of practical importance, and ought to be a link between the various parts of this book; unfortunately it is extremely difficult to get good answers.

The method of neutral pseudo-atoms does solve one of the basic problems; the problem of charge shift. It is not very difficult to prescribe the positions of the ions in some idealized model of a stacking fault, dislocation, or grain boundary, but the calculation of the effect of this arrangement upon a conduction electron requires a knowledge of the self-consistent field of the ' other ' electrons in the neighbourhood of the imperfection. Within the approximations of linear screening theory (§ 5.3 above) this field is automatically taken care of by assuming that every ion has just carried its screening cloud around with it. In other words, we may write, just as in (5.26)

$$\mathscr{V}(\mathbf{r}) = \sum_l \mathscr{U}_{\mathrm{N\Psi A}}(\mathbf{r} - \mathbf{R}_l), \qquad (5.38)$$

for the effective potential at any point in the metal, even though now the neutral pseudo-atoms are no longer on the sites of a translationally invariant lattice. This is, of course, only an approximation, but quite good enough for present practical purposes.

The only case that is well understood in principle is that of a stacking fault or twin boundary (Ziman, 1960, § 6.6). For such a system, elementary diffraction theory, based upon first-order perturbation arguments is not really appropriate (but see Harrison, 1956). We know how to set up Bloch functions in the two regions on either side of the plane of the fault as mixtures of plane waves as in the N.F.E. model (Ziman, 1964a, §§ 3.2, 3.3). It is not difficult to construct solutions corresponding to transmitted and reflected waves that match on the boundary, and hence to calculate the reflection coefficient (Howie, 1960; Lackmann, 1964). The geometrical set-up of a particular problem tends to be quite complicated, but the principles that emerge are relatively simple.

We find, for example, that the fault will not be seen by a Bloch function that could be described by a single 'pseudo-plane-wave' Such a wave would obviously be a solution on both sides of the boundary, and would join perfectly across it. It is only states with k-vectors near the zone boundaries that are scattered.

But a state on, say, the (111) neck on the Fermi surface of Cu would not be affected by a stacking fault in the (111) plane (Howie, 1960); the mixing of waves in such a state is really determined by the spacing and total scattering power of the successive lattice planes with that orientation, which is unaltered by the fault. Thus, the reflection tends to be strongest for electrons moving at a relatively large angle to the normal of the fault plane. It is also necessary, in some of the calculations, to include localized Tamm states (Ziman, 1964a, §6.8) dying away exponentially from the boundary plane, to match the propagating waves properly.

The result is that the scattering of an electron in some state $\mathbf{k}$ depends strongly on the position of $\mathbf{k}$ on the Fermi surface, and on its direction relative to the fault plane. It must also depend strongly on the magnitude of the Fourier component of the pseudo-potential on the nearest zone boundary, which governs the shape of the Fermi surface and the admixture of plane waves in the Bloch states (Ziman, 1959a). For the noble metals and possibly for Li (Dugdale & Gugan, 1961; Howie, 1961) these coefficients are thought to be large enough to give the observed quite small resistivity that has definitely been attributed to stacking faults—but the whole subject has by no means been thoroughly explored.

Scattering by dislocations is a more baffling problem. The diffraction method would have us take the Fourier transform of (5.38)

$$\mathscr{V}(K) = \mathscr{U}_{\text{N}\Psi\text{A}}(K) \cdot \frac{1}{N} \sum_l e^{i\mathbf{K} \cdot \mathbf{R}_l} \tag{5.39}$$

as the matrix element for scattering by this arrangement of ions. But we are not so much interested in the absolute structure factor of the dislocated lattice as in its difference from the corresponding function for the perfect crystal. The perturbing potential due to the dislocation ought to be written

$$\mathscr{U}_{\text{disl}}(K) = \mathscr{U}_{\text{N}\Psi\text{A}}(K) \cdot \frac{1}{N} \sum_l (e^{i\mathbf{K} \cdot \mathbf{R}_l} - e^{i\mathbf{K} \cdot l}). \tag{5.40}$$

When K is small, so that the exponential factors vary only slowly over substantial distances, each sum may be approximated to by an integral over a continuous medium. Let $N'(\mathbf{R})$ be the local density in the neighbourhood of $\mathbf{R}$ near the dislocation, then

$$\mathscr{U}_{\text{disl}}(\mathbf{K}) \approx \mathscr{U}_{\text{N}\Psi\text{A}}(K) \cdot \frac{1}{N} \int e^{i\mathbf{K}\cdot\mathbf{R}} \{N'(\mathbf{R}) - N\} \, d\mathbf{R}$$

$$\approx -\mathscr{U}_{\text{N}\Psi\text{A}}(K) \cdot \Delta(\mathbf{K}),\tag{5.41}$$

where $\Delta(K)$ is the Fourier transform of the dilatation corresponding to the variation of density $N'(\mathbf{R}) - N$.

As we saw in §5.3, $\mathscr{U}_{\text{N}\Psi\text{A}}(K)$ tends to the deformation potential as $K \to 0$. This result therefore justifies the continuum model for dislocation scattering (Hunter & Nabarro, 1953; Ziman, 1960, §6.5). Unfortunately it does not, as a general formula, agree with experiment. With all due allowances for difficulties in solving the Boltzmann equation for such anisotropic scattering, one arrives at a resistivity of about $1 \times 10^{-14} \mu\Omega\,\text{cm}^3$/dislocation line in a typical metal, whereas the experimental value is about $30 \times 10^{-14} \mu\Omega\,\text{cm}^3$ (Clareborough *et al.* 1962; Rider & Foxon, 1966).

Various attempts have been made to explain the discrepancy. The idea that all dislocations are extended into strips of stacking fault is not confirmed by experiment. The suggestion that the Born approximation to the scattering (implicit in the use of (5.14)) is not valid in the strong potential near the dislocation core (Seeger & Bross, 1960) is invalidated by an exact solution of the two-dimensional scattering problem for a 'Coulomb' field (Nabarro & Ziman, 1961). The effect of allowing a dislocation to have a hollow core—as if it were a row of vacancies—has been calculated by Harrison (1958), and shown to be large enough but not necessarily physically plausible.

The basic error must surely lie in the transition from (5.40) to a continuum description (5.41) which is only justifiable in the limit of long wavelengths. We ought therefore to calculate the sums in (5.40) properly. But this raises serious difficulties. The atoms in a dislocated crystal cannot be labelled as if derived from the sites of a perfect lattice by a continuous deformation; by the very definition of a dislocation, we have to make cuts, remove or add half-planes of atoms, make displacements of a lattice vector, etc. If we are not careful, we shall find ourselves computing the scattering power of the extra half-plane or treating one half of the dislocated crystal as a perturbation on the other half, which would be nonsensical.

Apart from some results on the diffraction patterns of dislocations by Wilson (1950) there is no evaluation of the structure factor suitable for a numerical calculation of the scattering cross-section and resistivity of dislocations. It is likely that the mixing of waves in Bloch states near zone boundaries plays a fundamental role in this phenomenon, as in the stacking fault problem, but this again has not been proved.

Nevertheless, there is hope that the observed resistivity could be explained as a diffraction effect due to the large displacements of the

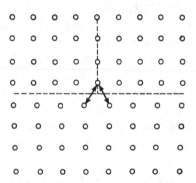

Fig. 5.1. Unusual interatomic vectors in an edge dislocation.

ions in the core of the dislocation line. Consider the square modulus of the potential (5.39); the scattering probability will be proportional to

$$\left| \frac{1}{N} \sum_l e^{i\mathbf{K}.\mathbf{R}_l} \right|^2 = \frac{1}{N^2} \sum_{l,l'} e^{i\mathbf{K}.(\mathbf{R}_l - \mathbf{R}_{l'})}. \qquad (5.42)$$

In the perfect lattice this sum reduces to a delta function at $\mathbf{K} = \mathbf{g}$. The dislocated lattice has almost the same distribution of interatomic vectors $\mathbf{R}_l - \mathbf{R}_{l'}$ as the perfect lattice, and therefore also gives nearly the same transition probabilities so the basic Bloch functions of the crystal are not changed.

But in an idealized edge dislocation there are at least two interatomic vectors that have no counterpart in the perfect crystal (see Fig. 5.1) and these must contribute large scattering effects away from the reciprocal lattice points in $\mathbf{K}$-space. Without detailed calculation we cannot work out exactly how strong this scattering would be, but hand-waving arguments suggest that it ought to be as large as if we had moved one atom in each plane normal to the dislocation line by a distance of about a lattice spacing (Ziman, 1964 b). We may thus give support to the heuristic argument of Basinski, Dugdale & Howie (1963) who show that for a number of metals the observed resistivity of dislocations is quite well predicted by either of the two formulae

$$\Delta\rho_D/N = \left\{ \begin{array}{l} \alpha b^2 V^{\frac{4}{3}} M \Theta^2 \rho_i(T)/T, \\ \beta b^2 \rho_L(T_m), \end{array} \right\} \tag{5.43}$$

where b is the Burgers vector, ρ_i the ideal resistivity of the solid metal, $\rho_L(T_m)$ the resistivity of the liquid metal at the melting point, and α, β are constants. These formulae are derived upon the very reasonable assumptions that the resistivity of a dislocation may be compared with the ideal (phonon-scattering) resistivity that would be produced by an equivalent amount of mean square lattice displacement in the solid (cf. Ziman, 1964a, §7.5) or with the resistivity of a cylinder of liquid metal replacing the core of the dislocation.

5.7 THE BOLTZMANN EQUATION: 'IDEAL' RESISTIVITY

For transport properties dominated by the electron–phonon interaction the setting up of the Boltzmann equation is not quite trivial. Let us write this in the form

$$\left(-\frac{\partial f_k^0}{\partial \mathscr{E}} \right) \mathbf{v_k} \cdot \mathbf{F(k)} = -f_k]_{scatt}, \tag{5.44}$$

where we write

$$\mathbf{F(k)} = e \left(\mathbf{E} - \frac{1}{e} \nabla \zeta \right) + \frac{1}{T} \{ \mathscr{E}(\mathbf{k}) - \zeta \} (-\nabla T), \tag{5.45}$$

for the thermodynamic force produced by the electric field and thermal gradient.

As we saw in §5.1, the state $\mathbf{k}$ may gain or lose electrons by transitions to and from some other state $\mathbf{k'}$ with the creation or destruction of a phonon. The matrix elements for such transitions are given by expressions such as $\mathscr{M}_{\mathbf{k'k}}$ in (5.8) with energy conditions such as (5.10). The square of the matrix element gives the transition probability, which will thus contain a factor n_q or $n_q + 1$ according as the phonon in state $\mathbf{q}$ is destroyed or created in the process. We must also allow for the intrinsic probability, f_k of the initial state containing an electron and $(1 - f_{k'})$ of a final state being available for it. Thus, the contribution to the right-hand side of (5.44) from the process considered in (8) is of the form

$$f_k (1 - f_{k'}) (n_q + 1) \delta \{ \mathscr{E}(\mathbf{k}) - \mathscr{E}(\mathbf{k'}) - \hbar\nu_q \} Q_{\mathbf{k}}^{\mathbf{k'q}}, \tag{5.46}$$

where the symbol $Q_{\mathbf{k}}^{\mathbf{k'q}}$ includes the other factors such as $|\mathscr{I}(\mathbf{k'},\mathbf{k})|^2$ which could readily be calculated by perturbation theory.

To construct the whole rate of change of $f_{\mathbf{k}}$ due to scattering by phonons, we must integrate (5.46) over all the final states $\mathbf{k'}$. But there are inverse processes also to account for, with statistical factor $(1-f_{\mathbf{k}})f_{\mathbf{k'}}n_{\mathbf{q}}$, and also the conjugate transitions in which the phonon is destroyed rather than created. The final expression for $-f_{\mathbf{k}}^0]_{\text{scatt}}$ is thus rather complicated. We recall, however, that the steady-state distributions are near the equilibrium distributions $f_{\mathbf{k}}^0, n_{\mathbf{q}}^0$, where, by the principle of detailed balance, the inverse rate exactly equals the forward rate of scattering between any two states. We therefore measure the 'out-of-balance' distribution,

$$g_{\mathbf{k}} \equiv f_{\mathbf{k}} - f_{\mathbf{k}}^0 \equiv (-\partial f_{\mathbf{k}}^0/\partial\mathscr{E}(\mathbf{k}))\,\Phi_{\mathbf{k}}, \qquad (5.47)$$

by a function $\Phi_{\mathbf{k}}$ which is, so to speak, the effective energy shift of the Fermi distribution caused by the thermodynamic force. With the help of a similar function, the out-of-balance of the phonon distribution, we get (Ziman, 1960, §§7.7, 9.5)

$$f_{\mathbf{k}}]_{\text{scatt}} = \frac{1}{kT}\iint[\{\Phi_{\mathbf{k}} - \Phi_{\mathbf{k'}} - \Phi_{\mathbf{q}}\}\mathscr{P}_{\mathbf{k}}^{\mathbf{k'q}} + \{\Phi_{\mathbf{k}} - \Phi_{\mathbf{k'}} + \Phi_{\mathbf{q}}\}\mathscr{P}_{\mathbf{kq}}^{\mathbf{k'}}]\,\mathbf{dk'}\,\mathbf{dq},$$

where now
$$\text{(5.48)}$$

$$\mathscr{P}_{\mathbf{k}}^{\mathbf{k'q}} = f_{\mathbf{k}}^0(1-f_{\mathbf{k'}}^0)\,(n_{\mathbf{q}}^0+1)\,\delta\{\mathscr{E}(\mathbf{k}) - \mathscr{E}(\mathbf{k'}) - \hbar\nu_{\mathbf{q}}\}\,Q_{\mathbf{k}}^{\mathbf{k'q}} \qquad (5.49)$$

is the actual transition rate in the equilibrium state, allowing for statistical factors.

The combination of (5.44) and (5.48) constitute our Boltzmann equation. But it is apparent that $\Phi_{\mathbf{q}}$, the out-of-balance of the phonon distribution, is undetermined in this equation; we need a further equation counting all scattering processes and thermodynamic forces to which the phonon system may be subject. For the moment we shall make the 'Bloch assumption' that $\Phi_{\mathbf{q}}$ is zero, although, as we shall later remark, this can cause serious error in some cases.

Unfortunately, we have still an inhomogeneous integral equation, to be solved for the unknown distribution $\Phi_{\mathbf{k}}$. Such equations, which turn up in many different branches of physics, cannot be solved by an elementary algorithm. But there is a general variational procedure (Kohler, 1948) for constructing a convergent sequence of approximate solutions, provided one has a reasonably good trial function to start from.

T

The algebraic derivation and thermodynamical significance of the variational method are adequately discussed in the literature (see, e.g. Ziman, 1960, §§7.7–7.10). But the formulae produced are rather complicated and do not have the intuitive physical significance that we have learnt to attach to the concept of the relaxation time. It is instructive, therefore, to try to bridge the gap between these two methods.

Consider, for example, the case of elastic scattering by impurities. The electrical resistivity is the minimum of the ratio (Ziman, 1960, §7.9)

$$\rho = \frac{1}{2kT} \iint \{\Phi_\mathbf{k} - \Phi_{\mathbf{k}'}\}^2 \mathscr{P}_\mathbf{k}^{\mathbf{k}'} \, d\mathbf{k} \, d\mathbf{k}' \Big/ \left| \int e v_\mathbf{k} \Phi_\mathbf{k} (-\partial f^0/\partial \mathscr{E}) \, d\mathbf{k} \right|^2. \tag{5.50}$$

Varying this integral gives the functional form of $\Phi_\mathbf{k}$, whose absolute magnitude, however, may be chosen at will. In the standard case of the elementary solution for an isotropic medium (Ziman, 1960, §7.4; Ziman, 1964a, §7.3) $\Phi_\mathbf{k}$ turns out to be proportional to the cosine of the angle between $v_\mathbf{k}$ and the electric field. If the constant of proportionality were unity, we should then get

$$\left| \int e v_\mathbf{k} \Phi_\mathbf{k} (-\partial f^0/\partial \mathscr{E}) \, d\mathbf{k} \right| = \tfrac{1}{3} e v_F \mathscr{N}(\mathscr{E}_F) \tag{5.51}$$

by the standard properties of the Fermi function (Ziman, 1964a, §4.5). Let us suppose that the distribution function $\Phi_\mathbf{k}$ is always normalized to make (5.51) true.

But according to the elementary kinetic formula, the resistivity is given by (Ziman, 1964a, §7.2)

$$\rho = \frac{3}{(e v_F)^2} \frac{1}{\mathscr{N}(\mathscr{E}_F)} \frac{1}{\tau}. \tag{5.52}$$

Comparing (5.50) and (5.51) and (5.52), we get a formula for the relaxation time

$$\frac{1}{\tau} = \frac{3}{\mathscr{N}(\mathscr{E}_F)} \frac{1}{2kT} \iint \{\Phi_\mathbf{k} - \Phi_{\mathbf{k}'}\}^2 \mathscr{P}_\mathbf{k}^{\mathbf{k}'} \, d\mathbf{k} \, d\mathbf{k}' \tag{5.53}$$

subject of course, to $\Phi_\mathbf{k}$ being, indeed the solution of the equation.

In the standard isotropic case, where the scattering probability is elastic and depends only on the angle θ between $\mathbf{k}$ and $\mathbf{k}'$, i.e.

$$\mathscr{P}_\mathbf{k}^{\mathbf{k}'} = f_\mathbf{k}^0 (1 - f_{\mathbf{k}'}^0) \, \delta\{\mathscr{E}(\mathbf{k}) - \mathscr{E}(\mathbf{k}')\} \, \mathscr{Q}(\theta), \tag{5.54}$$

it is easy to show (Ziman, 1960, §7.9) that the exact solution

$$\Phi_\mathbf{k} \propto \mathbf{E} . v_\mathbf{k} \tag{5.55}$$

leads at once to the familiar result

$$\frac{1}{\tau} = \int (1 - \cos\theta)\, \mathcal{Q}(\theta)\, d\Omega'. \tag{5.56}$$

But (5.53) is more powerful than this, being a prescription for the effective relaxation time when the elementary solution is not strictly correct.

When calculating the 'idea' resistivity each of the two integrals in (5.48) contributes the same to the total scattering probability in a variational integral such as (5.53); going back to (5.8), and the elementary theory of the mean square displacement $|U_q|^2$ in the qth mode, (Ziman, 1964a, §§ 2.1, 2.9) we may write

$$\mathcal{P}_{k}^{k'} = 2f_{k}^{0}(1 - f_{k'}^{0})\,\delta(\mathcal{E} - \mathcal{E}' + \hbar\nu_q)\, n_q^0\, Q_{kq}^{k'}$$

$$= f_{k}^{0}(1 - f_{k'}^{0})\,\delta(\mathcal{E} - \mathcal{E}' + \hbar\nu_q)\,\frac{2\pi}{\hbar}\,\frac{2\hbar n_q^0}{2MN\nu_q}\,|\mathcal{S}_q(k', k)|^2$$

$$= f_{k}^{0}(1 - f_{k'}^{0})\,\delta(\mathcal{E} - \mathcal{E}' + \hbar\nu_q)\,\frac{2\pi}{\hbar}\,|U_q|^2\,|\mathcal{S}_q(k', k)|^2. \tag{5.57}$$

In other words, the scattering of electrons by phonons is proportional to the square of the matrix element of the appropriate Fourier components of the potential produced by the lattice displacements, between electron states. This expression implies, of course, a selection rule for the phonon wave-vectors and an eventual summation over phonons of various polarization.

But the inelasticity of the scattering has some effect at low temperatures when the change of energy $\hbar\nu_q$ is comparable with the thickness kT of the thermal layer on the Fermi surface. By elementary analytical arguments (e.g. Ziman, 1960, §9.5), one can show that the fermion statistical factor behaves like the corresponding elastic scattering factor (in (5.54), say) in being effectively a delta function at the Fermi level multiplied by a function of $z = \hbar\nu_q/kT$; i.e.

$$f_{k}^{0}(1 - f_{k'}^{0}) \approx \frac{z}{1 - e^{-z}}\,kT\left(-\frac{\partial f^0}{\partial \mathcal{E}}\right). \tag{5.58}$$

If we were to go to the trouble of finding the 'best' variation of Φ_k over the Fermi surface (Bross & Holz, 1963, 1965), then (5.52), (5.53), (5.57) and (5.58) would be an exact formula for the ideal resistivity of a metal. In practice, only the alkali metals have been studied in detail, and it is usual to take as the approximate trial function for

Φ_k the standard 'isotropic' form (5.55), which simplifies all the algebra and leads to an expression similar to (5.56),

$$\frac{1}{\tau_i} \approx \iint (1 - \cos\theta)\, \mathcal{Q}(\mathbf{k}, \mathbf{k}')\, d\Omega\, d\Omega'. \tag{5.59}$$

Apart from the electron–phonon interaction factor $|\mathcal{I}_q(\mathbf{k}, \mathbf{k}')|^2$, the transition probability $\mathcal{Q}(\mathbf{k}, \mathbf{k}')$ is proportional to $|U_q|^2$, which at high temperatures always makes ρ_i proportional to T. At low temperatures the Bose–Einstein distribution governs the phonon density; as is well known, this usually gives a T^5 law, with or without the inelasticity factor in (5.58) (Ziman, 1964a, §7.5).

If we make further, poorly justifiable approximations, such as treating the Fermi surface as a sphere, using the diffraction model for the electron–phonon interaction, and ignoring U-processes we end up with a crude formula for the electron mean free path in terms of the differential scattering cross-section σ_a of a neutral pseudo-atom (Ziman, 1964a, §7.5).

$$\frac{1}{\Lambda_i} = N2\pi \int \sigma_a(\theta)\, N\, |\mathbf{K} \cdot \mathbf{U}_q|^2\, (1 - \cos\theta) \sin\theta\, d\theta \tag{5.60}$$

which is useful for back-of-envelope calculations.

But a realistic evaluation of (5.59) even for a simple metal such as Na is exceedingly complicated in practice. The inclusion of U-processes (Ziman, 1954; Ziman, 1960, §9.5) requires nasty geometrical averages. The difference between longitudinal and transverse modes, and the anisotropy of the lattice spectrum is most important in determining the temperature variation of ρ_i (Bailyn, 1960b; Darby & March, 1964; Hasegawa, 1964; Bross & Holz, 1963, 1965; Sharma & Joshi, 1965). Distortions of the Fermi surface from a sphere, and concomitant mixing of wave-functions in the electron–phonon matrix elements may also be important in the effect of U-processes (Collins, 1961; Collins & Ziman, 1961). And now, after all this effort, we have come to realize that ρ_i is very sensitive to the form factor $\mathcal{U}_{N\Psi A}$ in the electron–phonon interaction, and that the Bardeen formula used in most of the numerical calculations is not very accurate. Although Greene & Kohn (1965) have made a start in constructing a better empirical form factor for Na, their results are not yet precise (Wiser, 1966). It must be admitted, in the end, that although the whole theory of the ideal resistivity of metals, dating back in its essentials to the work of Bloch some 35 years ago, can give fair agreement with experiment, it has not been fully tested quantitatively even in its first term.

Moreover, apart from an unpublished calculation on Al (Ashcroft, thesis, 1964) numerical work has been confined to the alkali metals, mostly to Na.

For this reason, the volume dependence of resistivity in the alkali metals (Dugdale & Phillips, 1965) must be interpreted with caution. There is obviously an effect due to changes in the Debye temperature (Mott, 1934a) but the origin of the large coefficients found in K and Rb has not been isolated. It could be distortion of the Fermi surface; but variations of the pseudo-potential with volume, similar in principal to the variation of the Bardeen matrix element calculated by Hasegawa (1964) are quite possible, as has been shown in the context of the resistivity of liquid metals (Ziman, 1967). Another interesting experimental result still to be interpreted quantitatively is that of Dugdale & Gugan (1960) showing the difference between the resistivities of the hexagonal and cubic phases of Na at low temperatures. On a more optimistic note, it is worth recording that Dugdale, Gugan & Okumura (1961) have measured the ideal resistivity of Li[6], and shown that it follows the curve for natural Li, appropriately scaled to allow for the effect of the change of mass on the Debye temperature.

5.8 HEAT CONDUCTION BY ELECTRONS

The elementary theory (Ziman, 1964a, §§7.7, 7.8) gives the thermal conductivity in terms of a relaxation time τ_w by a formula which we may write

$$\kappa = \frac{\pi^2}{3} \frac{(kT\,v_F)^2}{3T} \tau_w \mathcal{N}(\mathcal{E}_F). \tag{5.61}$$

For elastic scattering by impurities, τ_w is the same as the relaxation time τ for electrical conductivity, so that (5.52) and (5.61) are equivalent to the Wiedemann–Franz law

For scattering by phonons, the whole problem is much more complicated and can only be solved efficiently by the variational method (Ziman, 1960, §§9.9, 9.10). To link this with (5.61) we may use a variational functional for the thermal resistivity, analogous to (5.50); when Φ_k is correctly chosen, it minimizes

$$\frac{1}{\kappa} = \frac{1}{2k} \iint \{\Phi_k - \Phi_{k'}\}^2 \mathscr{P}_k^{k'}\, d\mathbf{k}\, d\mathbf{k}' \bigg/ \left| \int \mathbf{v}_k \{\mathscr{E}(\mathbf{k}) - \zeta\}\, \Phi_k(-\partial f^0/\partial\mathscr{E})\, d\mathbf{k} \right|^2. \tag{5.62}$$

The denominator looks almost the same as in (5.50) but the numerator is different, because it measures heat currents rather than electric currents. The standard solution (5.55) for the electrical conductivity carries almost no heat; we must construct a distribution function that carries the maximum heat with the minimum charge. In the standard elastic scattering case (Ziman, 1964a, §7.8) this is of the form

$$\Phi_{\mathbf{k}} = \frac{\mathscr{E}(\mathbf{k}) - \zeta}{kT} \cos(\mathbf{v_k}, \mathbf{E}), \qquad (5.63)$$

which leads to a normalization condition, analogous to (5.51),

$$\left| \int \mathbf{v_k} \{\mathscr{E}(\mathbf{k}) - \zeta\} \Phi_{\mathbf{k}} (-\partial f^0 / \partial \mathscr{E}) \, d\mathbf{k} \right| = \tfrac{1}{3} v_F \tfrac{1}{3} \pi^2 kT \mathscr{N}(\mathscr{E}_F), \qquad (5.64)$$

as if now each electron on the Fermi surface were carrying its contribution to the electronic specific heat rather than its electrical charge.

From (5.61) to (5.64) we get a formula for the relaxation time to be used in (5.61),

$$\frac{1}{\tau_w} = \frac{3}{\mathscr{N}(\mathscr{E}_F)} \frac{3}{\pi^2} \frac{1}{2kT} \iint \{\Phi_{\mathbf{k}} - \Phi_{\mathbf{k}'}\}^2 \mathscr{P}_{\mathbf{k}}^{\mathbf{k}'} \, d\mathbf{k} \, d\mathbf{k}', \qquad (5.65)$$

which is exact if $\Phi_{\mathbf{k}}$ is chosen to minimize the integral, subject to (5.64).

It is very easy to generalize (5.63) a little, and use the standard theorem for integrals involving Fermi functions (Ziman, 1964a, §§ 4.5, 7.7) to show that if the scattering is elastic this reduces exactly to the corresponding functional (5.53) for the relaxation time in electrical resistivity; this proves the Wiedemann–Franz law in its most general form.

But for phonon scattering the inelasticity of the electron scattering cannot be ignored. Suppose we put into the variational integral (5.65) the standard approximation (5.63). We then have from the energy condition in (5.57)

$$\{\Phi_{\mathbf{k}} - \Phi_{\mathbf{k}'}\}^2 = (1/kT)^2 \{(\mathscr{E} - \zeta) \cos(\mathbf{v}, \mathbf{E}) - (\mathscr{E}' - \zeta) \cos(\mathbf{v}', \mathbf{E})\}$$
$$\approx (1/kT)^2 \{(\mathscr{E} - \zeta) \cos(\mathbf{K}, \mathbf{E}) + \hbar \nu_{\mathbf{q}} \cos(\mathbf{v}', \mathbf{E})\} \qquad (5.66)$$

where $\mathbf{K}$ is the scattering vector, in the direction $(\mathbf{v}' - \mathbf{v})$.

At high temperatures, the first term is dominant; its energy dependence cancels the factor $(3/\pi^2 k^2 T^2)$, whilst its angular variation produces the usual factor $(1 - \cos\theta)$, leading to a result like (5.59); in other words, when the phonon scattering is quasi-elastic, $\tau_w \to \tau_i$, and the Wiedemann–Franz law holds.

At low temperatures the second term, which is not weighed against small-angle scattering, comes into its own. Physically this means that the energy of the phonon is large enough to carry away the extra heat carried by the electron, so that even small-angle scattering is a serious hindrance to thermal conduction. It is easy to show (Ziman, 1964a, §7.8) that this effect makes τ_w smaller than τ_i by a factor of the order of $(T/\Theta)^2$, and leads to a T^{-2} law for κ at low temperatures.

The actual evaluation of (5.65) in a specific case is even more tedious than the calculation of τ_i, and again depends sensitively on the form assumed for the lattice spectrum, electron–phonon interaction, Fermi surface, etc. Apart from some calculations on the alkali metals by Collins (1961; Collins & Ziman, 1961) nothing much has been done on this problem in the past decade (see Ziman, 1960, §9.10).

5.9   THERMO-ELECTRIC POWER: PHONON DRAG

It is easy, though tedious (Ziman, 1960, §9.12) to carry through a variational calculation of the thermo-electric power for free electrons scattered by phonons; but such an analysis is scarcely necessary.

At high temperatures the result is to reproduce the elementary formula first given by Mott & Jones (1936, chapter VII; see Ziman, 1964a, §7.9),

$$Q = \frac{\pi^2}{3}\frac{k}{e}kT\left[\frac{\partial \ln \sigma(\mathscr{E})}{\partial \mathscr{E}}\right]_{\mathscr{E}=\zeta}. \tag{5.67}$$

We calculate the conductivity as in §5.7, ignoring the inelasticity of electron scattering, and then vary the Fermi energy.

At low temperatures the effects of inelasticity on the electronic contribution to the thermopower are almost always masked by phonon drag (Ziman, 1964a, §7.11). In going from (5.48) to (5.50) and (5.62), we made the 'Bloch assumption', $\Phi_\mathbf{q} = 0$, as if the phonon distribution were always in equilibrium. This cannot be true, the phonons must be influenced to some extent by any change in the distribution of the electrons with which they are in interaction.

When we have allowed for direct heat conduction by the phonons, the thermal conductivity is scarcely affected by phonon drag. The electrical resistivity at very low temperatures could, in principle be considerably reduced (Ziman, 1960, §9.13), but this does not seem to have been observed experimentally. But the effect on the thermopower at low temperatures is striking.

Here again, we need a variational solution of the coupled Boltzmann equations for electrons and phonons (Ziman, 1960, §9.13). This can be linked with the elementary kinetic formulation (Ziman, 1964a, §7.11) as follows:

Suppose we assume that the 'drift velocity', of the phonon system is some fraction, α, of the drift velocity of the electron system under the influence of an electric field. This will give rise to a heat current which will, in turn contribute to the Peltier effect. The lattice thermopower may then be expressed in the form.

$$Q_L = \alpha \frac{C_L}{3ne}, \qquad (5.68)$$

where C_L is lattice specific heat (the factor $\frac{1}{3}$ being as usual a consequence of averaging over directions).

The coefficient α evidently measures the ratio of the rates at which phonons receive momentum from the electrons and dissipate it in various scattering processes (including scattering by the electrons). We may therefore express α in the form

$$\alpha = \tau_L/\tau_{e-p} \qquad (5.69)$$

where τ_L is the relaxation time governing the phonon system— as might be used in a theory of thermal conductivity by the lattice— and τ_{e-p} stands for a characteristic 'transfer time' between the two systems, to be calculated from our knowledge of electron–phonon interactions.

Here again the variational principle may be invoked. It turns out that $1/\tau_{e-p}$ is a weighted average of transition probabilities, just as in the formula (5.53) for $1/\tau_i$; for the distribution functions Φ_k, Φ_q that actually satisfy the coupled Boltzmann equations, (5.68) and (5.69) are exact if

$$\frac{1}{\tau_{e-p}} = \frac{3}{\mathcal{N}(\mathscr{E}_F)} \frac{1}{2kT} \iint \{\Phi_{k'} - \Phi_k\} \Phi_q \mathscr{P}_k^{k'} \, dk \, dk'. \qquad (5.70)$$

Of course, we still have to find these functions. But the general behaviour of Q_L may be calculated by assuming that the electron distribution is of the standard 'isotropic' form (5.55), and that the phonon distribution, also is of the similar simple form (but see Bailyn, 1960a)

$$\Phi_q = q \cos(\mathbf{q}, \mathbf{E}). \qquad (5.71)$$

The result is that the integrand in (5.70) behaves as if proportional to $\mathbf{K} \cdot \mathbf{q}$, where $\mathbf{K}$ is the 'scattering vector', $\mathbf{k'} - \mathbf{k}$.

If we allow only electron–phonon N-processes, where $\mathbf{q} = \mathbf{K}$, we shall find nothing more interesting than that $\tau_{e-p} = \tau_i > \tau_L$. The lattice thermopower will then be negative (because of the sign of e) and proportional to the lattice specific heat, until (presumably) the effects of phonon–phonon interaction reduce τ_L to a negligible fraction of τ_i.

But the geometry of electron–phonon U-processes can make $\mathbf{q}$ and $\mathbf{K}$ take opposite directions, which would make τ_{e-p} negative, and thus reverse the sign of Q_L (Ziman, 1964a, §7.11). As we have learnt from the calculation of other transport coefficients, the ratio of N-processes to U-processes in an integral such as (5.70) is very sensitive to the lattice spectrum, the electron–phonon matrix element and the shape of the Fermi surface, and can vary rapidly with temperature. Thus, by allowing the Fermi surface to approach a zone boundary, and providing a low-velocity lattice mode in the appropriate direction, one can so enhance the U-processes as to make the magnitude of $1/\tau_{e-p}$ somewhat greater than $1/\tau_L$, and hence explain the very large positive thermopower found in Cs at low temperatures.

In principle, the effects of phonon drag should be an excellent index of the accuracy of our ideas concerning the various factors that influence it. Unfortunately, the models that have been used to compute Q_L for the alkali metals (Ziman, 1959b; Bailyn 1960a; Collins & Ziman, 1961), although quite capable of reproducing the different experimental curves, were not sufficiently accurate in all the variable parameters for any reliable conclusions to be drawn from the results.

Nevertheless, the sensitivity of this phenomenon to the geometry of the reciprocal lattice and the Fermi surface must encourage further work on the theory. For example, Klemens, van Baarle & Gorter (1964) have observed that Q_L in Sn is anisotropic, being positive in one crystal direction and negative in another. This was interpreted semiquantitatively as evidence for different proportions of U-process being involved, according to the direction of the electric current relative to the Brillouin zone.

But one must be careful not to draw simple qualitative conclusions from the sign and magnitude of the lattice thermopower. In the noble metals Q_L shows only a small positive peak even though the Fermi surface is known to touch the zone boundaries. The distinction between 'N-processes' and U-processes' is then meaningless. But we may study the sign of the phonon drag effect by using the exact electron distribution function (Ziman, 1964a, §7.3)

$$\Phi_{\mathbf{k}} \propto \Lambda(\mathbf{k}) . \mathbf{E}, \tag{5.72}$$

where $\Lambda(\mathbf{k})$ is a vector mean free path, which is approximately in the direction of the local electron velocity at the point $\mathbf{k}$ on the Fermi surface, i.e.

$$\Lambda(\mathbf{k}) \approx \tau(\mathbf{k})\, \mathbf{v_k}, \tag{5.73}$$

where $\tau(\mathbf{k})$ would be the local value of the relaxation time (see §10 below). Each transition would then contribute to the integrand of (5.70) a term like

$$\mathbf{q} \cdot \{\tau(\mathbf{k}')\,\mathbf{v_{k'}} - \tau(\mathbf{k})\,\mathbf{v_k}\}, \tag{5.74}$$

which would usually be positive or negative according as the chord passed through occupied or unoccupied regions of the Fermi surface

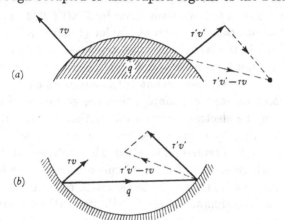

Fig. 5.2. (a) 'Electron-like'; (b) 'hole-like' transitions on a Fermi surface.

(Fig. 5.2). Thus, the sign of the contribution of this transition to $1/\tau_{e-p}$ would depend upon whether the states $\mathbf{k}$ and $\mathbf{k}'$ were 'electron-like' or 'hole-like' with respect to one another (Ziman, 1964 a, §9.3). The implications of this analysis for the noble metals have been discussed qualitatively (Ziman, 1961 b) but there is no quantitative analysis of phonon drag in this important group of metals, in spite of considerable experimental work on the pure metals and their alloys (see, e.g. Crisp & Henry, 1965; Weinberg, 1966; Huebner, 1966 for the most recent references).

5.10 THERMO-ELECTRIC POWER AT HIGH TEMPERATURES: ANISOTROPY OF τ

Although in principle the thermo-electric power of an 'ideal' metal at ordinary temperatures (i.e. $T > \Theta_D$) is given by (5.67), there is some controversy over the significance of the observed positive values of Q for the noble metals Cu, Ag, Au, and for Li and Cs. If we substitute for $\sigma(\mathscr{E})$ a formula like (5.25), say, then we can attribute the effect to a rapid decrease, with increasing $\mathscr{E}_F$, of any of the factors v_F^2, $N(\mathscr{E}_F)$, or τ, or some combination of them, like the area of the Fermi surfaces, or the mean free path.

The difficulty is that the factors governed by the geometry of the Fermi surfaces do not vary with energy in the way we would like (Ziman, 1961 b; Bross & Häcker, 1961). We must blame the dependence of τ on $\mathscr{E}$; but why should this vary so much from one metal to another? Qualitative arguments have been given by Blatt (1963), in favour of electron–electron scattering, and by Taylor (1963 a), in favour of the Kohn anomaly in the lattice spectrum as causes of the phenomenon, but these have not been worked out in detail. Nor does it seem likely that phonon drag is still significant in these conditions.

The explanation may be, simply, that the geometry of the Fermi surface and of the electron–phonon interaction allows stronger and stronger scattering by U-processes (or their geometrical equivalent, as in Fig. 5.2) as the Fermi level is raised. This is obvious, for example, in the standard geometry of a Fermi sphere near a zone boundary; increasing the radius of the sphere decreases the minimum wave-vector for U-processes and enhances their contribution to the total resistivity. On the other hand, the contributions of N-processes will be decreased because a longer phonon wave-vector is required to turn the electron through a given angle. The relative importance of these two effects must, as ever, depend upon various factors and must be worked out in detail, but we can certainly understand how a positive 'diffusion thermopower' at high temperatures ought to go with a positive phonon drag hump at low temperatures, which is what is observed. The argument is, in fact, much the same as the interpretation of the positive thermopower in these same metals in the *liquid* state (Ziman, 1961 a; see §6.6); the rapid rise in the structure factor near $K = 2k_F$ causes $\tau(\mathscr{E})$ to decrease rapidly, provided that the pseudo-potential $\mathscr{U}_{N\Psi A}(K)$ is large enough for this sort of scattering to be an important part of the total resistivity.

Unfortunately, the evaluation of this effect requires just those detailed calculations of ideal resistivity which we still lack. Moreover, the whole result may be quite sensitive to the weight to be given to various scattering processes in the total expression for the conductivity or resistivity; the anisotropy of the relaxation time $\tau(\mathbf{k})$ over the Fermi surface cannot be neglected.

The solution of the Boltzmann equation under these circumstances is a major analytical and computational task. It is not difficult to set up formal algorithms for constructing successive approximations to $\tau(\mathbf{k})$ (Sondheimer, 1962; Taylor 1963b) but these require long computations in practice. Thus, for example, Taylor had to assume a cut-off in the electron–phonon matrix element which automatically nullified the effect of U-processes on the diffusion thermopower discussed above.

As a rough approximation, a formula akin to (5.56) and (5.59) suggests itself (Deutsch, Paul & Brooks, 1961; Ziman, 1961b), i.e.

$$1/\tau(\mathbf{k}) \approx \int \{1 - \cos(\mathbf{v}_{\mathbf{k}}, \mathbf{v}_{\mathbf{k}'})\}\, \mathscr{Q}(\mathbf{k}, \mathbf{k}')\, d\Omega', \qquad (5.75)$$

which goes over into the usual elementary 'isotropic' case and whose behaviour for some simple models is similar to the solution obtained by an exact procedure (Sondheimer, 1962; Taylor, 1963b). It is important here to use the electron velocity rather than the wave-vector $\mathbf{k}$ to define a local direction of propagation on the Fermi surface, to avoid ambiguities and serious errors in extreme cases.

Again, the evaluation of even such a formula as (5.75) requires us to set up simplified models of the factors governing the scattering probability $\mathscr{Q}(\mathbf{k}, \mathbf{k}')$, but some crude calculations (Ziman, 1961c) show that $\tau(\mathbf{k})$ may indeed be very anisotropic, especially in the noble metals, where conditions on the 'necks' are quite different from those on the 'belly' of the Fermi surface. For phonon scattering, the anisotropy depends quite strongly on the temperature and also, of course, on the assumptions made about the lattice spectrum, etc. The effect of this anisotropy on the thermo-electric power has not been properly calculated, but it does go some way towards explaining the peculiarities of the Hall effect in the monovalent metals (Ziman, 1961c; Deutsch, Paul & Brooks, 1961). There is not space in the present chapter to give an account of low-field galvanomagnetic phenomena, but it can easily be seen that the relative weights of various regions of the Fermi surface in the conductivity must also determine the relative contributions of 'hole-like' and 'electron-like' orbits to these phenomena (see Ziman, 1960, §12.5; Ziman, 1961b).

5.11   IMPURITY AND IMPERFECTION SCATTERING

It is often assumed that the theory of the transport coefficients in a metal where scattering by impurities is dominant is much easier than the theory of the 'ideal' transport properties. This is only true for a free-electron metal such as Na, with a nice spherical Fermi surface; in other cases, especially for the noble metals, we have no warrant for ignoring the shape of the Fermi surface, mixing of plane waves, etc. Calculations of electrical resistivity may not be seriously affected by these complications, but other more subtle coefficients, such as thermo-electric power and Hall effect may be quite sensitive to them.

The free-electron theory is, of course, perfectly familiar; we simply feed the effective potential of the impurity or defect, as calculated above in §§5.4–6, into the usual formula for the scattering cross-section, and then calculate the relaxation time as in (5.56). If, as in the case of a dislocation or stacking fault, the scattering object is not spherically symmetrical, then we may have to use some averaging procedure based upon the variational method; it seems that (5.59) is then a fair approximation (Ziman, 1960, §9.4; Lackmann, 1964). In all such calculations, the main interest is in the behaviour of the effective potential as a function of scattering angle and energy, rather than on the underlying electronic structure of the parent metal.

Nevertheless, one must now have doubts whether such a model is adequate for the interpretation of the familiar phenomenon as the effects of alloying upon the thermo-electric power of the noble metals. It is clear that the relaxation time may be very anisotropic in such an alloy. For example, an impurity such as Ag, in Au, whose whole effect is concentrated in the core region of the ion, can scarcely be seen by a p-like Bloch state in the neck of the Fermi surface; this has very significant effects upon the Hall coefficient (Ziman, 1961c, 1961b; Hurd, 1965; Heine, 1965b) and is within range of direct experimental verification using 'Fermi surface' effects (Deaton & Gavenda, 1963). Until the magnitude of these corrections to the free-electron model have been computed, it is unprofitable to pursue lengthy analyses of the transport coefficients in alloys. Here again, we are not much further advanced in our quantitative theories than Mott and Jones were, some 30 years ago.

CHAPTER 6

ELECTRONIC TRANSPORT PROPERTIES OF LIQUID METALS

by T. E. FABER†

6.1 SOME SIMPLE FACTS...

As a general rule the resistivity of a metal rises when it melts, often by a factor of about 2. The main exceptions to the rule (see Table 6.1) are Sb, Bi and (for certain orientations of the crystalline solid) Ga, but these are 'semi-metals' in the solid phase, which crystallize with rather open structures of low co-ordination number and have unusually high resistivities in consequence. It is well known that they contract on melting and to judge by the results of X-ray diffraction studies their structure in the liquid phase is no longer one of low co-ordination, so it is hardly surprising that melting causes the resistivity of these three elements to fall, to values which are characteristic of a metal proper. Si and Ge show the same sort of behaviour in a more extreme way; they are semiconductors in the solid phase but can be classified when liquid as ordinary metals.‡ Otherwise the only exception seems to be Mn; but the data available for liquid transition and rare earth metals are still very meagre and no attempt is made to discuss them in this chapter.

The temperature coefficient of the resistivity is generally smaller above the melting point than it is below. In the solid phase the resistivity is usually regarded as made up of a 'residual' term contributed by impurity atoms and lattice defects which is independent of temperature and a term due to thermal vibrations of the lattice which is proportional to T; in a reasonably pure specimen near its melting point the residual resistivity is negligible so that $(T/\rho)(\partial\rho/\partial T)_p$ is close to unity. A typical value for this quantity in a polyvalent liquid metal, however, is only about 0·3 and in liquid Zn and Cd it is even slightly negative.

---

† Dr Faber is University Lecturer in Physics at the Cavendish Laboratory, Free School Lane, Cambridge, and Fellow of Corpus Christi College, Cambridge.

‡ Te is another semiconductor whose resistivity drops markedly on melting. But several aspects of the electrical behaviour of liquid Te distinguish it from the liquid metals proper (Tièche & Zareba, 1963; Hodgson, 1963), so it is not included in Table 6.1.

If one looks at the coefficient at constant volume rather than constant pressure (liquid metals have sufficiently high coefficients of expansion for there to be substantial differences betwen the two) the contrast between the behaviour of the two phases is still more striking. Some values for $(T/\rho_l)(\partial\rho_l/\partial T)_V$ just above the melting point are shown in Table 6.1, though some of them include an element of guess-work.[†] It appears that this quantity is distinctly negative for all the divalent metals and surprisingly small for those with more than two valence electrons per atom; only the alkali metals show values comparable with those found for solid metals.

Other data are of course available concerning the electronic transport properties of liquid metals. We shall have occasion below to discuss the dependence of the resistivity on volume, the thermo-electric power, the effect of adding impurity, the Hall effect, and the nature of the response to an oscillating electric field of high frequency. Information relating to some of these points is included in Table 6.1. But most of this is of fairly recent origin. The simple facts summarized above, which have been known for fifty years or more, are enough to set us on the right lines in establishing a framework of theory.

6.2 ...AND SIMPLE THEORIES

Theorists are easily fascinated by dimensionless ratios and there have been many attempts to explain the observed values of ρ_l/ρ_s or at least to correlate them with other properties of the metals concerned (for references to the early literature the reader should consult a useful review article by Cusack, 1963). Nowadays we prefer to focus attention on ρ_l and ρ_s separately. Nevertheless, it is appropriate in this volume to record one of the first and most constructive theories, which had a lot of influence on subsequent thinking about liquid metals: it was put forward by Mott (1934b).

Mott suggested that ρ_l/ρ_s could be correlated with the latent heat of melting. He pictured a liquid metal as an essentially solid-like array of ions, but with each ion vibrating with a lower characteristic frequency ν and therefore a greater amplitude than it would have in the true solid at the same temperature—the result, presumably, of a

---

[†] Direct measurements of $(\partial\rho/\partial T)_V$ have been made for all the liquid alkali metals except Li and for Ga and Hg. Where it has not been measured directly it can be deduced from $(\partial\rho/\partial T)_p$ if $(\partial V/\partial T)_p$ and $(\partial\rho/\partial V)_T$ are known. If $(\partial\rho/\partial V)_T$ has not been measured this can sometimes be estimated from the thermo-electric power (see §6.6).

Table 6.1. *Some electrical properties of liquid metals*

Metal	Valency	$\dfrac{\rho_l}{\rho_s}$	ρ_l ($\mu\Omega$ cm)	$\dfrac{l}{d}$	$\dfrac{T_m}{\rho_l}\left(\dfrac{\partial\rho_l}{\partial T}\right)_V$	$\dfrac{V}{\rho_l}\left(\dfrac{\partial\rho_l}{\partial T}\right)_V$	ξ	$\dfrac{R}{R_0}$
Li	1	1·59	24·7	13	(0·6)	⩽0	−9·3	—
Na	1	1·45	9·6	158	0·85	4·8	2·8	0·98
K	1	1·58	13·0	177	0·76	5·1	3·5	—
Rb	1	1·60	22·5	118	0·70	4·2	1·7	—
Cs	1	1·67	37	85	0·69	2·1	−1·4	—
Cu	1	2·1	21	13	(0·4)	—	−3·6	1·0
Ag	1	2·1	17·2	18	(0·4)	—	−1·8	1·0
Au	1	2·3	31·2	10	(0·4)	—	−0·8	1·0
Mg	2	1·8	27	8·1	—	—	—	—
Ba	2	1·6	(134)	(2·3)	—	—	—	—
Zn	2	2·2	37	4·6	(−0·24)	—	−0·1	1·01
Cd	2	2·0	34	6·7	(−0·22)	—	−0·2	0·99
Hg	2	3·7–4·9	91	2·3	(−0·10)	8·0	4·1	0·99
Al	3	2·2	24	6·0	—	—	1·0	—
Ga	3	0·45–3·1	26	6·3	0·14	2·9	0·55	0·97
In	3	2·2	33	5·0	(0·16)	—	0·8	0·93
Tl	3	2·1	73	2·5	—	—	0·3	0·96
Si	4	—	(80)	(1·5)	—	—	—	—
Ge	4	—	(70)	(2·4)	—	—	—	1·06
Sn	4	2·1	48	3·1	(0·13)	—	0·4	1·0
Pb	4	1·9	95	1·6	−(0·02)	—	2·1	0·88
Sb	5	0·6	114	1·1	—	—	—	0·92
Bi	5	0·35–0·47	128	1·0	0·02	3·1	0·55	0·95
Mn	—	0·6	(40)	—	—	—	—	—
Fe	—	1·01	(140)	—	—	—	—	—
Co	—	1·09	(100)	—	—	—	—	—
Ni	—	1·33	(85)	—	—	—	—	—
Pt	—	1·4	—	—	—	—	—	—
W	—	1·1	—	—	—	—	—	—
Mo	—	1·2	—	—	—	—	—	—

Here ρ_l and ρ_s denote the resistivities of the liquid and solid phases; l/d is the ratio of electronic mean free path in the liquid to the mean interatomic spacing; ξ is the thermoelectric power parameter defined in the text; R/R_0 is the ratio of the observed Hall coefficient to the free-electron value, calculated on the assumption that the number of conduction electrons per unit volume is equal to the number of valence electrons. All the data refer to the melting temperature, T_m. Where a range of values is quoted for ρ_l/ρ_s it is because ρ_s is sensitive to crystal orientation.

References to most of the original papers from which these data have been collected may be found in review articles by Cusack (1963) and Wilson (1965). More recent sources which have been consulted are Faber (1967a) (Li); Marwaha (1967) (ξ); Busch & Güntherodt (1967) (R/R_0). Many of the quantities listed have been measured by more than one experimenter, with results that are not necessarily identical, and authors are liable to differ as to which is the 'best' value to quote. Entries which seem especially uncertain are enclosed in parentheses in the above table.

rather looser packing all round. In the elementary theory of the resistivity of a solid metal at high temperatures ρ is proportional to the mean square amplitude of vibration and hence to ν^{-2}. Mott's model therefore suggests
$$\rho_l/\rho_s = (\nu_s/\nu_l)^2. \tag{6.1}$$

The thermal entropy of an Einstein solid, on the other hand, is given by
$$S = 3Nk\ln(kT/h\nu) + O(h\nu/kT)^2, \tag{6.2}$$

so that if a similar formula can be applied to the liquid phase we may expect
$$\rho_l/\rho_s = \exp(2(S_l - S_s)/3Nk) = \exp(2L/3RT_m). \tag{6.3}$$

This simple prediction does work out quite well, though its success lies more in the fact that both sides of the equation are roughly 2 for a large number of metals (L/RT_m is usually about unity) than in a detailed correlation from one metal to the next. Mott attempted to explain away the relatively poor agreement for the alkali metals, but his argument at this point is fallacious.

There are a number of difficulties in Mott's theory. It requires, first, that we ignore all contributions to S_l arising from the disorder of the liquid structure, i.e. the so-called *configurational* entropy. This in itself is implausible, and the consequent deduction that ν_l differs from ν_s by 40 % or more is almost equally so. A typical value of Grüneisen's constant γ is about 2, so that a 5 % increase of volume for a solid metal decreases ν_s by about 10 %; why should the 5 % increase in volume which typically occurs when a metal melts affect the vibrational frequencies to any greater extent than this? In any case, can the hypothesis of a 40 % change in ν be reconciled with the direct evidence that is now available from the inelastic scattering of neutrons concerning the ionic motions in solid and liquid metals? It seems not. On a different tack, the theory does not make it easy to understand the behaviour of ρ as a function of temperature above the melting point; it suggests that it should continue to rise in proportion to T, which is far from what is observed in polyvalent metals. Mott did point out that there might be some residual band structure affecting the conduction electrons in a liquid metal and limiting the effective number of carriers; if this band structure were to dissolve on heating one might find that in the elementary free-electron formula
$$\rho = (m/ne^2)(1/\tau) \tag{6.4}$$

the factor (mn/e^2) decreased with temperature. It is difficult to believe, however, that it decreases fast enough to outweigh an increase

U

286   T. E. FABER

of the scattering probability $1/\tau$ proportional to T. At any rate if one wants to argue that (m/ne^2) is so sensitive to temperature above the melting point it is scarcely consistent to make the assumption, which Mott's theory requires, that it remains unchanged during the drastic upheaval of the melting process itself.

An alternative approach to the whole problem, advocated by MacDonald (1959) and others, is to suppose that virtually all the entropy of melting is configurational, and that the part of the resistivity which increases on melting is the 'residual' rather than the thermal term—as though lots of vacancies and dislocations had been introduced. This implies a somewhat less rapid increase of ρ with T in the liquid phase than Mott's theory, but it makes it no easier to understand values of $(\partial\rho/\partial T)_V$ which are zero or even negative.

The weakness of both the theories outlined above is that they use the familiar idea that the resistivity is composed of separate residual and vibrational terms, without questioning whether this is appropriate in the case of a liquid. The ionic configuration in simple liquids may well be so random that comparison with a gas is more meaningful than with a solid. The magnitude of the entropy of melting, indeed, is one of the arguments in favour of adopting a model distinctly different from that of a polycrystalline solid: Mott & Gurney (1939) and others have shown that it is hard to explain a configurational entropy as large as R on a micro-crystallite model unless the micro-crystals are only one or two ions across. Nowadays it is more fashionable to picture simple liquids as being 'random close-packed' structures in the sense defined by Bernal (1964), which implies that the characteristic symmetry of the solid is not to be discerned, even in the smallest unit of an ion and its immediate neighbours.

As soon as one starts to think in terms of a gas-like situation the problem concerning $(\partial\rho/\partial T)_V$ in the polyvalent metals disappears, for if the ions are in effect completely disordered just above the melting point no amount of further heating can increase the disorder or increase the resistivity. The explanation provided by the quantitative theory to be discussed below lies on just these lines. It turns out that the gradual disappearance on heating of such residual order as must persist in the liquid due to the short-range repulsive forces between the ions is sufficient to explain the behaviour of the monovalent and divalent metals also.

6.3 THE N.F.E. MODEL

If the ionic arrangement in a liquid metal is highly disordered it seems unlikely that much memory of the Brillouin zones and aniso-tropic Fermi surfaces which are characteristic of the solid phase can persist on melting. A plausible model to use for the conduction elec-trons as a first approximation is therefore one in which their states are represented by ordinary plane waves rather than Bloch waves, with an isotropic relation between energy E_k and wave-number k. The plane waves must of course be *pseudo*-wave-functions, in the sense defined in Chapter 1, for the true wave-functions oscillate rapidly inside each ion core. They will be acted upon by a relatively weak *pseudo-potential* $\mathscr{V}(\mathbf{r})$. To first order in this pseudo-potential the potential energy of an electron in a state k is given in the notation of Chapter 1 by

$$\langle -\mathbf{k}|\mathscr{V}|\mathbf{k}\rangle \tag{6.5}$$

and because $\mathscr{V}$ is a non-local operator this is not necessarily indepen-dent of k; there may also be significant second-order energy shifts (cf. (1.62)) not necessarily independent of k. Hence it is not necessarily sound to equate E_k to $\hbar^2 k^2/2m$, and in particular at the Fermi radius k_F we should set

$$\partial E_k/\partial k = \hbar^2 k_F/m^* \tag{6.6}$$

and be prepared to find that the effective mass m^* thus defined differs significantly from the free-electron mass m. Even if one ignores the effects of the interaction parameter η introduced in §1.7, therefore (and they will be ignored throughout this chapter), the electrons cannot be treated as entirely free carriers. It is for this reason that the model to be used below is referred to as the *nearly* free electron, or N.F.E., model.

The pseudo-potential is not only responsible for a shift of E_k, it is also liable to scatter electrons from one value of k to another. If, as a first approximation, we ignore the motion of the ions the scattering can be treated as elastic and we need only consider processes in which an electron initially on the Fermi surface, at $\mathbf{k}_1$ say, moves to another state on the Fermi surface, $\mathbf{k}_2$, the angle between these two vectors being θ (see Fig. 6.1). The matrix element responsible for the scattering is

$$\langle -\mathbf{k}_2|\mathscr{V}|\mathbf{k}_1\rangle = V^{-1}\int^V \exp(-i\mathbf{K}.\mathbf{r})\,\mathscr{V}(\mathbf{r})\,d\mathbf{r} = V^{-1}\mathscr{V}(\mathbf{K}), \tag{6.7}$$

where K is the scattering vector, $(\mathbf{k}_2-\mathbf{k}_1)$. Hence, by application of the 'golden rule', the lifetime τ of an electron at the Fermi surface is given to second order in $\mathscr{V}$ by

$$\frac{1}{\tau} = \frac{m^* k_F}{2\pi\hbar^3 V} \int_0^\pi |\mathscr{V}(K)|^2 \sin\theta\, d\theta. \tag{6.8}$$

The ensemble average of $|\mathscr{V}(K)|^2$ is taken to be independent of the direction of $\mathbf{K}$ because the liquid is isotropic. The scattering probability depends upon the density of states (for one spin orientation only, since the chance of an electron reversing its spin during a scattering event is very remote) and it is for this reason that the effective mass m^* appears in (6.8).

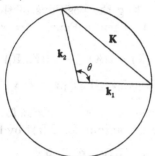

Fig. 6.1. The scattering vector $\mathbf{K}$ links two states on the Fermi surface, $\mathbf{k}_1$ and $\mathbf{k}_2$.

The relaxation time which is needed for insertion into (6.4) is not quite the same as the lifetime described by (6.8). Solution of the Boltzmann equation shows (see §5.7) that in calculating the relaxation time for current decay we should weight the large-angle scattering events by a factor $(1-\cos\theta)$, which expresses the proportion of the original momentum of the electron which is lost during the scattering. If we write

$$\frac{1}{\tau_z} = \frac{mk_F}{2\pi\hbar^3 V} \int_0^\pi \overline{|\mathscr{V}(K)|^2} \sin\theta(1-\cos\theta)\, d\theta$$

$$= \frac{mk_F}{\pi\hbar^3 V} \int_0^1 \overline{|\mathscr{V}(K)|^2}\, 4\left(\frac{K}{2k_F}\right)^3 d\left(\frac{K}{2k_F}\right)$$

$$= mk_F \langle\overline{|\mathscr{V}|^2}\rangle/\pi\hbar^3 V, \tag{6.9}$$

where brackets thus $\langle\,\rangle$ are used to denote an average over the range of $\mathbf{K}$ from 0 to $2k_F$ heavily weighted towards the upper end, then we have

$$\rho = (m^*/m)^2 (m/ne^2)(1/\tau_z). \tag{6.10}$$

One factor of (m^*/m) enters as in (6.8) because of the density of states; the second would seem to be required if we think of the current carried by an electron near the Fermi surface as determined in the usual way by its group velocity, i.e. by $\hbar^{-1}\partial E_k/\partial k$ (see Chapter 1). We should presumably take n in (6.10) to be the number of valence electrons per unit volume, which determine the size of the Fermi sphere according to the equation

$$k_F^3 = 3\pi^2 n. \tag{6.11}$$

On the assumption that there is no serious overlap between the ion cores in a liquid metal we may represent the total pseudo-potential $\mathscr{V}(\mathbf{r})$ as the sum of (screened) pseudo-potentials $v(\mathbf{r} - \mathbf{R}_l)$ due to individual ions, $\mathbf{R}_l$ denoting the positions of the N ions which are present in the volume V. Then in Fourier transform for $K \neq 0$ (see §§ 1.2, 5.2 and 5.3)

$$\mathscr{V}(\mathbf{K}) = v(\mathbf{K}) \sum_l \exp(i\mathbf{K} \cdot \mathbf{R}_l) \tag{6.12}$$

and

$$|\mathscr{V}(K)|^2 = |v(K)|^2 Na(K), \tag{6.13}$$

where $a(K)$ is the so-called *interference function* for the liquid structure, related to the pair correlation function $P(r)$† by the equations

$$a(K) = \overline{\sum_m \exp i\mathbf{K} \cdot (\mathbf{R}_m - \mathbf{R}_l)}$$
$$= 1 + \frac{N}{V} \int_0^\infty 4\pi r^2 \, dr (P-1) \frac{\sin Kr}{Kr}. \tag{6.14}$$

In terms of this quantity, then, the equation for τ_z may be written

$$1/\tau_z = (N/V) mk_F \langle |v|^2 a \rangle / \pi\hbar^3. \tag{6.15}$$

Were the liquid structure truly gas-like, P would equal unity for all values of r so that a would equal unity for all values of K. Equation (6.15) would then express the well-known result that the intensity of a wave scattered by a random array of scatterers can be calculated as the sum of the intensities scattered individually. The interference function allows for the effects of any short-range order that may exist.

The theory for the scattering of say X-rays by a liquid (e.g. Fournet, 1957) is essentially identical to the above, v being replaced by the X-ray form factor for the individual atoms. In the case of X-rays, or for that matter neutrons, the ratio between the intensity observed in a particular direction of scattering and the intensity that would be

† $NP(r)/V$ is the probability, given that an ion is centred on the point $\mathbf{r} = 0$, of finding another one per unit volume at a radius r.

observed for a perfectly random array is something that can be measured directly. Hence if we want to apply (6.15) to a given liquid metal we may obtain the appropriate values of $a(K)$ from an X-ray or neutron experiment.

This method for calculating the resistivity is essentially due to Ziman and the suffix in τ_z is a discreet acknowledgement of this. Several previous authors (Shubin, 1934; Bhatia & Krishnan, 1948; Gerstenkorn, 1952) had approached the problem along similar lines, but were hampered by a lack of understanding of the modern pseudo-potential concept, which is so useful in this context. In Ziman's first paper (1961 a) the factor $(m^*/m)^2$ in equation (6.10) was not included and in most applications of his theory it has either been omitted or set equal to unity. More sophisticated arguments, which are touched on briefly below, suggest that it ought indeed to be dropped. The fact is that the deviations of m^* from m are due essentially to the perturbing effects of the pseudo-potential, and if the whole of this pseudo-potential is included in the calculation of τ_z we should be allowing for the effects twice over if we added a factor of $(m^*/m)^2$ as well.

For the sake of simplicity the fact that the ions are in motion has been neglected in the above calculation and the scattering treated as elastic. Strictly, the quantity $\sum_l \exp(-i\mathbf{K} \cdot \mathbf{R}_l)$ which occurs in (6.12) is time-dependent and so is $\mathscr{V}(K)$. This means that an electron or an X-ray or a neutron which is scattered by the vector $\mathbf{K}$ is liable to suffer a change of energy in the process. One may describe the probability of a particular energy gain $\hbar\omega$ in terms of a function $S(K, \omega)$, such that

$$\int_{-\infty}^{\infty} S(K, \omega)\, d\omega = a(K). \qquad (6.16)$$

A more careful argument then shows (e.g. Baym, 1964) that in the final expression for τ_z $a(K)$ should be replaced by

$$\int_{-\infty}^{\infty} S(K, \omega)\, \frac{\beta\omega\, d\omega}{1 - \exp(-\beta\omega)} = \int_{-\infty}^{\infty} S(K, \omega)\,(1 + \tfrac{1}{2}\beta\omega + \tfrac{1}{12}\beta^2\omega^2 + \ldots)\, d\omega,$$
$$(6.17)$$

where β is an abbreviation for $\hbar/kT$; the factor $\beta\omega/(1 - \exp(-\beta\omega))$ arises out of an integration of $f(E)(1 - f(E + \hbar\omega))$ across the Fermi surface, $f(E)$ being the Fermi–Dirac distribution function. $S(K, \omega)$ can be determined for liquid metals, in principle at any rate, by resolving the energy spectrum after diffraction of a beam of slow neutrons, but we do not need to invoke any experimental data for it in order to see that the ionic motion makes a negligible difference to τ_g. There exist sum rules which tell us the first and second moments of the function $S(K, \omega)$ (Placzek, 1952), and from these one can evaluate (6.17) as

$$a(K) - \hbar^2 K^2/6MkT + ..., \qquad (6.18)$$

where M is the ionic mass. In the lightest metal, Li, the correction term amounts to about $0 \cdot 015 \, (K/2k_F)^2$ which is just big enough to be significant (see Fig. 6.3) In heavier metals it can safely be forgotten.

6.4 CORRECTIONS TO ZIMAN'S THEORY

The main object of this chapter is to discuss the application of the formulae derived above, but it may be desirable to spend a little time first in pointing out their limitations. It tends to be assumed that if Ziman's theory fails to give quite the right answer for the resistivity of a liquid metal then the fault lies in the values assumed for $v(K)$; indeed, such faith is reposed on the form of the equations that it is sometimes suggested that observed values for the resistivity and perhaps thermo-electric power of the liquid should be used, by turning the handle of Ziman's theory backwards, to generate a semi-empirical $v(K)$ to be applied subsequently in quite a different context. It therefore needs emphasis that even if there are no errors in $v(K)$ the theory is capable of giving the wrong answer for ρ; discrepancies of up to 50 % do not seem out of the question.

One assumption of the theory which has already been mentioned is that the ions do not overlap; to be more precise, what is required is that the usual linear screening theory should everywhere be valid, which means that in regions where the pseudo-potentials of two neighbouring atoms overlap both should be weak. No calculations to justify this assumption seem to have been reported. It could lead to serious errors in the evaluation of $\mathscr{V}(K)$ for small K, though perhaps near $K = 2k_F$, which tends to be the important region as far as the resistivity is concerned because of the weighting factor in (6.9), they may indeed be negligible.

Secondly, the fact that the plane waves of the N.F.E. model are only *pseudo*-wave-functions means that certain normalization corrections are required (Faber, 1966). They are likely to *reduce* the answer for ρ, by 10 % or more.

Thirdly, a rather important feature of the integral in (6.9) seems to be the sharp cut-off at an upper limit corresponding to $K = 2k_F$,

292   T. E. FABER

i.e. to scattering processes which take an electron from a state on the Fermi sphere to the state diametrically opposite. Can the Fermi surface in a liquid metal really be sufficiently well-defined to justify so abrupt a cut-off? The thermal broadening, kT, is admittedly still small compared with the Fermi energy E_F, even in the higher melting point metals; but one would expect some additional lifetime broadening for each plane wave of order $\hbar/\tau$ and the observations suggest that this may be as much as $\frac{1}{2}E_F$ in some polyvalent metals. At first sight this seems to undermine the whole structure of Ziman's theory, making the application even of the Boltzmann equation a rather perilous matter.

There is an alternative approach to the theory of electrical conduction, discussed in a variety of forms by Kubo, Greenwood, Edwards and several others (Faber, 1966). One starts by trying to discover the nature of the eigenstates for the electrons in the absence of any electric field; simple plane waves do not qualify, of course, because they are subject to scattering by the pseudo-potential. It turns out that in a liquid metal the eigenstates may be represented by wave-groups of the form

$$\Phi(E) = \Sigma a_{\mathbf{k}} \phi(\mathbf{k}) \qquad (6.19)$$

with

$$\overline{|a_k|^2} = \frac{1}{\pi \mathcal{N}(E)} \frac{\hbar/2\tau_k}{(E_k - E)^2 + (\hbar/2\tau_k)^2}, \qquad (6.20)$$

where $\mathcal{N}(E)$ is the density of states for one spin orientation and the prescription for the calculation of τ_k is such that in the limit when $\hbar/\tau_k$ is small it corresponds exactly with the lifetime defined by (6.8), though with a general value for k instead of k_F.† The Lorentzian character of (6.20) is obviously connected with the fact that a single plane wave $\phi(k)$ suffers exponential attenuation due to scattering, but since E_k and τ_k may both be functions of k the Lorentzian is not necessarily perfect. All directions of k must, on the average, be equally represented in the group, so that the eigenstates carry no net current.

Current arises when a field is switched on because transitions are excited from one eigenstate to another. Its magnitude is easily calculated by Kramers–Heisenberg dispersion theory, supposing the field to be oscillatory with an angular frequency ω which is subse-

† To simplify the presentation the notation used by Faber (1966) has been slightly modified here. The lifetime τ_k includes a factor (m^*/m)—see (6.8)—which is omitted from the definition of τ_k in Faber (1966); the first and second-order energy shifts due to $\mathscr{V}$ are here lumped in with E_k instead of being exposed in a separate term A_k^*; and some complications that may arise if A_k^* varies with E are here ignored.

quently allowed to go to zero. The result for the d.c. conductivity
takes the form

$$\sigma = \frac{2\pi e^2 \hbar^3}{m^2 V} \mathcal{N}(E)_F^2 \overline{|D|_F^2}, \tag{6.21}$$

where D is a matrix element between two adjacent states; i.e.

$$D = \left\langle \Phi_1^* \frac{\partial}{\partial x} \Phi_0 \right\rangle = \sum_{\mathbf{k}} ik \cos \alpha_{\mathbf{k}} a_{\mathbf{k}} b_{\mathbf{k}}^*, \tag{6.22}$$

where $\alpha_{\mathbf{k}}$ is the angle between the directions of $\mathbf{k}$ and of the applied
field and $b_{\mathbf{k}}$, the coefficient which describes Φ_1, is such that $|b_{\mathbf{k}}|^2$
and $|a_{\mathbf{k}}|^2$ have the same ensemble average, though there must be
subtle differences of phase between $b_{\mathbf{k}}$ and $a_{\mathbf{k}}$ to ensure that Φ_1 and Φ_0
are orthogonal. Careful evaluation of the mean square matrix element
in the limit when $\hbar/\tau$ is small then gets us back to (6.10) as expected,
though without the $(m^*/m)^2$ factor; all the terms which include the
effective mass cancel from the final answer for σ or ρ. It is the cross-
terms in $|D|^2$, i.e.

$$\sum_{\mathbf{k}} \sum_{\mathbf{k}' \neq \mathbf{k}} kk' \cos \alpha_{\mathbf{k}} \cos \alpha_{\mathbf{k}'} a_{\mathbf{k}} a_{\mathbf{k}}^* b_{\mathbf{k}'} b_{\mathbf{k}}^*, \tag{6.23}$$

which are responsible for the apearance of a $(1 - \cos \theta)$ correction and
hence for the fact that it is the relaxation time τ_z which determines
the answer, rather than the lifetime τ.

The advantage of this approach is that it is easy to see in principle
how the calculation needs to be extended when $\hbar/\tau$ is not small. Detailed computations have been carried through only for an ideal situation in which $\mathcal{V}(K)$ is independent of K,† and it turns out then that
the correction needed to Ziman's formula is a factor

$$(1 + O(\hbar/2\tau E_F)^4).$$

It seems to be something of an accident that the terms of order
$(\hbar/2\tau E_F)^2$ cancel, and there is no reason to suppose that they do so
for any real liquid metal. In cases where $\hbar/\tau$ is as large as $\frac{1}{2}E_F$ corrections of up to say 10 % (positive or negative) may be required on
this account. It is some consolation, however, to find from the analysis
that no terms of order $(h/2\tau E_F)$ are ever likely to appear.

---

† And incidentally independent of $\mathbf{k}$ too. It has been emphasized in §1.1. that the
pseudo-potential, being a non-local operator, depends on the energy and wave-
number of the electron upon which it acts and also upon the angle of scattering θ.
So long as $\hbar/\tau$ is small it is not necessary to allow for this dependence explicitly
because one is only concerned with values of $\mathbf{k}_1$ near k_F, and θ is fixed (for given K)
by the consideration that $\mathbf{k}_2$ lies near k_F also. When $\hbar/\tau$ is comparable with E_F
the situation is no longer so simple.

Finally, the most serious limitation of Ziman's theory may well be its reliance on the Born approximation in the evaluation of scattering probabilities. It is well known that the matrix element which determines the probability of scattering from a state ψ_1 to a state ψ_2 due to a potential $V(\mathbf{r})$ is

$$\langle \psi_2^* | V | \psi \rangle, \qquad (6.24)$$

where ψ is the *whole* wave-function at the point $\mathbf{r}$, i.e. it includes all the scattered waves, which have the effect of modulating to some extent the amplitude and phase of the incident wave ψ_1. The Born approximation amounts to the neglect of this modulation, ψ being replaced by ψ_1 in (6.23); clearly if ψ_1 and ψ_2 are plane waves proportional to $\exp(i\mathbf{k}_1 . \mathbf{r})$ and $\exp(i k_2 . \mathbf{r})$ respectively (6.24) then reduces to $V(K)$. Suppose, however, that

$$\psi(\mathbf{r}) = (1 + \gamma(\mathbf{r})) \psi_1(\mathbf{r}) \qquad (6.25)$$

and that the factor γ is not negligible. It is clear by substitution of (6.25) into (6.24) that we may if we wish retain the machinery of the Born approximation, provided that we replace the true potential V by an effective one V' such that

$$V'(\mathbf{r}) = (1 + \gamma(\mathbf{r})) V(\mathbf{r}), \qquad (6.26)$$

so that $V'(K)$ rather than $V(K)$ is the appropriate matrix element. It may be helpful to note here that $\gamma(\mathbf{r})$ may depend upon $\mathbf{k}_1$ and moreover that if $V(\mathbf{r})$ is spherically symmetric about the point $\mathbf{r} = 0$ it does *not* follow that $\gamma(\mathbf{r})$ is spherically symmetric; hence $V'(K)$ may be complex and is liable to depend both on the magnitude of $\mathbf{k}_1$ and on the scattering angle θ.

Inside an ion core in a metal the potential is so strong compared with the Fermi energy that a wave-function proportional to

$$\exp(i\mathbf{k}_F . \mathbf{r})$$

must suffer very substantial modulation; $\gamma(\mathbf{r})$ may be much greater than unity and may oscillate rapidly. Hence if we wish to calculate the scattering by a single ion the particular recipe suggested by equation (6.26) may not be of great help to us. There should be an infinite variety of other effective potentials, however, all of them liable to be non-local operators, which can be used in the same sort of way. The aim in calculating a pseudo-potential for a metal is to find one which, for convenience, will not have the large oscillations displayed by the V' of (6.26) but will, nevertheless, describe the scattering from a single ion correctly when used in conjunction with the Born approximation.

We are concerned, however, with scattering by a whole assembly of ions and it is quite possible that around any one of them there may be appreciable modulation, say by a factor $\gamma_n(\mathbf{r})$, due to its neighbours. Should we not then use for the central ion an effective pseudo-potential

$$v'(\mathbf{r}) = (1 + \gamma_n(\mathbf{r}))\, v(\mathbf{r}) \qquad (6.27)$$

in place of v?

It is easy enough (Faber, 1966) to write down an expression for the mean value of γ_n in terms of the interference function. Numerical computations for liquid Na and Li suggest that actually at an ion's centre γ_n is about -0.2 in both cases. Since it is the central region of the ion which determines the large K Fourier components of the pseudo-potential it follows that for $K \gg k_F$ $v'(K)$ is about 0.8 times $v(K)$. Further computations (which will be laborious to perform and very susceptible to errors in the raw data for $a(K)$ and $v(K)$) may show that in the range of K between 0 and $2k_F$, which is what matters as far as the resistivity is concerned, the difference between v' and v is less marked. Clearly, however, we must take calculations in which this difference is entirely ignored with a pinch of salt.

In a really complete calculation we should have to allow for the fact that γ_n may fluctuate from one ion to the next about its mean value $\overline{\gamma_n}$, so that v may also fluctuate. By and large it is probably true to say that all the neighbours within a distance of the electronic mean free path contribute to the local value of γ_n and when, as in liquid Na the number of these is large (i.e. when the mean free path l is much greater than the inter-ionic spacing d) it is a reasonable guess that the fluctuations are unimportant. But they may be quite significant in some of the liquid polyvalent metals, for which l/d is not much more than unity.†

In principle it is straightforward enough to develop a complete expression for the resistivity as a perturbation series in powers of the pseudo-potential v. Acceptance of the Born approximation amounts to neglect of all terms beyond the one in v^2. The use of a mean effective pseudo-potential v' in place of v, provided that it is computed by an iterative procedure so as to ensure self-consistency, corresponds to the summation of a good proportion of all the subsequent terms. But

† Values of l/d, derived from the measured resistivity using the free-electron theory in its simplest form are included in Table 6.1. For a discussion of the corrections which are necessary in the light of the more sophisticated theory outlined above see Faber (1966).

to sum the residue requires a knowledge of the three-body and higher correlation functions for the ions in the liquid, which no experiments on X-ray or neutron diffraction can provide.

6.5 CALCULATION OF THE RESISTIVITY

Let us now suspend criticism and see how well in practice the simple formula

$$\rho = \frac{m}{ne^2\tau_z} = \frac{3\pi m^2}{\hbar^3 e^2} \frac{N}{Vk_F^2} \langle |v|^2 a \rangle \tag{6.28}$$

is capable of explaining the available data.

It is necessary to appreciate that, although the dependence of $v(K)$ and $a(K)$ on K must vary in detail from one metal to the next, the common pattern of behaviour for these two functions is as shown in Fig. 6.2. The pseudo-potential near $K = 0$, for reasons outlined in Chapter 5, behaves like the real screened Coulomb potential round a point charge ze, where z is the valency; that is to say,

$$v(K) = -4\pi ze^2/(K^2 + K_s^2), \tag{6.29}$$

where K_s is a screening parameter determined by the density of states at the Fermi level. If E_F denotes the free-electron Fermi energy $\hbar^2 k_F^2/2m$, then the limit of $v(K)$ as K tends to zero is

$$-\tfrac{2}{3}E_F(m/m^*)(V/N). \tag{6.30}\dagger$$

The components of the pseudo-potential for large K, however, are determined by the nature of the ion core and are not necessarily proportional to z, though a tendency for $v(K)$ to increase systematically with valency may be discerned within each period of the table of elements. As a general rule $v(K)$ seems to change sign when $K/2k_F$ is about 0·8 and to reach about $-0·1\,v(0)$ when $K/2k_F$ is 1. Rb and Cs have such relatively small cores that $v(K)$ remains negative up to $K/2k_F \sim 2·5$, while $v(2k_F)$ is exceptionally small in Na and K; but most of the polyvalent metals follow the pattern described. As for the interference function, $a(K)$, this is small initially because long-range density fluctuations are limited by the repulsive forces between ions; the classical theory of Ornstein & Zernike (1914) shows that

$$a(0) = NkT\beta/V, \tag{6.31}$$

† The factor (V/N) does not occur in a similar expression quoted in Chapter 5; it reflects a different procedure for normalizing the Fourier components of the pseudo-potential, which is more convenient especially for the discussion of alloys.

where β is the isothermal compressibility, and a typical value for this in a metal just above its melting point would be only 10^{-2}. As K increases $a(K)$ rises to a sharp peak and subsequently oscillates about the value unity which, as explained above, describes a completely random arrangement of ions. In monovalent liquid metals the point at which $K = 2k_F$ lies to the left of the main peak in $a(K)$; but in polyvalent liquid metals the Fermi sphere has to be bigger in order to

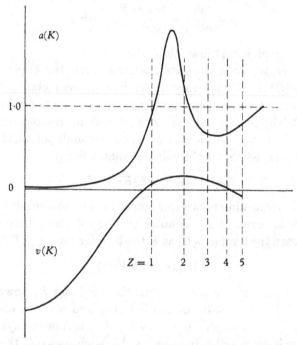

Fig. 6.2. Typical behaviour of interference function (upper curve) and ionic pseudopotential (lower curve) as functions of K. The vertical lines marked $Z = 1$, 2, etc., show where the limit $K = 2k_F$ would lie for different valencies. The position of the first node in the $v(K)$ curve relative to the first peak in the $a(K)$ curve varies from metal to metal.

accommodate more electrons (6.11) and this critical point, which marks the upper limit to the range of integration in (6.9), comes to the right of the peak.

It is roughly true, therefore, that the product $|v|^2 a$ is the same near both limits to the range of integration, but it is liable to fall to zero and perhaps to pass through a maximum in between. The region near the upper limit is of course accentuated in importance by the $(K/2k_F)^3$ weighting factor in (6.9). In Fig. 6.3 $|v|^2 a$ is plotted against

$(K/2k_F)^4$—though the abscissa axis is still labelled in terms of $(K/2k_F)$
—for three different metals, and it is the area under each of these
curves which determines the appropriate value of $\langle |v|^2 a \rangle$. It may be

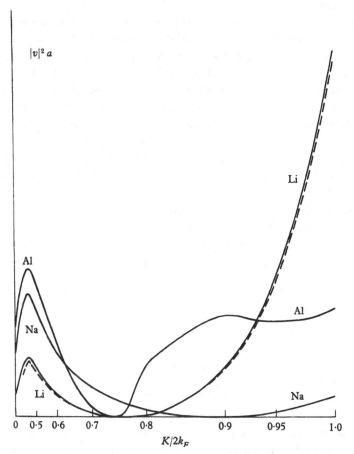

Fig. 6.3. The variation of $|v|^2 a$ with $K/2k_F$ for three different metals. The values of
the ordinate, which are in arbitrary units, have been scaled by a factor proportional
to vk_F^2 and the scale for the abscissae is linear in $(K/2k_F)^4$. The area under each
curve determines the resistivity of the metal concerned and its distribution shows
whether the resistivity is due primarily to large-angle or to small-angle scattering
events. The broken curve for Li shows the effect of correcting for ionic motion using
(6.18). The curves are based on Animalu's pseudo-potentials. Reliable data for the
interference function are not available for small values of $K/2k_F$ and the curves are
therefore somewhat schematic in this region.

seen that in trivalent Al the area is contributed mainly by the region
$0.8 < K/2k_F < 1$ within which $a(K)$ is close to unity, so in this case
the resistivity is due mainly to large angle scattering processes and

takes more or less the value characteristic of a random, 'gas-like', array of ions. Large angle processes still dominate small angle ones in liquid Li. It is only for liquid Na that the region $K/2k_F < 0.8$ is really significant.

The numerical results obtained for ρ depend to a considerable extent upon whose pseudo-potential one prefers and whose data for $a(K)$. The variation is greatest in the monovalent metals where a small shift in the zero of the potential (say 0.2 eV or so) is liable to change the answer for ρ by a factor of 2 or more, and where it is also sensitive to the precise shape of the $a(K)$ curve to the left of the main peak, which is often not determined accurately in diffraction experiments. But the sort of agreement with experiment which is commonly obtained is illustrated by the figures in Table 6.2; these are due to Animalu (1965) and are based upon the so-called 'model' potential of Heine & Abarenkov (1964), modified to take account of non-local screening. More extensive calculations using the same potential have been reported by Ashcroft & Lekner (1966), but these authors prefer to derive $a(K)$ from a theoretical model rather than to trust the diffraction data, and it might confuse the reader to quote them here.

Table 6.2. Comparison of theory with experiment

	Li	Na	K	Rb	Cs	Zn	Al	Pb
ρ_l(expt) ($\mu\Omega$ cm)	24·7	9·6	13·0	22·5	37	37	24	95
ρ_l(theory) ($\mu\Omega$ cm)	25	7·9	23	10	10	37	27	64

In a few cases the agreement is very good indeed—for example Li, Zn and Al—and one is naturally reluctant to suppose that this is fortuitous. When it is shown, as in the case of Al (Ashcroft & Guild, 1965), that the same model potential which gives such a good answer for the resistivity of the liquid is capable of describing the de Haas–van Alphen data for the solid with equal success (see §2.2.1), one's faith in (6.28) is further strengthened. It is therefore commonly supposed that the discrepancies which do appear in other cases are attributable solely to errors in the raw data for $v(K)$ and $a(K)$. A note of caution about this attitude has been sounded in the previous section.

The only simple metals for which calculations based on (6.28) seem likely to go seriously astray are Cu, Ag and Au. De Haas–van Alphen measurements and complementary band-structure calculations for these noble metals in the solid phase imply an ionic pseudo-potential with a rather large Fourier component when $\mathbf{K} = \mathbf{G}$, the (111) reciprocal lattice vector. If it were equally large in the liquid phase near $K = 2k_F$ (which is close to G) the answer for ρ would come out to be something approaching 10 times too big. It is probable that the whole pseudo-potential concept needs scrutiny in the case of the noble metals, because of the proximity of the d-band to the Fermi level; this is liable to make the dependence of $v(k)$ on k and θ unusually rapid, in which case it may be rash to apply the same pseudo-potential to two rather different problems. The d-band is fairly close ($\sim$ 8 eV) to the Fermi level in Hg also, and we shall meet some anomalies below which may perhaps be related to this fact.

Whether or not the agreement between the two rows of figures in Table 6.2 is as good as could be expected, it is better than has yet been achieved for solid metals. Except perhaps for Na and K, the N.F.E. model is a poor starting point for the calculation of ρ_s; one has to work with Bloch waves, to distinguish between normal and umklapp scattering processes, between longitudinal and transverse phonons, and so on. That is why it is not possible to quote here any reliable predictions concerning ρ_l/ρ_s for comparison with experiment. Ziman (1961a) has argued that since the resistivity of Na is influenced by small angle scattering processes, i.e. by the low K region, we might expect

$$\frac{\rho_l}{\rho_s} \sim \frac{a(0)_l}{a(0)_s} \sim \frac{\beta_l}{\beta_s} \qquad (6.32)$$

in this metal. In fact the ratio of the compressibilities in liquid and solid Na seems to be only about 1·2, whereas ρ_l/ρ_s is 1·45. The agreement is not impressive, but then the argument is greatly over-simplified.

6.6 DERIVATIVES OF THE RESISTIVITY

The elegance of Ziman's theory really emerges when one comes to consider the temperature coefficient of the resistivity. At constant volume the only quantity on the right-hand side of (6.28) which depends upon T is the interference function. Hence.

$$\frac{T}{\rho}\left(\frac{\partial\rho}{\partial T}\right)_V = \frac{\langle |v|^2 a(\partial\ln a/\partial\ln T)_V\rangle}{\langle |v|^2 a\rangle}. \qquad (6.33)$$

The complications that arise theoretically when thermal expansion has to be taken into account will be discussed as a separate topic below. If $a(K)$ were everywhere proportional to T, as it is at $K = 0$, this coefficient would be unity. In fact the effect of heating must surely be to make $a(K)$, for all values of K, tend towards the value of unity that characterizes a gas-like array. This means that

$$(\partial \ln a / \partial \ln T)_V$$

must decrease to zero as K increases and become slightly negative in the region where $a(K)$ has its main peak; beyond this peak it should oscillate in sign but is likely to remain quite small in magnitude. Fig. 6.4 shows schematically the behaviour to be expected. Hence

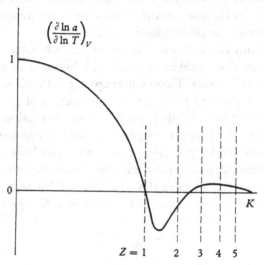

Fig. 6.4. Schematic diagram to illustrate the variation of $a(K)$ with temperature at constant volume as a function of K. The broken vertical lines have the same significance as in Fig. 6.2.

it is only in a monovalent liquid metal, and particularly in Na and K where the low K region is significant, that values approaching unity are to be expected for $(\partial \ln \rho / \ln T)_V$. In a divalent metal the region where $(\partial \ln a / \partial \ln T)_V$ is negative comes just below $K = 2k_F$ and must largely determine the averages on the right-hand side of (6.33), so a negative value for $(\partial \ln \rho / \partial \ln T)_V$ seems almost inevitable. It could be negative also in a metal with more than two valence electrons but whatever its sign it should be small in such a case. Hence the figures quoted in column 6 of Table 6.1 and discussed already in §6.2 are provided with a straightforward qualitative explanation.

It is not possible to check the explanation quantitatively because measurements of $(\partial a / \partial T)_V$ are difficult to make and none have yet

x

been reported. It will be apparent from an examination of the sodium curve in Fig. 6.3 that $(\partial \ln a / \partial \ln T)_V$ must stay quite close to unity over most of the range of K between 0 and $2k_F$ if a value as big as $0·85$ is to be obtained for $(\partial \ln \rho / \partial \ln T)_V$; it is bound to fall off to zero near $K = 2k_F$ but it must not fall off too soon. Ashcroft ($1966b$) has calculated a curve for $(\partial \ln a / \partial \ln T)_V$ in liquid Na using a theoretical model, and it does have the required form. However, if Ashcroft's model is used to calculate $(\partial \ln a / \partial \ln T)_p$ it gives results which are in rather serious disagreement with those which have been measured experimentally by Greenfield (1966); the latter fall off a good deal more quickly with K. Hence the quantitative success of Ziman's theory for the temperature coefficient of the resistivity in liquid Na is far from assured.

Of course if the Born approximation is not to be trusted in sodium this is bound to affect the temperature coefficient to some extent. The mean effective pseudo-potential v' which should replace v in the theory is determined by $a(K)$ and may therefore be temperature-dependent even at constant volume, and any contribution to the resistivity due to fluctuations in v' may also be temperature-dependent. It seems rather unlikely, however, that these effects can be large enough in Na to remove the discrepancy which Greenfield believes to exist.

Another type of derivative of the resistivity is revealed by measurements of the thermo-electric power, S. The standard theoretical expression for S is

$$S = \frac{\pi^2 k^2 T}{3e} \frac{1}{\rho} \left(\frac{\partial \rho}{\partial E_F} \right) = \frac{\pi^2 m k^2 T}{3e\hbar^2 k_F} \frac{1}{\rho} \left(\frac{\partial \rho}{\partial k_F} \right), \qquad (6.34)$$

where the partial derivative expresses the rate of change that would be observed for the resistivity if the Fermi sphere could be expanded, without altering the volume of the specimen or, e.g., the screening of the pseudo-potential. It is customary to discuss the results in terms of a dimensionless parameter

$$\xi = -\frac{3eE_F}{\pi^2 k^2 T} S = -\frac{k_F}{2\rho} \left(\frac{\partial \rho}{\partial k_F} \right).$$

According to (6.28) we should be able to express this as

$$\xi = 3 - 2 \frac{|v(2k_F)|^2 a(2k_F)}{\langle |v|^2 a \rangle} - \frac{\langle k_F (\partial |v|^2 / \partial k_F) a \rangle}{2 \langle |u|^2 a \rangle}. \qquad (6.35)$$

The final term is necessary because, even though the screening is not supposed to alter when the Fermi sphere suffers its notional expansion and the real potential round each ion remains quite unchanged, the *pseudo*-potential is sensitive to the energy of the electron on which it acts and to the angle of scattering (for given K), both of which depend on the Fermi radius.

One can understand most of the results for ξ which are quoted in Table 6.1, in a qualitative fashion at any rate, on the assumption that this final term is negligible. In a polyvalent metal we do not expect $|v|^2 a$ to vary rapidly over the important region close to $K = 2k_F$ in which case $|v(2k_F)|^2 a(2k_F)/\langle|v|^2 a\rangle$ should be roughly unity. Hence ξ should be roughly unity, as observed. In a monovalent metal such as Li it is evident from Fig. 6.3 that $|v|^2 a$ is rising sufficiently fast in the neighbourhood of $K = 2k_F$ to make

$$|v(2k_F)|^2 a(2k_F)/\langle|v|^2 a\rangle$$

quite large and hence to make ξ negative. While in Na $v(2k_F)$ should be small enough to let ξ approach 3, as it does.

Attempts to verify this line of explanation by quantitative calculations (Sundström, 1965; Marwaha, 1967) have so far been only moderately successful. One could blame the discrepancies onto errors in the pseudo-potentials used—replacement of v by the appropriate effective pseudo-potential v' to allow for breakdown of the Born approximation might improve the agreement. But no alteration of the pseudo-potential or of the interference function can make (6.35) yield an answer for ξ which is greater than 3 unless the third term is included. Hence it seems that this third term *must* be important in K and more particularly in Hg and it may well contribute for other metals also. The value of ξ for Hg gets even higher as the temperature is increased (ξ is slightly temperature-dependent in Zn and Tl also; the values quoted apply just above the melting point) but this is an effect of thermal expansion; there is no change of ξ if Hg is heated at constant volume (Bradley, 1963).

Data for the derivative of resistivity with respect to volume are limited; such as are available are listed in Table 6.1. A change of volume implies, of course, a change in the size of the Fermi sphere and there is therefore a link between the volume coefficient and the thermo-electric power. The N.F.E. model and (6.28) suggest

$$\frac{V}{\rho}\left(\frac{\partial\rho}{\partial V}\right)_T = \tfrac{2}{3}\xi - 1 + \frac{\langle|v|^2 V(\partial a/\partial v)_T\rangle}{\langle|v|^2 a\rangle} + \frac{\langle V(\partial|v|^2/\partial V)a\rangle}{\langle|v|^2 a\rangle}, \quad (6.36)$$

the final term being to allow for the effect on the pseudo-potential of any change in the screening properties of conduction electron gas.

304   T. E. FABER

Figure 6.5 shows a plot of $(\partial \ln \rho / \partial \ln V)_T$ against ξ. There does appear to be some correlation between the two, as (6.36) would suggest, and the slope of the best line through the points (those for Hg and Li excluded) is about 2/3.† Its intercept implies that the sum of the last two terms on the right-hand side of (6.36) is roughly 4.

One can explain this result in a semiquantitative way for the liquid monovalent metals, though the details of the explanation depend on whether (as in Na) weight is attached to the low K region or

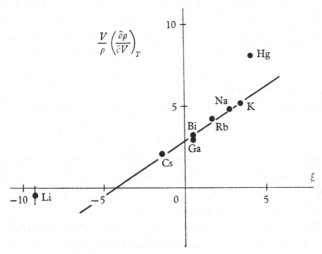

Fig. 6.5. The semi-empirical relation between the volume dependence of resistivity and the thermo-electric power parameter ξ. The line drawn through the points has a slope of 2/3. The values used for $(V/\rho)(\partial\rho/\partial V)_T$ are based on Bridgman's data (1931); Endo's measurements (1963) suggest a much larger value in the case of K.

(as in say Cs) to the region near $K = 2k_F$. At low K the magnitude of $a(K)$ is determined according to (6.31) by the compressibility, which varies with volume in such a way that $(\partial \ln a/\partial \ln V)_T$ is probably about 3. According to (6.30) the pseudo-potential is proportional to $V^{\frac{1}{3}}$ so that $(\partial \ln |v|^2/\partial \ln V)_T$ should be 2/3. Hence in so far as the low K region is important in determining the resistivity a total of about 4 for the last two terms in (6.36) is not surprising. Near $2k_F$ the pseudo-potential is little affected by screening and $(\partial \ln |v|^2/\partial \ln V)_T$ may be negligible. Moreover, it seems unlikely that the compressibility matters much, so that the shape of the peaks in the interference function may remain unaffected by small volume changes.

† The correlation revealed in Fig. 6.5 enables one to guess a plausible value for $(\partial \ln \rho/\partial \ln V)_T$ in metals where only ξ is known directly (see §6.1).

However, the values of K at which these peaks occur must surely scale like $V^{-\frac{1}{3}}$ and on this account $(\partial \ln a / \partial \ln V)_T$ may be given approximately by $\frac{1}{3}(K/a)(\partial a / \partial K)_V$. Examination of a typical $a(K)$ curve for a liquid monovalent metal suggests that this quantity is at least 4 near $2k_F$.

It is not easy to extend this line of argument to cover the polyvalent metals Ga and Bi, which also lie on the line in Fig. 6.5. The trouble is, of course, that although $\frac{1}{3}(K/a)(\partial a / \partial K)_V$ may rise to 4 or more on the low K side of the main peak of the interference function it becomes almost strongly negative on the high K side; it does not seem possible, therefore, for $\langle |v|^2 V(\partial a / \partial V)_T \rangle / \langle |v|^2 a \rangle$ in a liquid polyvalent metal to be anything like as big as 4. One is forced to conclude that the pseudo-potential in the large K region is more sensitive to changes of volume than is implied by the simple argument above. Granted that the effects of a change in the screening are more or less negligible, what about a shift of the conduction band as a whole relative to the core levels? A change of volume is obviously likely to cause such a shift, and it might easily have repercussions on $v(K)$; the pseudo-potential originally proposed by Phillips & Kleinman (1959) depended rather directly on the spacing between the Fermi level and the core levels, though the relevance of this spacing is not so apparent in (1.4). Here is a complicated problem on which further calculations are required.

The position of the point for Hg in Fig. 6.5 emphasizes the anomalous character of this metal, to which reference has already been made. Not only is ξ unexpectedly large (and thermal expansion makes it larger still) but $(\partial \ln |v|^2 / \partial \ln V)$ seems to be very large too.[†] It does not seem out of the question that all this is a fairly straightforward consequence of the proximity of the d-band to the Fermi level in Hg (and that the somewhat anomalous behaviour of Li on the other side of Fig. 6.5 is related to the absence of any d-electrons in that metal), but here too more calculations are required. Interesting information is now becoming available about the resistivity of Hg near its critical point, where its volume is 2 or 3 times its volume at room temperature (Franck & Hensel, 1966). Whether Ziman's theory will still be useful in this regime is as yet undetermined; Mott (1966, 1967) has argued in favour of a rather different approach.

[†] It may be of interest to note that Bridgman's (1931) data for Hg at high pressures (see also Postill et al. 1967) suggest that $(\partial \ln \rho / \partial \ln V)_T$ remains about 8, even when the volume has been reduced by 4 %. This degree of compression would seem sufficient, to judge by Bradley's results (1963), to restore ξ to a value quite typical of other polyvalent liquid metals.

6.7. EXTENSION OF ZIMAN'S THEORY TO LIQUID ALLOYS

If there are two species of ion present in a liquid the structure will be characterized by three pair correlation functions, P_{AA}, P_{BB} and P_{AB}, and correspondingly one may define three interference functions, a_{AA}, a_{BB} and a_{AB} by relations similar to (6.14). Faber & Ziman (1965) have showed that the equation which replaces (6.13) for a pure liquid is

$$|\mathscr{V}(K)|^2 = N\{(\overline{|v|^2} - |\bar{v}|^2) + \sum_\alpha \sum_\beta c_\beta c_\alpha v_\beta v_\beta^* a_{\alpha\beta}\}, \qquad (6.37)$$

where the suffices α, β can take the values A or B; c_A and c_B, v_A and v_B are the atomic concentrations and the pseudo-potentials of the two species.† The rest of Ziman's theory goes through as before.

Just as $a(K)$ in a pure liquid can be determined by a single diffraction experiment, the three quantities a_{AA}, a_{BB} and a_{AB} for a binary alloy can in principle be determined by three diffraction experiments, if some way can be found of altering the ratio between the two form factors involved. Some alteration is possible in the case of neutron diffraction by isotopic replacement, and the first attempt at a structural analysis of a liquid binary alloy using this technique (Cu_6Sn_5 was chosen) has been reported by Enderby et al. (1966). When such data become more plentiful we shall be in a stronger position to discuss the resistivity of liquid alloys in a quantitative fashion. For the present we are obliged to make some assumption about the 'partial' interference functions if we wish to arrive at any definite predictions.

The simplest assumption is that they are all three equal, which is justified so long as A and B ions are able to replace each other without distorting the surrounding structure, and so long as there is no tendency for ions of either species to cluster together or to form ordered arrays. This is the so-called 'substitutional' model. It means that

$$|\mathscr{V}(K)|^2 = N\{\overline{(v-\bar{v})^2} + a|\bar{v}|^2\} \qquad (6.38)$$

(an almost self-evident result) and hence that

$$\rho = \rho' + \rho''$$

† Equation (6.37) is still valid if there are more than two components to the alloy, but the dummy suffices α and β must then be supposed to stand for more than two possibilities.

where $$\rho' = \left(\frac{3\pi m^2}{\hbar^3 e^2}\right) \frac{N}{Vk_F^2} \langle c_A |v_A|^2 a + c_B |v_B|^2 a \rangle \qquad (6.39)$$

and $$\rho'' = \left(\frac{3\pi m^2}{\hbar^3 e^2}\right) \frac{N}{Vk_F^2} \langle c_A c_B |v_A - v_B|^2 (1-a) \rangle; \qquad (6.40)$$

these expressions follow from (6.9) and (6.38) after a simple rearrangement of terms.

Many of the gross features displayed by the curves of resistivity versus concentration for liquid binary alloys can be explained by this extension of Ziman's theory. If both components are polyvalent, then over the important range of K near $2k_F$ the interference function should be close to unity, in which case ρ'' (which is determined by $(1-a)$) should be fairly small compared to ρ'. Hence to a first approximation the resistivity should vary linearly with concentration between the values appropriate for pure A and pure B at the temperature concerned. In systems such as Sn–Pb, or Sb–Bi, this is just the behaviour observed (see Fig. 6.6). If both components are monovalent, then a is not necessarily much greater than $(1-a)$ over the whole of the important range, and ρ'' is liable to make a significant contribution. A curve with distinct convex hump is the result (see Fig. 6.6), not too dissimilar to the curves which are familiar for disordered alloys in the solid phase. If one component is polyvalent and the other monovalent, then for a given value of $(c_A c_B)$ ρ'' is liable to be bigger near the monovalent than near the polyvalent end, so that the hump in the curve for ρ is liable to be asymmetric (see Fig. 6.6 yet again).

The only liquid alloys whose behaviour is difficult to square with this theory, even at a qualitative level, are the Hg amalgams. It has been known for many years that the addition to Hg of almost any metallic impurity (some of the alkali metals are exceptions) lowers the resistivity quite sharply; the curve for the Hg–Pb system which is reproduced in Fig. 6.6 is a case in point. Many explanations for the anomaly have been put forward, the most recent by Mott (1966). Obviously it must be linked in some way with the anomalies for pure Hg which have already been mentioned, and one possibility which is worth further study is that it is a consequence of the large volume coefficient of ρ. To explain the form of the Hg–Pb curve in Fig. 6.6, for example, one might postulate that the volume occupied by each Hg atom decreases by some 4 % when 30 % of Pb is added, but is unaffected by further addition. Density measurements do rather suggest that a shrinkage of this kind occurs (Kleppa *et al.* 1961) though it would not seem to be quite abrupt enough, at very small concentrations of Pb to explain the very sharp drop in the resistivity curve.

For detailed quantitative calculations the substitutional model is unlikely to be good enough; the simple observation that the mean atomic volume of an alloy varies with concentration (a phenomenon

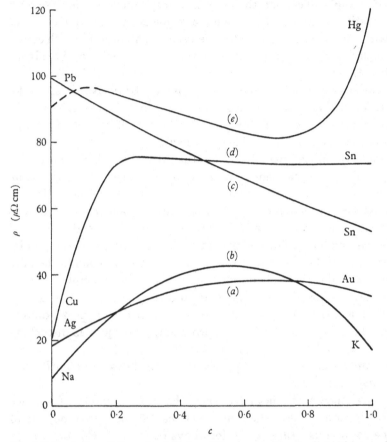

Fig. 6.6. Typical variation of resistivity with concentration for liquid binary alloys: (a) Ag–Au at 1200 °C; (b) Na–K at 100 °C; (c) Pb–Sn at 400 °C; (d) Cu–Sn at 1200 °C; (e) Pb–Hg at 250 °C. For sources of data see Wilson (1965), Some authors have reported curves with small anomalous bumps in them for Ag–Au, Pb–Sn, Cu–Sn and other systems, but these are believed to be spurious.

which is certainly not confined to Hg amalgams) warns us that *some* distortion must be occurring when an A ion replaces a B ion. In the absence of experimental data for a_{AA}, a_{BB} and a_{AB} one may be able to go a step further by using the sort of theoretical models which Ashcroft & Lekner (1966) have applied to pure liquids (Toombs 1965).

Another approach is to assume that a_{AA}, a_{BB} and a_{AB} are independent of concentration, in which case they may be deduced from a single type of diffraction experiment carried out for at least three different values of c; this hypothesis has been applied to the Ag–Sn system with satisfactory results (Wagner & Halder, 1967). Yet a third idea, which is applicable when the alloy is a very dilute one, is to extend the substitutional model by a simple recipe suggested by Faber and Ziman; it involves using an effective pseudo-potential for each solute ion, including a term proportional to the local dilatation of the solvent (see §5.4). This recipe has been used by Dickey et al. (1966) in their calculations for dilute solutions of Li, K and Rb in Na. It suggests, rather surprisingly, that the dilatation adds little to the resistivity in these particular systems, despite the large differences of atomic volume that exist between the components.

Serious complications arise in any quantitative calculation for an alloy (unless it so happens that the two components have the same volume and the same valency) from the fact that the Fermi radius is a function of concentration. This means that v_A and v_B are liable to vary with concentration, and in particular that the pseudo-potential for a solute ion at infinite dilution in a foreign solvent may be quite different from the one that is appropriate for the same ion in the pure solute. Few detailed calculations have yet been attempted. Those of Dickey et al. referred to above are probably the most careful and they are at least as successful as one is entitled to expect—doubts about the validity of the Born approximation are just as disturbing in the case of alloys as for pure metals (Faber, 1967b).

Nothing is known about $(\partial\rho/\partial T)_V$ in liquid alloys, but $(\partial\rho/\partial T)_p$ has of course been measured in a number of systems. It is of interest that this tends to be negative when the mean number of valence electrons per atom is in the range between 1·5 and 2 (see the data of Bornemann et al. (1912, 1913) for alloys of Cu with Cd, Zn, Al, Sn and Sb). The explanation may be similar to the one put forward above for the negative temperature coefficients observed in pure divalent metals. Negative coefficients are also observed for liquid alloys such as Zn–Sb (Matuyama, 1927), but this is a semiconductor when solid and perhaps should be classified as a semiconductor in the liquid state as well.

It is noticeable that $(\partial\rho/\partial T)_p$ tends to increase with concentration c in many dilute liquid alloys, or in other words that $(\partial\rho/\partial c)_T$ tends to increase on heating; Matthiessen's rule is not obeyed as accurately as it is in solid alloys. The Li–Mg system provides an example of this tendency which has been analysed in detail (Faber, 1967 b) to see whether (6.39) and (6.40) can account for it. It turns out that $(\partial \rho''/\partial c)_T$, when the concentration of Mg is very small, should decrease rather than increase on heating, mainly because of the rise in $a(K)$ over the relevant range of K. At first sight $(\partial \rho'/\partial c)_T$ should be negligible because the resistivities of pure Li and pure Mg are so similar that a straight line drawn between the two on a plot of ρ versus c would be virtually horizontal. It must not be forgotten, however, that the addition of small quantities of Mg to Li is liable to swell the Fermi sphere; according to the N.F.E. model

$$dk_F/dc = k_F/3 \qquad (6.41)$$

and there is some evidence from experiments on positron annihilation to support this relation. Now we know that the resistivity of pure Li is unusually sensitive to changes of k_F, because $|\xi|$ is so large, and in so far as ρ' is contributed mainly by the Li$^+$ ions when c is small it should be equally sensitive. Taken at their face value equations (6.34) and (6.41) imply that at infinite dilution

$$\left(\frac{\partial \rho'}{\partial c}\right)_T \simeq \frac{dk_F}{dc}\left(\frac{\partial \rho_0}{\partial k_F}\right)_V = -\tfrac{2}{3}\xi_0 \rho_0 = 6 \cdot 2\rho_0, \qquad (6.42)$$

where the suffix zero indicates a property of pure Li. If this is right $(\partial \rho'/\partial c)_T$ should increase with temperature quite fast enough to outweigh any decrease of $(\partial \rho''/\partial c)_T$.

Actually, there is a flaw in this argument—which is just as well, since the experimental data cannot support the interpretation that $(\partial \rho'/\partial c)_T$ is as big as $6 \cdot 2\rho_0$, or anything like it. In the theory of the thermo-electric power one is concerned with changes of k_F *while the bottom of the conduction band* (relative say to the core levels) *is kept fixed*. When k_F is changed by the addition of impurity, however, it is the *Fermi level* which stays fixed (Friedel, 1954); the change in the kinetic energy E_F is compensated by a change of potential energy. Hence the partial differential with respect to k_F does not stand for quite the same thing in (6.34) and (6.42). The difference between the two should only be significant in cases where the pseudo-potential is appreciably dependent upon energy. The implication of the Li–Mg resistivity data, therefore, is that the pseudo-potential *is* energy-dependent in pure Li and that attempts such as those of Sundström (1965) to explain its large value of $|\xi|$ with a model potential which does not depend upon energy, ignoring the final term in (6.35), are doomed to failure. How complex the theory becomes!

The theory should, of course, be capable of explaining the variation of thermo-electric power with concentration as well as the variation of resistivity. It fits some of the observations well enough (see Bradley (1962) for a discussion of dilute alloys of the alkali metals, and Howe & Enderby (1967) for a discussion of liquid Ag–Au) but the Hg amalgams are once again anomalous (Cusack *et al.* 1964; Takeuchi & Noguchi 1966; Fielder, 1967).

6.8   THE EFFECT OF A MAGNETIC FIELD

Application of the Boltzmann equation to the simple N.F.E. model leads to quite unambiguous predictions about the effects of a magnetic field: whatever the value of m^*/m there should be *no* magneto-resistance and the Hall coefficient should be given by the elementary free-electron formula

$$R_H = -1/nec \qquad (6.43)$$

(see, for example, Ziman, 1964a, p. 213).

It is harder experimentally to study these effects in a liquid metal than a solid, because the Lorentz force exerted on the measuring current by the applied magnetic field tends to set up circulating currents which can lead to spurious results. Finite values have been reported for the longitudinal magneto-resistance in liquid K and Na–K alloys (Kikoin & Fadikov, 1935; Armstrong, 1935) but these are generally attributed to experimental error. Recent values obtained by a number of experimenters for the Hall coefficient, however, seem to be trustworthy. The best results are summarized in table 1, where values for $-(nec R_H)$ are listed. They are quite remarkably close to unity for a number of liquid metals, but in others, e.g. Pb, Sb and In, they fall significantly below this expected level; the experimental errors seem unlikely to exceed about 5 %.

The various criticisms which have been levelled against Ziman's theory in §6.4 above apply equally strongly in the presence of a magnetic field and cast considerable doubt on the simple prediction expressed by (6.43). In particular it seems that some correction must be required on account of the blurring of the Fermi surface over a range of energy of order $\hbar/\tau$. A number of attempts have been reported (Kubo, 1964; Springer, 1964; Evans, 1966; Banyai & Aldea, 1966) to extend the Kubo–Greenwood–Edwards theory so as to enable one to calculate a value for the Hall coefficient in cases where $\hbar/\tau E_F$ is no longer small, but no very concrete answer emerges from them. Intuitively one expects a correction proportional to $(\hbar/\tau E_F)^2$—just as thermal blurring of the Fermi surface is responsible for a correction proportional to $(kT/E_F)^2$ according to orthodox transport theory—but the correlation between the values of $-(necR_H)$ in Table 6.1 and the corresponding values for $(\hbar/\tau E_F)^2$ is not particularly impressive. We may need to consider the normalization corrections mentioned in §6.4 and perhaps to scrutinize once more the validity of the Born approximation before the data for the Hall coefficient can be fully understood.

6.9   THE OPTICAL PROPERTIES

From a study of the way in which electromagnetic radiation is reflected from a clean metal surface it is possible to deduce how the conductivity $\sigma(\omega)$ varies with frequency in the infra-red, visible and ultra-violet regions of the spectrum. Complications due to the anomalous skin effect may arise in the infra-red, but these are most troublesome for pure solid metals at low temperatures in which the mean free path is long; in most liquid metals they can safely be ignored. A dielectric constant $\epsilon(\omega)$ can also be deduced, but since, according to the Kramers–Kronig relations, the value of ϵ is a mere corollary of $\sigma(\omega)$, this quantity does not require a separate discussion; it is σ which determines the absorption and which is therefore of greater physical significance.

It is to be expected that σ should fall off at high frequencies due to relaxation effects, and if we are to write

$$\sigma(0) = ne^2 \tau_z/m \qquad (6.44)$$

for the d.c. conductivity, then the classical theory of Drude would suggest that

$$\sigma(\omega) = ne^2 \tau_z/m(1 + \omega^2 \tau_z^2). \qquad (6.45)$$

There are few solid metals for which (6.45) can be fitted satisfactorily to the data, if only because most solid metals show absorption edges in the visible corresponding to excitation of electrons from the Fermi level to empty bands above it. Absorption edges are not observed in liquid metals until the frequency is high enough to excite the core electrons (Otter, 1961; Wilson & Rice, 1966)—it is of course consistent with the N.F.E. model that no signs of band structure above the Fermi level should be detectable—and (6.45) has been found to fit the data rather well. The details of the best fit obtainable depend upon whether or not one insists that the curve for σ should extrapolate to the d.c. conductivity. Hodgson (1959, 1960, 1961, 1962) has allowed himself some freedom in this respect, arguing that the effective value of $\sigma(0)$ in a surface layer of thickness say 10^{-6} cm, which is the depth to which a light wave penetrates, may well be less than the bulk conductivity by a few parts per cent. He has fitted his data for $\sigma(\omega)$ and also for $\epsilon(\omega)$ by Drude expressions, using an 'effective' density of conduction electrons n^* whose relation to the number of valence electrons n is shown in Table 6.3. The fit is not always perfect in the infra-red where the measured $\sigma(\omega)$ tends to fall below the theoretical curve, but one could attribute the discrepancies to experimental error.

Table 6.3. *Values of n^*/n deduced from optical properties*

Cu	K	Ag	Cd	Hg	Ga	In	Ge	Sn	Pb	Sb	Bi
(0·8)	1·0	1·1	1·06	(1·25)	0·98	1·07	1·07	1·17	1·17	1·22	1·06

(The figures quoted for K and Ga are respectively from Mayer & El Naby (1963) and Schulz (1957) rather than Hodgson.)

It will be no surprise to the reader to learn that the only liquid metal for which the Drude theory seems to be really unsatisfactory is Hg. A rough value of n^*/n is included for Hg in Table 6.3, but it has been derived using the data in a narrow range of the visible only. In the infra-red Hodgson's curve for $\sigma(\omega)$, which is reproduced in Fig. 6.7, rises to a maximum and there are corresponding anomalies in $\epsilon(\omega)$. His results have been disputed by Schulz (1957), but the careful experiments of Smith (1967) together with an argument based upon the sum rule (Faber, 1966) make it almost certain that the anomaly is genuine. Is it possible that for Hg the effective value of $\sigma(0)$ in the surface layer is some 20% higher than the bulk value, perhaps as a result of the unusual sensitivity of its pseudo-potential to the density of the ions? This hypothesis would imply an absorption curve of the form observed, apparently extrapolating to a high d.c. value but dropping towards the bulk value at frequencies low enough to make the skin depth greater than the thickness of the anomalous surface layer. There are other ways of explaining a maximum in $\sigma(\omega)$ (e.g. Mott, 1966) but none of them seems to be applicable to a maximum which occurs at an energy $\hbar\omega$ which is much smaller than $\hbar/\tau_z$.

Drude's theory is a primitive one by modern standards and the validity of (6.45) requires a careful check. In particular one may question the grounds for using τ_z in the denominator rather than the lifetime τ which (6.8) defines. According to the philosophy outlined in §6.4 the conductivity $\sigma(\omega)$ is determined by the matrix element D between two eigenstates which are separated in energy by $\hbar\omega$. If the eigenstates are regarded as wave-groups, as in (19), the magnitude of

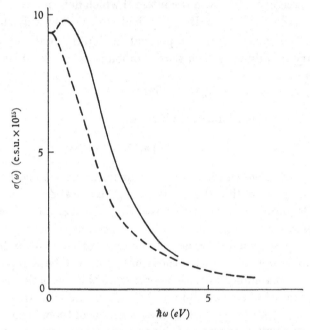

Fig. 6.7. Absorption of Hg as a function of frequency. The broken curve follows the simple theoretical curve of Drude (6.45).

D may be seen to depend on the extent to which they overlap in k-space. Obviously the overlap must diminish as $\hbar\omega$ increases and this is sufficient to account in a general way for the fall-out in $\sigma(\omega)$. But since, according to (6.20), the spread of each eigenstate in k-space is determined by $\hbar/\tau$ rather than $\hbar/\tau_z$, one might well expect $\omega\tau$ rather than $\omega\tau_z$ to control the scale of the σ curve.

This situation is borne out by a fuller analysis (Faber, 1966); in so far as $|D|^2$ is determined by

$$\sum_k k^2 \cos^2 \alpha_k a_k a_k^* b_k b_k^* \qquad (6.46)$$

it should diminish with frequency like $(1+\omega^2\tau^2)^{-1}$. Complications arise, however, from the fact that the cross-terms in $|D|^2$ (cf. (6.23)) diminish more rapidly than this and the result is a rather clumsy expression for $\sigma(\omega)$ in general. The fact that the Drude expression usually works so well is probably an indication that the $(1 - \cos\theta)$ factor has less effect upon the magnitude of τ_z than one might suppose, so that the only real difference between τ_z and τ lies in the fact that one involves m and the other m^* (see (6.8) and 6.9)). This $(1 - \cos\theta)$ changes the weighting factor in the integral which determines τ_z from $2(K/2k_F)$ to $4(K/2k_F)^3$ and it is not difficult to see that this should make little odds, when, as in the polyvalent liquid metals, $\overline{|\mathscr{V}(K)|^2}$ does not vary rapidly with K in the neighbourhood of $2k_F$. If indeed we may write

$$\tau_z = (m^*/m)\,\tau \qquad (6.47)$$

then the full expression for $\sigma(\omega)$ reduces to

$$\sigma(\omega) = (1+\lambda)^2\,(m^*/m)\,ne^2\tau/(1+\omega^2\tau^2), \qquad (6.48)$$

which has exactly the Drude form; λ is the normalization correction referred to in §6.4. Within the limits of experimental error, (6.48) is probably adequate for all the metals listed in Table 6.3. The $(1 - \cos\theta)$ term should make an appreciable difference for one or two monovalent metals, e.g. it should decrease τ_z by a factor of about 0·7 in liquid Li, but reliable optical data for these have not yet been reported.

The point of real importance to emerge from this discussion is that Hodgson's values for n^*/n should provide us with an indication of the magnitude of $(1+\lambda)^2\,(m^*/m)$. Since they are none of them bigger than one can hope to explain in terms of the normalization correction alone, it is not unreasonable to suppose that for those liquid metals which feature in Table 6.3 (m^*/m) is not appreciably different from unity.

6.10 CONCLUSIONS

The theory of electronic transport properties in liquid metals is by no means complete. However, the emphasis which has been placed throughout this chapter on the discrepancies between theory and experiment should not be allowed to distract the reader's attention from the very real advances that have been made during the last five years. The concept of the pseudo-potential and the idea that the scattering of electrons is governed by the same structural properties that determine the scattering of X-rays and neutrons have enabled us to explain a whole range of phenomena in much greater detail than was hitherto possible.

The success of the theory may be regarded as evidence in favour of the N.F.E. model, which was chosen as the starting point in §6.3. If the analysis of the optical properties in §6.9 is to be trusted, one may go further and state that in many liquid metals the conduction electrons behave as though they were genuinely free carriers; there seem to be little sign of residual band structure after melting. From the point of view of those theoreticians who have worked on the fascinating problem of how to describe electrons in a disordered lattice (e.g. Edwards, 1962; Beeby & Edwards, 1963; Phariseau & Ziman, 1963; Lloyd, 1967 and many others) this may be a disappointing conclusion; but it will not be altogether a surprise (Ballentine, 1966). It would alas, require a separate chapter to do justice to their work.

CHAPTER 7

EXPERIMENTAL STUDIES OF THE STRUCTURES OF METALS AND ALLOYS

by P. J. BROWN AND W. H. TAYLOR†

7.1 INTRODUCTION

This chapter treats the development, in recent years, of experimental studies of the structures of metals and alloys by diffraction methods, with particular reference to attempts to determine electron density and spin density distributions. As will be seen, it is very difficult to obtain accurate and reliable information on electron configuration, partly on account of the nature of the experimental measurements themselves, partly as a consequence of uncertainties encountered in their interpretation. It is for this reason that relatively little success has so far been achieved in the comparison of structural detail with theoretical treatments of the metallic state, although at first sight this might seem likely to offer the obvious and direct means of providing a factual basis on which mathematical developments might rest. It will be shown, however, that information obtained by combining X-ray diffraction and neutron diffraction studies is now beginning to be used in the discussion of theoretical models in much the same way as are the measured physical properties.

Diffraction studies aimed at this primary objective of contributing to the theoretical treatment of metals must nevertheless rest on a broad understanding of the systematic structural geometry of pure metals and alloys and of the factors governing the stability of the various structural types. Section 7.2 indicates briefly the importance of these factors, and is followed by a review of recent advances in systematic alloy chemistry (§7.3).

The study of magnetic structures, now in a very active stage of development, is considered in §7.4 in relation to the magnetic moments of the atoms and their relative orientations in the crystal; in §7.5 there is an account of information obtained about total and unpaired

† Dr Brown is Assistant Director of Research at the Crystallographic Laboratory, Free School Lane, Cambridge, and Fellow of Newnham College, Cambridge. Dr Taylor is Reader in Crystallography at the Crystallographic Laboratory, Free School Lane, Cambridge, and Fellow of Clare Hall, Cambridge.

318   P. J. BROWN AND W. H. TAYLOR

electron distributions. The earlier sections may be regarded as providing the background for the researches described in §§ 7.4 and 7.5.

In a short chapter, of a few thousand words, only an outline treatment of this mass of material can be attempted. The choice of examples for detailed consideration must in these circumstances be somewhat arbitrary. That they are taken very largely from the researches of the group with which the authors have been associated in the Cavendish Laboratory, may not be inappropriate in a volume dedicated to the head of that laboratory.

7.2 STABILITY OF ALLOY STRUCTURES

Three main factors are recognized as governing the stability of alloy structures, and thus providing the basis for a semi-empirical classification of structure types: they are the sizes of the atoms, the ratio (e/A) of the number of free (valency) electrons to atoms in the structure, and electrochemical (bonding) effects. The separate discussion of each factor in turn, which follows, is convenient since for a given material one factor may be dominant, but is in principle physically unrealistic since for any particular structure the effects due to the dominant factor are likely to be modified to a significant extent by the others.

7.2.1 Atomic size

The size factor is important in relation to the range of composition over which solid solution may occur in an alloy, in its influence on the production of long range (superlattice) order, and in considering the formation of interstitial compounds and the occurrence of Laves phases and other 'packing structures'. The differences between the sets of atomic radii proposed by different workers (see, for example, Pauling, 1945; Wells, 1950) are relatively unimportant since in any case allowance must be made for the possible influence of the other factors (discussed below) when interatomic distances are to be predicted, or measured values explained, with any accuracy. Thus arguments based on the concept of atomic size, though valuable when expressed in general terms, must not be applied too rigidly (see, for example, Geller, 1956, 1957; Pauling, 1957; Shoemaker & Shoemaker, 1964).

7.2.2 The e/A ratio

The empirical classification of a large number of alloy phase structures in terms of the numerical values of the e/A ratio, and the discussion of these 'electron compounds' in terms of Brillouin Zone theory, marked important steps towards the understanding of alloy structures. (For a short review, see Raynor, 1949.) Attempts to apply Brillouin Zone methods to obtain information about electron distributions in complex alloy structures involving transition metals were, in some cases, much too naïve but led to more critical interpretations of the relevant X-ray diffraction measurements (Taylor, 1954; Black, 1956) and thus made clear the essentially inexact nature of the Brillouin Zone (e/A) criteria for stability. More recently this whole approach has acquired much more precise physical significance since it has become possible to obtain detailed information about the Fermi surface from measurements of various physical properties. (For short reviews, see Chapter 2 and Ziman, 1963.)

7.2.3 Electrochemical effects

The cohesive forces in an alloy may assume a partly ionic character—when one component is more electropositive and the other more electronegative—or may show partly covalent characteristics—as, for example, with B subgroup metals. These electrochemical or bonding effects necessarily modify any conclusions based on size factor or e/A ratio which would apply to purely metallic structures.

An important application is in the classification of alloys between two transition metals of which one is an A-type (Sc, Ti, V, Cr groups—i.e. to the left of Mn in the Periodic System), the other a B-type (Mn, Fe, Co, Ni, Cu groups—i.e. to the right of, and including, Mn). (For a comprehensive discussion, see Nevitt, 1963.) Here again a detailed study shows that flexibility is needed in applying the broad generalizations derived from a large amount of structural data.

7.3 ALLOY CHEMISTRY

In recent years the most important advances have resulted from the application of single-crystal X-ray diffraction techniques to the study of alloys containing transition metals, many of the structures being too complicated for analysis by powder methods. Some of the alloys studied are of direct importance for their technological properties, but in most cases the aim has been to obtain a better understanding of the role of d-electrons in metallic structures in general, since it is known that these electrons play a major role in the technologically important refractory and magnetic alloys. Some general trends observed in transition metal alloys are described in the following paragraphs, and provide a basis for the more detailed studies.

7.3.1   Alloys of transition metals with B subgroup metals

For these alloys the atomic sizes, the e/A ratio and bonding effects may all be important, so that many different and often complex structures occur; the range of homogeneity is usually small.

Alloys with compositions not far from the equiatomic TM:B ratio or with more TM than B atoms, often form structures of one of the common types such as CsCl, γ brass, Cu_3Au, $CuAl_2$ or the Laves phases.

By contrast, alloys relatively rich in the B metal form many phases with very complex structures. The first of a number of Al-rich transition metal alloys to be studied in the authors' laboratory was Co_2Al_9 (Douglas, 1950), in an attempt—now recognized as rather naïve—to decide whether or not electron transfer took place from Al to unoccupied d-levels of neutral Co atoms. The subsequent examination of more alloy structures of this type, though providing no reliable information about electron distribution, led to the critical examination of the significance of the purely geometrical application of Brillouin Zone methods (§ 7.2.2 above) and also established the geometrical features characteristic of the structures. In particular, certain polyhedral complexes were recognized as basic building blocks, the different phase compositions representing different ways of packing the blocks together; the icosahedral group of 12 Be atoms around a central TM atom, seen in many TM–Be alloys, may be quoted as an example. The polyhedral units are not, in general, regular in form—thus in the icosahedron there are usually 6 long and 6 short bonds, and in some polyhedral groups some of the bonds may be very short. The effect of size factor is seen on comparing alloys of TM's with various B metals—for example, Be, Zn, Al; alloys very rich in the B metal tend to form when the size factor is large as in Be compounds, but not in the Al compounds—the effect being particularly marked with TM atoms at the beginning of the long Periods, less obvious with the later TM atoms (see, for example, Brown, 1959, 1962).

While it is obviously convenient to visualize these structures as linked polyhedral complexes, it is rather doubtful whether the polyhedral complex is in fact a physically stable unit and it seems more probable that it is the framework formed by the complexes which is truly stable; Black & Cundall (1966) have shown that the complexes do not exist in the liquid state of some of the alloys. It may be supposed that compound formation at a particular composition depends in the first instance on packing considerations which control the formation of the polyhedral complexes, and secondly on the bonding interaction between electrons in the first half of the d-band in the TM and the s- and p-electrons of the B metal, which in turn leads to the formation of stable frameworks of polyhedra.

7.3.2 Alloys of two transition metals

The classification of these alloys is based on the division of the transition metals into A- and B-types (§ 7.2.3, above), to the left and to the right of the Mn–Tc–Re group: the division is not rigid, since both Mn and Tc behave as either A- or B-type, though Re is definitely B-type.

While alloys of types A–A and B–B frequently show considerable ranges of homogeneous solid solution, alloys of type A–B are best regarded as compounds in which electrochemical (bonding) effects are prominent and (in general) the range of homogeneity is small. Of the A–B compounds, the σ-phase structures in the FeCr and CoCr systems were among the first to be determined and the state of order in the latter was examined (Dickins, Douglas & Taylor, 1956). More recently, structures of this and related types have again been studied with the improved techniques now available and have provided new information about size factor and electrochemical effects. For more detailed discussions Nevitt (1963), da Veiga (1963) and Hall & Algie (1966) may be consulted.

7.3.3 Long-range order

A further advance in understanding alloy structures has been made possible by the increased sensitivity of modern techniques for X-ray diffraction studies and by the development of electron diffraction methods, which have led to the recognition of long-period regularities in an increasing number of systems. These long-range phenomena provide a particularly favourable field for theoretical interpretation since, of the three factors which may influence stability, the size factor and electrochemical effect are essentially short-range interactions and can be assumed to be largely responsible for the primary structure but not to influence the long-range regularities which must then be attributed almost entirely to interaction between electrons and the Brillouin Zone (e/A).

Long-period regularities of this kind have been shown to occur in a number of simple alloys, including CuAu, $PdCu_3$, $PtCu_3$, Ag_3Mg, in addition to the known simple superlattice. The long-range periodicity is best regarded as a regular long-range antiphase domain structure; since this is a truly equilibrium configuration, with fixed domain size, it is unlikely to be due to random nucleation. Sato & Toth (1962) have proposed an explanation in terms of the Brillouin Zone.

Another example of long-range interaction is provided by the variety of stacking sequences, variants of the usual simple close-packed structures, seen in the rare earth structures and in a number of alloys. Sophisticated theoretical treatments relate the observed long-period repeats to the form of the Fermi surface, and to the long periodicities of the observed spiral spin magnetic structures.

The position reached at this stage in the account of the experimental study of metals and alloys by diffraction methods is that much has been established, empirically, about the factors which influence the stability of many structures and that systematic classifications of structural types are available, which rest on accurate determinations of their geometry. What is completely lacking is any *direct* correlation between structure type and stability on the one hand, and the electron wave-functions on the other.

Direct information about the spatial variation of the electron wave-functions should be obtainable from X-ray determination of the total electron density distribution, but for various technical reasons this is extremely difficult and progress remains slow. A further direct method is applicable to alloys in which the d-electrons have a magnetic moment; this magnetic moment can interact with the neutron magnetic moment and under ideal conditions measurement of the neutron magnetic scattering will yield the distribution of (vector) magnetic moment in the unit cell. Thus, ideally, measurement of X-ray and neutron scattering from the same crystal could be used to derive separately the spatial distributions of electrons in each of the two spin states. However, just as the derivation of the total electron density distribution from the X-ray scattering depends on accurate knowledge of the geometric structure so the interpretation of neutron scattering in terms of spin density requires knowledge of the basic magnetic structure of the crystal. In §7.4 some examples of the wide variety of magnetic structures exhibited by transition metal alloys are described as a necessary preliminary to the more fundamental studies described in §7.5.

7.4 MAGNETIC STRUCTURES

In this section magnetic ordering in metals and alloys is discussed. The experimental study of magnetic structures depends upon the use of neutron diffraction methods which permit the measurement of the individual atomic moments and the determination of their orientations. No account will be given of the experimental techniques involved in obtaining diffraction measurements. These may be carried out over a wide range of temperatures, with unpolarized or polarized neutrons, and from specimens which may be subjected to strong magnetic fields. It is assumed that the atomic structure of the material is determined accurately by X-ray methods, and that the magnetic properties of the material are known—usually over a wide range of temperature. The Mössbauer effect offers an important ancillary means of examining certain aspects of the magnetic state of some metals, but will not be discussed here.

The details of the distribution of the (unpaired) spin density which constitutes the magnetic moment of an atom will be considered in §7.5. In the present section the treatment is, in fact, an extension of the familiar treatment of ferromagnetic materials in terms of the parallel alignment of localized atomic or ionic moments. The ordered structures revealed by neutron studies range from the simple (collinear) ferromagnetic, antiferromagnetic and ferrimagnetic arrays, through fairly complex non-collinear arrays such as triangular groups, single and double spirals, to spin density waves in the Cr structures, complex spirals or modulated spin structures in the rare earths, and—a recent example—a three-component modulated non-collinear structure observed in Mn_5Si_3.

The effects under discussion are observed principally in transition metals of the first long period and in the rare earth metals, with a few special cases from the second and third long periods and the actinides U and Pu. The moment associated with a given chemical atom varies from one alloy structure to another and even between different sites in the same structure, is not (in general) an integral number of Bohr magnetons, and in a given alloy is not usually the same as is derived from a Curie–Weiss temperature dependence of the paramagnetic susceptibility.

Examples selected from among the less complex magnetic structures which have been determined by single-crystal methods are described in the following paragraphs (§ 7.4.1). Information obtained from these and similar studies is then reviewed (§ 7.4.2) and it is shown that it is becoming possible to set up tentative generalizations which may serve as a basis for theoretical work.

7.4.1 Examples of ordered structures

The magnetic structures described in the following paragraphs are, roughly, in order of increasing complexity. In some cases the complexity arises from the nature of the atomic arrangement in the crystal structure—the ideal in this respect is the structure in which all atomic positions are fixed by symmetry—in others it is the distribution of magnetic moments which is complex, even though the pattern of atomic sites may be relatively simple.

7.4.1a *Ferrimagnetic* Mn_2Sb (Wilkinson, Gingrich & Shull, 1957; Alperin, Brown & Nathans, 1963). The structure is tetragonal $a = 4.08$ Å, $c = 6.56$ Å, space group P4/nmm with atomic positions

>2 Mn (I) in 2 (a) (0, 0, 0), etc.,
>
>2 Mn (II) in 2 (c) (0, $\frac{1}{2}$, z), etc., with $z = 0.2897 \pm 0.0006$,
>
>2 Sb in 2 (c) (0, $\frac{1}{2}$, z), etc., with $z = 0.7207 \pm 0.0002$.

Magnetic structure factors were obtained from polarized neutron diffraction measurements at room temperature using nuclear scattering lengths -0.37×10^{-12} cm for Mn and $+0.54 \times 10^{-12}$ cm for Sb. Fourier projections of the magnetic scattering were prepared for [110] and [001] zones, and by integration of the density about each Mn atom the (oppositely directed) magnetic moments at room temperature were deduced as Mn (I) $-1.48 \pm 0.15 \mu$B, Mn (II) $+2.66 \pm 0.15 \mu$B, assuming that the total net moment 1.18μB, obtained from saturation magnetization measurements, is to be interpreted as the algebraic sum of the moments of these atoms. From the measured temperature variation of the net moment the atomic moments at 0 °K, Mn(I) -1.77 μB and Mn (II) $+3.55 \mu$B were deduced.

Other interesting features of the magnetic structure are discussed in § 7.5 (see Fig. 7.4).

7.4.1*b*   *Ferromagnetic* $Fe_{1.67}Ge$ (Forsyth & Brown, 1964). The structure approximates to a partially-filled $B8_2$ (NiAs) type, with space group $P6_3/mmc$, and atomic positions

$$2\,Fe_I \text{ in } 2\,(a)\,(0, 0, 0), \text{ etc.,}$$

$$1{\cdot}33\,Fe_{II} \text{ in } 2\,(c)\,(\tfrac{1}{3}, \tfrac{2}{3}, \tfrac{1}{4}), \text{ etc.,}$$

$$2\,Ge \text{ in } 2\,(d)\,(\tfrac{1}{3}, \tfrac{2}{3}, \tfrac{3}{4}), \text{ etc.}$$

In fact the apparently hexagonal crystal represents random twinning, on $(11{\cdot}0)$, of a structure with trigonal space group $P\bar{3}m1$, in which the atomic positions are

$$1\,Fe'_I \text{ in } 1\,(a)\,(0, 0, 0),$$

$$1\,Fe''_I \text{ in } 1\,(b)\,(0, 0, \tfrac{1}{2}),$$

$$1{\cdot}33\,Fe_{II} \text{ in } 2\,(d)\,(\tfrac{1}{3}, \tfrac{2}{3}, z), \text{ etc., with } z \sim 0{\cdot}225,$$

$$2\,Ge \text{ in } 2\,(d)\,(\tfrac{1}{3}, \tfrac{2}{3}, z), \text{ etc., with } z \sim 0{\cdot}730.$$

In addition, the X-ray diffraction pattern shows evidence of a further distortion of the structure which moves the Fe_I atoms off the hexagonal axis to fill, at random, any one of six positions (lying around the ideal positions) at $\pm(x, 0, z) \pm (0, x, z) \pm (x, x, z)$ with $x \sim 0{\cdot}03$ and $z = 0$ or $\tfrac{1}{2}$.

Clearly the interpretation of neutron diffraction measurements must be less direct, as a consequence of these structural complications, than for Mn_2Sb; in principle, however, the procedure is much the same. The extrapolated saturation magnetization at $0\,°K$ is $1{\cdot}59\,\mu B$ per Fe atom, and if it is assumed that the moments of Fe'_I and Fe''_I are identical then the (parallel) moments at $0\,°K$ are Fe_I $1{\cdot}4 \pm 0{\cdot}1\,\mu B$ and Fe_{II} $1{\cdot}9 \pm 0{\cdot}1\,\mu B$. These measurements contradict the values (Fe_I $2{\cdot}0\,\mu B$, Fe_{II} $1{\cdot}0\,\mu B$) predicted by Kanematsu (1962) from the magnetic measurements and a theoretical model of a bond and spin scheme for this structure.

7.4.1*c*   *Ferrimagnetic* K–Cu–Mn–Al (Wilkinson, 1965). The basis of this structure is the very simple CsCl type, with atomic sites fixed by symmetry at $(0, 0, 0)\,(\tfrac{1}{2}, \tfrac{1}{2}, \tfrac{1}{2})$ in a small cubic unit cell, but a serious complication arises from the need to determine the partially-ordered distribution of Cu, Mn and Al atoms on both sites. An earlier ordering scheme (Tsuboya, 1961) derived from magnetic and X-ray measurements was proved incorrect and a new scheme proposed which was derived from combined X-ray and neutron measurements.

These and additional polarized neutron measurements were then used in deducing, for the particular alloy crystal studied, moments $+1\cdot4 \pm 0\cdot1\,\mu\mathrm{B}$ for a Mn atom on site $(0, 0, 0)$ which is relatively rich in Mn, and $-2\cdot4 \pm 0\cdot4\,\mu\mathrm{B}$ for a Mn atom on site $(\frac{1}{2}, \frac{1}{2}, \frac{1}{2})$ which is rich in Al and contains rather little Mn (moments correspond to temperature $295\,^{\circ}\mathrm{K}$).

Other features of this structure are discussed in §7.5.

7.4.1 d *Antiferromagnetic* NiMn (Kasper & Kouvel, 1959). The crystal structure of the ordered alloy NiMn is tetragonal, with unit cell $a \sim 3\cdot7\,\text{Å}$, $c \sim 3\cdot5\,\text{Å}$, and atomic planes parallel to (001) containing alternately all Ni and all Mn.

In fields up to 8 kGs, over the temperature range $1\cdot8\,^{\circ}\mathrm{K}$ to room temperature, the volume susceptibility is constant and field-independent, the value being 6×10^{-5} e.m.u., which is consistent with an antiferromagnetic state with Néel temperature well above room temperature.

Neutron diffraction patterns at $77\,^{\circ}$, $298\,^{\circ}$ and $600\,^{\circ}\mathrm{K}$ from a polycrystalline material contained peaks due to nuclear scattering, nuclear superlattice scattering and magnetic scattering; only 4 magnetic reflections were measurable. For these reflections $(h + k)$ is odd, whence it is deduced that atomic moments in planes parallel to (001) are antiparallel to their nearest neighbours in the same plane. From the measured intensities, it is possible to exclude one model in which moments are parallel to the c-axis, and various models in which moments are perpendicular to the c-axis, leaving as the most probable magnetic structure one with moments parallel to [100] directions and of magnitudes $4\cdot0\,\mu\mathrm{B}$ for Mn and $< 0\cdot6\,\mu\mathrm{B}$ for Ni. The magnetic unit cell for this structure is the same as the chemical cell.

7.4.1 e *Non-collinear structures.* A completely systematic treatment would now deal with progressively more complicated cases of non-collinear arrays of moments, perhaps in the following order:

triangular arrays: Mn_3Sn, Mn_3Ge, Mn_3Rh (Kouvel & Kasper, 1964)

simple spirals: Mn_2Au (Herpin & Meriel, 1961)

double spirals: MnP (Forsyth, Pickart & Brown, 1966)

spin density waves: Cr (Bacon, 1961; Brown, Wilkinson, Forsyth & Nathans, 1965) modulated spin structures and complex spirals: the rare earths (many papers, especially in *Phys. Rev.* and *Jnl Appl. Phys.*, 1961 onwards).

Such a full account would be inappropriate, even if space were available; instead, two examples will be described—the double spiral MnP and a three-component modulated non-collinear structure Mn_5Si_3.

Metamagnetic MnP (Forsyth, Pickart & Brown, 1966). The structure is orthorhombic, type B31, with unit cell $a \sim 5 \cdot 2$ Å, $b \sim 3 \cdot 2$ Å, $c \sim 5 \cdot 9$ Å, space group Pnma, atomic positions

$$4 \text{ Mn in } 4(c) \ (x, \tfrac{1}{4}, z), \text{ etc., with } x = 0 \cdot 0049 \pm 0 \cdot 0002,$$

$$z = 0 \cdot 1965 \pm 0 \cdot 0002,$$

$$4 \text{P in } 4(c) \ (x, \tfrac{1}{4}, z), \text{ etc., with } \quad x = 0 \cdot 1878 \pm 0 \cdot 0005,$$

$$z = 0 \cdot 5686 \pm 0 \cdot 0005.$$

At 50 °K, there is a magnetic transformation from ferromagnetic (above 50°) to metamagnetic (below 50°) in which state the structure exhibits an antiferromagnetic-ferromagnetic transition which is field- and temperature-dependent. This metamagnetic phase is discussed.

With unpolarized neutrons at 4·2 °K no extra reflections due to a simple collinear antiferromagnetic array of moments were observed, but near (nuclear) peaks h0l pairs of satellites were seen, with displacements $\pm 0 \cdot 112 c^*$ from the nuclear peaks. Only one pair of satellites was associated with each nuclear reflection, and the intensities were usually markedly different for the two members of a pair.

The interpretation in general terms is that the magnetic structure, i.e. the pattern of atomic moments, is modulated in the c^* direction, that the modulation is closely sinusoidal, and that the structure cannot be represented by a simple single spiral. The detailed interpretation of the intensities of the satellites leads to a model in which all the atomic moments are parallel to the (001) plane, those of the 2 Mn atoms at $(x, \tfrac{1}{4}, z)$ are parallel, so are those of the 2 Mn atoms at $\pm (\tfrac{1}{2} + x, \tfrac{1}{4}, \tfrac{1}{2} - z)$, and the angle between the first pair of moments and the second pair is $\sim 20°$. Each pair of moments is thus coupled ferromagnetically across a symmetry centre of the space group, while the moments of a pair are rotated 20° for each operation of the screw diad axis parallel to c, which thus propagates the spiral of moments through successive unit cells along the c-direction (Fig. 7.1). The envelope of the the spiral is probably not perfectly cylindrical but has unequal axes $1{\cdot}73 \pm 0{\cdot}01\,\mu B$ parallel to b and $1{\cdot}41 \pm 0{\cdot}10\,\mu B$ parallel to a, with mean moment per Mn atom $1{\cdot}58 \pm 0{\cdot}10\,\mu B$, significantly larger than the value $1{\cdot}29\,\mu B$ determined from the saturation magnetization in the ferromagnetic state above 50 °K.

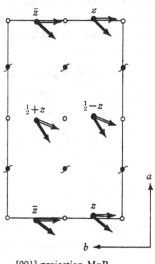

[001] projection MnP

Fig. 7.1. The spin arrangement in the metamagnetic phase of MnP. The Mn atoms in two unit cells are projected down the direction of spin propagation (c). The z-coordinates refer to the origin in each unit cell and the arrows show how the spin directions are related in the upper cell (black arrow) to those in the lower (white arrow). (We thank the editors of *Proc. Phys. Soc.* for permission to publish this figure.)

The double spiral structure proposed also explains satisfactorily details of the effects observed on applying a field to the metamagnetic phase at 4·2 °K.

Antiferromagnetic Mn_5Si_3 (G. H. Lander, 1966, private communication). The structure of Mn_5Si_3 is hexagonal, type DO_{19}, with unit cell $a \sim 6{\cdot}9\,\text{Å}$, $c \sim 4{\cdot}8\,\text{Å}$, space group $P6_3/mcm$, and atomic positions

$4\,Mn_I$ in $4\,(d)$ $(\frac{1}{3}, \frac{2}{3}, 0)$, etc.,

$6\,Mn_{II}$ in $6\,(g)$ $(x, 0, \frac{1}{4})$, etc., with $x = 0{\cdot}2358 \pm 0{\cdot}0006$,

$6\,Si$ in $6\,(g)$ $(x, 0, \frac{1}{4})$, etc., with $x = 0{\cdot}5992 \pm 0{\cdot}0015$.

Below the Néel temperature 68 °K the structure is antiferromagnetic, with a (magnetic) unit cell in which the a-axis of the chemical cell is doubled (the c-axis remaining the same) and the full hexagonal symmetry is preserved.

Attempts to explain the measured magnetic intensities in terms of a simple series of ferrimagnetic sheets, stacked perpendicular to the c-axis, were unsuccessful.

The observed reflections can be explained successfully in terms of a non-collinear magnetic structure based on an orthorhombic cell having twice the volume of the chemical cell, i.e. the smallest ortho-hexagonal cell, which may be set up with its (short) a-axis parallel

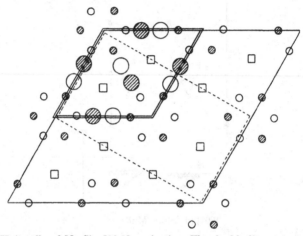

Fig. 7.2. Unit cells of Mn_5Si_3. [00·1] projection. The double lines, single lines and broken lines show the chemical, magnetic and c-face centred orthorhombic cells. Open circles denote atoms at $z = \frac{1}{4}$, shaded circles atoms at $z = \frac{3}{4}$. Squares denote Mn_I atoms at $z = 0$ and $z = \frac{1}{2}$, small circles Mn_{II} atoms, large circles Si atoms (shown only in the chemical unit cell) (G. H. Lander, 1966, Thesis).

to any one of the three equivalent a-axes of the hexagonal chemical cell (Fig. 7.2). To each of the three possible orthorhombic cells so described there corresponds a different reciprocal lattice and the three reciprocal lattices overlap only at points corresponding to the reciprocal lattice of the chemical cell. Hence each magnetic reflection corresponds to a particular orthorhombic cell, and the fact that hexagonal symmetry of the reflections is preserved indicates that equal volumes of the crystal can be described by each of the orthorhombic cells. Consideration of the observed reflection intensities, which are characterized by some very striking systematic absences, enables magnetic moments to be assigned to all the Mn atoms in the orthorhombic cell, but it is not found possible to assign equal moments to all Mn atoms which are crystallographically equivalent in the chemical cell.

A model which maintains equivalence of structurally identical Mn atoms, and is thus more physically reasonable, is one in which each of the orthorhombic magnetic cells corresponds to a sinusoidal modulation of the magnetic structure with propagation vector parallel and equal in length to the longest axis of that orthorhombic cell. It must be postulated that all three modulations can co-exist in the same volume of crystal, and the model then has the further merit of introducing hexagonal symmetry of the reflection intensities without the necessity of invoking equality of volume of three types of domain. It is unfortunately, difficult to visualize the model of the magnetic structure in this unexpectedly complicated case.

7.4.2 Systematic study of magnetic moments

Information about the effective magnetic moments of various atoms in different structural environments is beginning to accumulate to an extent which permits the formulation of tentative generalizations, on an empirical basis, which may serve to guide theoretical discussions.

The influence of magnetic order on the moments of the individual atoms is most clearly demonstrated when the atomic moment in an (ordered) magnetic structure is compared with the moment deduced from a Curie–Weiss temperature dependence of the measured paramagnetic susceptibility.

Two trends are apparent in the influence of alloying atoms on the atomic moments. First, alloying a magnetic atom with a B subgroup metal reduces the moment by an amount which is greater the lighter the B metal and the greater the number of B atoms in the group around the magnetic atom. Secondly on alloying a magnetic atom with a transition metal the moment of the magnetic atom is greater for a B-type alloying transition metal (to the right of Mn—§7.2.3) than for an A-type (to the left of Mn).

If the experimental measurements of atomic moments are very far from being comprehensive, attempts to link the available measurements with fundamental ideas about electron states in metals are even more 'patchy'. Thus, it is true that some success has been achieved in discussing the modulated structures (of ordered moments) observed in Cr and rare earth metals with reference to the Fermi surface and features of the electron distribution (e.g. Tachiki & Nagamiya, 1963; Koehler, Cable, Child, Moon & Wollan, 1966; Lebech & Mikke, 1966). Yet, at a much less sophisticated level, little, if any, solid progress can be claimed in the detailed interpretation of the atomic moments observed even in simple alloy structures, or in the theoretical investigation of the factors controlling the sign of the magnetic interactions in these structures.

The most obvious need is for more measurements, but coupled with this is the need for a more detailed examination of the nature of the electronic systems which confer the magnetic moments on the atoms and control their interactions in the crystalline structure.

7.5 ELECTRON CONFIGURATION

This section deals first with the information available about the total electron configuration, secondly with measurements of the unpaired spin density distribution which constitutes the magnetic moment of the atom.

In principle, the total electron distribution may be derived directly from measurements of the X-ray scattering. In practice, formidable difficulties are encountered; in measuring diffraction intensities on an absolute scale, in order to avoid the introduction of arbitrary constants in the Fourier conversion of intensities to electron densities; in securing a set of intensity measurements sufficiently complete to avoid the introduction of false detail in the Fourier synthesis; and, perhaps most difficult of all, in making allowance for extinction effects, in order to arrive at diffraction intensities truly characteristic of the unit cell structure. In the following paragraphs (§ 7.5.1) published measurements are reviewed critically.

The unpaired electron spin density distribution has been determined for a number of ferromagnetic and ferrimagnetic metals and alloys, using polarized neutron diffraction methods. These methods avoid some of the difficulties inherent in X-ray studies so that in general the results obtained are rather more accurate, though extreme care must still be taken in their interpretation. The relationship between the atomic moments and the unpaired spin density distributions derived from neutron studies may be regarded as analogous to that between atomic positions and total electron density distributions derived from X-ray measurements. Selected examples of spin density distributions are discussed in § 7.5.2.

7.5.1 Total electron density distribution

Most of the experimental measurements published before about 1960 may be ignored in favour of the rather limited amount of information obtained from more recent work which is probably free from the sources of error mentioned above. Some of these measurements have been made with powders, some with single crystals. The salient points which now seem to be established with reasonable certainty are as follows:

(i) The atomic scattering factor (for X-rays) in a metal powder is approximately that expected for an electron distribution (including $3d$-electrons) similar to that of the free atom configuration (Batterman, 1959; Batterman, Chipman & DeMarco, 1961). This finding supersedes those based on single-crystal measurements by Weiss & DeMarco (1958), now known to be in error due to extinction effects, from which it was concluded that (for example) in metallic Fe the number of $3d$-electrons was about 2, compared with 6 in the free atom, and similarly for some other transition metals.

(ii) Comparison of the accurate absolute measurements of Batterman *et al.* (1961) with calculated scattering factors based on the free atom shows that for Fe, Cu and Al powders the observed values lie about 4 % below the calculated values for reflections at low angles.

In view of the known difficulties in obtaining absolute measurements of intensities with the accuracy needed to prove the reality of this discrepancy, caution is necessary. Nevertheless, the evidence is now very strong that this is an effect characteristic of the metallic state of these and some other materials. Thus:

(a) measurements of scattering factor for the rare gases Ne, Ar, Kr and Xe showed no such discrepancy with calculation (Chipman & Jennings, 1963);

(b) totally independent measurements in another laboratory led to precisely the same (4 %) discrepancy for Cr powder (Cooper, 1962) while measurements using the same procedures of specimen preparation and X-ray measurement showed no such difference between observation and calculation for powders of ordered alloys CoAl and NiAl (Cooper, 1963), (except for reflections {200}, see below);

(c) measurements from highly perfect single crystals of Cu (Jennings, Chipman & De Marco, 1964), and from an imperfect single crystal of Al (DeMarco, 1967), which involve the use of different techniques, both differing from those applicable to power measurements, confirm the 4 % discrepancy in these two materials. By contrast, the effect is not found in perfect single crystals of Si (DeMarco & Weiss, 1965).

The importance of the measurements summarized above is obvious.

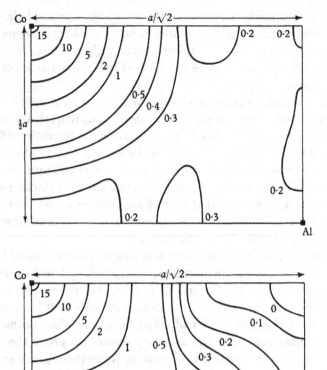

Fig. 7.3. The outer electron distribution in CoAl. The upper diagram shows the outer electron distribution calculated from theoretical structure factors, the lower that derived from experimental structure factors (Cooper, 1963). Contours are in electrons $Å^{-3}$ in a [110] section through the centres of the atoms Co and Al. (We thank the editors of *Phil. Mag.* for permission to publish this figure.)

In the first place, they show that— in the metals and alloys examined —the electronic distribution in the neighbourhood of the atom centres in the metal is, to a first approximation, that of the Hartree–Fock model for the free (isolated) atoms. Secondly, they provide very strong evidence that in the pure metals examined—Al, Cr, Fe, Cu—there is a reduction, of amount about 4%, in the scattering contributed by the outer electrons of the atom by comparison with that predicted on the

z

basis of the Hartree–Fock model, A direct interpretation of this effect for Cr, Fe and Cu may be proposed in terms of slight radial expansion of the electron distributions. For Al the reduction in scattering is greater than can be accounted for by assuming that the three valence electrons do not contribute at all to the scattering in the Bragg peaks; it must therefore be inferred that the charge density in the neon core is significantly different from that given by Hartree–Fock calculations. It must therefore be supposed *either* that the 4 % discrepancy represents some feature of the metallic binding which may be responsible for modification not only of the outer electron distribution but also of the core distribution, *or* that the Hartree–Fock method of calculation is in error. The latter explanation seems, perhaps, less likely in view of the excellent agreement between theory and experiment for the rare gases, for Si and for CoAl and NiAl, already mentioned.

The possible relevance, to this discussion, of the following observations should perhaps be noted.

(i) In the ordered CoAl and NiAl alloys (Cooper, 1963) reference has already been made to the difference between structure factors {200} measured experimentally and those calculated for ground-state free atoms. This difference receives its most direct interpretation in terms of a small excess of electron density at the centre of the shortest contact between Co (or Ni) and Al atoms (Fig. 7.3).

(ii) The antiferromagnetic alloy Mn_5Si_3 (G. H. Lander, 1966, private communication) contains two groups of Mn atoms, Mn_I and Mn_{II} (7.4.1). The highly accurate analysis of the structure has provided some evidence that the scattering factor for the atoms in group Mn_I is low by comparison with that for group Mn_{II}. It is not impossible that this may be a consequence of the same feature of metallic binding which results in low scattering factors for Al, Cr, Fe and Cu.

Too much weight must not be attached to these isolated observations though the second, in particular, is intriguing, and the implication that metallic bonding in Fe, Cu and Al introduces more disturbance of the free atom electron wave-function than does the covalent interaction in Ge and Si is interesting. The need for more experimental measurements is clear.

7.5.2 Spin density distribution

A review of the information obtained in the last ten years from polarized neutron studies of metals and alloys would be out of place here, and, since our concern is primarily with alloys, no details can be given of the outstanding series of experiments by Shull and co-workers at M.I.T. which have established the spin density distributions in Fe, Co, and Ni (Shull, 1963; Moon, 1964; Mook & Shull, 1966). Instead, by reference to suitable examples, attention is directed to the following important features of unpaired electron spin density distributions, observed in ferromagnetic and ferrimagnetic alloys.

(i) The spin density distribution associated with atomic magnetic moment is aspherical in all cases, though the degree of asphericity may be small. The departure from spherical symmetry is shown most clearly on subtracting the spherically-averaged density from the actual density so as to display regions of excess or deficit of density, often in the form of + or − lobes.

(ii) The observed (spherically-averaged) radial density distribution is more compact for the magnetic atom than the calculated value for a free atom or ion, the form factor falling off more slowly as a consequence. One atom in a structure may be more compact than another chemically identical but crystallographically distinct atom in the same structure.

(iii) The centre of gravity of the unpaired spin density may be displaced relative to the centre of gravity of the total electron density, i.e. relative to the centre of the atom as determined from X-ray diffraction measurements. This can occur only when the atomic site is non-centrosymmetric, as in Mn_5Ge_3 and Mn_2Sb.

7.5.2a *Ferrimagnetic* Mn_2Sb (Alperin, Brown & Nathans, 1963). The crystal structure and the magnetic structure have already been described (p. 324); atom Mn(I) has moment $-1\cdot48\,\mu B$, atom Mn(II) $+2\cdot66\,\mu B$, at room temperature. Figure 7.4 shows the projection of the unpaired electron density on the (110) plane.

The most conspicuous feature of the map is the displacement of the spin density, with respect to the Mn(II) nucleus which is at a non-centrosymmetric site, by $0\cdot025\,$Å in the c-direction towards the nearest Sb atom. A similar displacement of the spin density is observed in Mn_5Ge_3 (P. J. Brown & J. B. Forsyth, unpublished) at a (non-centrosymmetric) Mn site and towards the nearest Ge neighbour. The significance of this special feature remains uncertain until further examples are discovered.

The following features of the structure, *not* obvious in the map of Fig. 7.4, emerge on detailed study of the spin density distribution. First, the unpaired electron distributions are aspherical for both

Mn(I) and Mn(II) atoms, asymmetric lobes pointing from Mn(II) towards Mn(I) and towards Sb atoms. Secondly, both Mn atoms are compact by comparison with the distribution calculated for the free ion Mn^{2+}, the (spherically-averaged) spin density for Mn(I) with the smaller moment more compact than that for Mn(II).

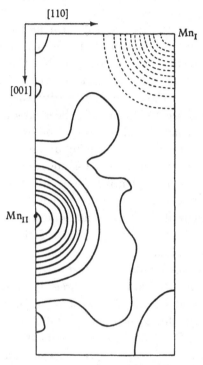

Fig. 7.4. Fourier projection on (110) of the unpaired spin density in Mn_2Sb. The contour interval is 0·40 μB $Å^{-2}$. Positive contours are shown as solid lines, negative contours as broken lines. The position of the Mn(II) atom obtained from X-ray diffraction is marked ●.

It may be supposed that the ferromagnetically aligned Mn(II) atoms, with the large moments, interact via the neighbouring Sb atoms and that the antiparallel coupling of equal numbers of Mn(II) and Mn(I) atoms is direct. The overlap of the (larger) positive spin of Mn(II) and the (smaller) negative spin of Mn(I) may result in the more compact spin density distribution of Mn(I); the possibility is not excluded that this effect would not be present in the true three-dimensional spin density distribution, but is a fictitious effect introduced in the two-dimensional projection.

7.5.2*b*   *Ferrimagnetic* K–Cu–Mn–Al (Wilkinson, 1965). The simple crystal structure (of CsCl type) and the magnetic structure have been described (p. 325). For the particular alloy crystal studied the moments (at room temperature) are $+1\cdot4$ μB for Mn(0) (atom at 0, 0, 0) and $-2\cdot4\mu$B for Mn($\frac{1}{2}$) (atom at $\frac{1}{2}, \frac{1}{2}, \frac{1}{2}$). In this partly-ordered material site (0) is occupied by $0\cdot69$ Mn, $0\cdot31$ Cu and site ($\frac{1}{2}$) by $0\cdot14$ Mn, $0\cdot06$ Cu, $0\cdot80$ Al.

The observed spherically-averaged form factors for both Mn (0) and Mn ($\frac{1}{2}$) fall off more slowly than the form factors calculated for the free atom Mn or the free ion Mn^{2+}, so that the unpaired electron densities for both atoms are more compact than those calculated for the free atom or ion; the atom Mn ($\frac{1}{2}$) with the *larger* moment is more compact than Mn (0). Again this effect may arise from overlap since the $-$ spin density at Mn ($\frac{1}{2}$) would be overlapped by $+$ spin density from $8 \times 0\cdot69 = 5\cdot5$ atoms Mn (0) of moment $1\cdot4$ μB (counting nearest neighbours only), whereas the $+$ spin density at Mn (0) would be overlapped by $-$ spin density from only $8 \times 0\cdot14 = 1\cdot1$ atoms Mn ($\frac{1}{2}$) of moment $2\cdot4\mu$B. By contrast with the case of Mn_2Sb, however, the overlap (if it occurs) is not a fictitious effect due to two-dimensional projection, but is a feature of the three-dimensional spin density distribution.

The spin density distributions are aspherical for both Mn (0) and Mn ($\frac{1}{2}$) atoms. A detailed study, in three dimensions, shows well-defined lobes of positive spin density for Mn (0) and negative spin density for Mn ($\frac{1}{2}$); the projection on (100) of this complicated pattern is shown in Fig. 7.5 which represents the difference between actual and spherically-averaged spin densities for the asymmetric unit. Though the details of the asymmetries cannot be deduced from this two-dimensional projection, it serves to demonstrate the nature and scale of the effects.

The alloy also shows another feature common to a number of magnetic structures. The atom Mn ($\frac{1}{2}$), with the larger magnetic moment, has $5\cdot5$ Mn (0) atoms as neighbours on sites 0, 0, 0 while Mn (0), with smaller moment, has only $1\cdot1$ Mn ($\frac{1}{2}$) atoms as neighbours on sites $\frac{1}{2}, \frac{1}{2}, \frac{1}{2}$. The correlation of larger moment with larger number of (magnetic) neighbours is maintained if second nearest neighbours are included (for Mn ($\frac{1}{2}$) $5\cdot5 + 6 \times 0\cdot14 = 6\cdot3$, for Mn (0) $1\cdot1 + 6 \times 0\cdot69 = 5\cdot3$), but comparison with the case of Fe_3Al suggests that nearest neighbours are much more important than next-nearest (Pickart & Nathans, 1961).

No *direct* lead into the interpretation of measured spin density dis-

tributions is provided by the generalizations illustrated by the structures of Mn_2Sb and K–Cu–Mn–Al and of a considerable number of
other alloys not mentioned here. At the present time, however, the
amount of factual information available is so small that no other
approach seems possible, unless by chance the following very tentative
suggestion resting on the comparison of spin density and total electron
density distributions were to prove fruitful.

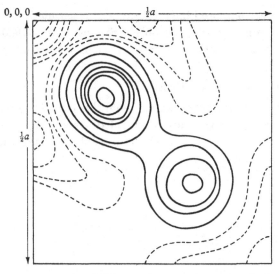

Fig. 7.5. Projection on (100) of difference between observed and spherically averaged
spin densities in K–Cu–Mn–Al. Solid contours denote positive spin density; dashed
contours negative spin density; the zero contour is omitted. Contours at equal arbitrary
intervals (Wilkinson, 1965, Thesis).

In all cases so far studied—for example, in antiferromagnetic
Mn_5Si_3 already discussed in §§ 7.4.1 and 7.5.1—a lower magnetic moment seems to be associated with a more symmetrical spin density
distribution. This, if firmly established, may be an indication of a
greater crystal field splitting of the d-band at sites of low symmetry,
it being assumed that the moment arises from holes at the top of the
d-band. Alternatively, it may merely arise as a consequence of a more
symmetrical total electron distribution, which could be checked by
X-ray measurements in a suitable case. It is even possible that the
antiferromagnetic Mn_5Si_3 may provide a comparison on these lines.
For, in a structural model admittedly not yet worked out in full detail,
the unpaired electron density of the spin component with the larger

moment is extended in the c-axis direction, and the total electron density is also extended in the same direction.

This section must end with a repetition of the warning (against too much reliance on isolated observations such as these) and of the appeal (for more experimental measurements) which concluded §7.5.1

Acknowledgments

The authors are indebted to G. H. Lander for permission to include an account of unpublished work on Mn_5Si_3 and to J. J. DeMarco for permission to refer to work on Al in advance of publication.

Note added in proof, August 1968

At the time of writing (November 1966) this chapter presented an up-to-date account of the topics included. In a subject which is developing extremely rapidly it is not possible to make that claim nearly two years later.

P. J. B.

W. H. T.

CHAPTER 8

TRANSITION METALS. ELECTRONIC STRUCTURE OF THE d-BAND. ITS ROLE IN THE CRYSTALLINE AND MAGNETIC STRUCTURES

by J. FRIEDEL†

8.1 BAND STRUCTURE OF PURE TRANSITIONAL METALS

The existence in the periodic table of three well-defined long periods, ending with Ni, Pd and Pt and corresponding to the filling respectively of the $3d$-, $4d$- and $5d$-shells strongly suggests that the peculiar properties of these metals correspond to a *strong d-character* in their valence states.

The peculiar originality of these states is a result of two factors. First, their orbits are fairly small, compared with the other (sp) valence states of comparable energy; their much smaller number of spherical nodal surface allows them, with comparable curvature, to decrease exponentially as $\exp(-qr)$ at a shorter distance from the origin (Fig. 8.1). As a result, the d-states are fairly localized; they are not strongly perturbed by the lattice potential and cannot overlap very strongly the atomic states of other atoms. Secondly, because of their parabolic increase with distance near the origin, the d-electrons screen the nuclear charge within an atom badly: in Slater's approximation, for instance (Slater, 1930), the effective screening charge of the d-electrons for each other is only 0·3, as compared with an effective charge of 1 of a d-electron for an s one.

As a result, the d-states of an atom fill successively in preference to other (sp) valence states when going through a transitional series.

8.1.1 Band structure versus localized states

The first question is to choose for the d-electrons between a localized 'atomic orbitals' description, as usually taken for the even more

† Dr Friedel is Professeur de Physique des Solides at the Faculté des Sciences de l'Université de Paris, 91-Orsay (France).

localized 4*f*-shells of rare earth metals, and an extended 'molecular orbital' approach, as for the other (*sp*) valence electrons.

The question would be very difficult to decide from purely theoretical grounds; one has to strike a delicate balance between the strength of the intra-atomic *dd*-correlation terms, and either the interatomic *dd* overlap integrals or the intra-atomic *d-sp* promotion terms. The correlation terms tend to give a fixed integral number of *d*-electrons in each *d* shell. The overlap or promotion terms tend to give to

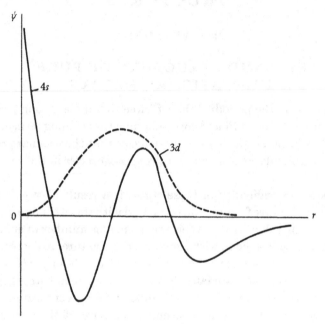

Fig. 8.1. Radial extension of a 4*s*- and a 3*d*-states for a transitional atom (schematic).

any electron in such an atomic *d*-state a finite lifetime, by inducing interatomic charge transfers and thus intra-atomic charge fluctuations in each *d*-shell. In the first case, it is certainly better to start from an atomic orbital picture of electrons in localized *d*-states, while the second case is better described by a gas of uncorrelated electrons running through the crystal in extended states. It is also obvious that each of these are only first-order descriptions: one must correct the localized picture for interatomic correlations by introducing interatomic exchange, van der Waals polarization and eventually charge transfer terms; and also the band picture for at least intra-atomic Coulomb and exchange correlations. In either case, some complication would arise from sd correlations. But, as emphasized by Mott (1949), these corrections, as long as they lead to a convergent series of approxima- tions, probably preserve an essential difference in the spectrum of low energy excitations corresponding to *charge* fluctuations: it is only the second type of approximation that allows a continuous spectrum of excited states, with a finite density of states of vanishingly small energy; these correspond to charge transfers where d-electrons take part, thus a d-part in the metallic conductivity and in the low tem- perature electronic specific heat.

It seems likely, therefore, that the most reasonable starting point, i.e. the 'molecular' or the 'atomic', will depend upon the presence or absence of metallic d-conduction.

It is therefore no surprise that the classical fight between the 'atomic' and 'molecular' approaches has had a long and distinguished career in the field of transitional metals. First defended by Pauling against Mott (Coll. Int. 1939), the 'atomic' orbital approach has more recently been revived by various authors in the context of magnetism (Mott & Stevens, 1957) or of cohesion (Lomer, 1962). For several reasons it is on the whole no longer fashionable (Friedel, 1955, 1958, 1962; Mott, 1964; Herring, 1966):

(1) Many transitional metals have large electronic specific heats, which would be difficult to explain without a strong d-part in the density of states at the Fermi level (cf. §8.1.6).

(2) Some transitional metals have permanent magnetic moments which are very far from integral values in Bohr magnetons per atom, although the measured g values are near to 2. These deviations from integral values cannot therefore be due to a partial lifting of the quenching of the orbital moment; it would be very difficult to assign them to a spin polarization of the sp-band (cf. §8.3.3).

(3) There is now evidence, in some transitional metals, for fairly complex Fermi structures, which could not be explained in terms of a nearly free sp-band only (cf. §8.1.7)

(4) There is some evidence for d-band conduction, and much evi- dence for $s \rightarrow d$ scattering at the Fermi level, through phonons or im- purities (Mott & Jones, 1936; Sondheimer, 1948; Campbell *et al.* 1967).

(5) The regular variation of cohesion through the transitional periods, with the very large maxima observed for half-filled d-bands, indicates clearly that the d-states dominate the cohesion, and that they are fairly strongly perturbed when the atoms are condensed into the metallic state (cf. §8.2.1).

(6) The absence of permanent moments in all transitional metals except those at the end of the first transitional series, and the fairly weak energies of magnetic coupling observed there, compared with the cohesive energies, clearly show that the main interaction between neighbouring d-shells in a metal is not of magnetic origin: it cannot be related to an interatomic exchange term, as would be the case in an 'atomic' orbital picture (cf. §8.3.1).

(7) It is true that, in many insulating transitional compounds such as the halides or most oxides, there is no dd conduction, and the atomic orbital approach must be used. But dd conduction is clearly present in the oxides at the beginning of the transitional series (Morin, 1959), despite a distance between transitional ions larger than in transitional metals; and many compounds with higher valencies (sulphides, selenides, tellurides, nitrides, phosphides, carbides, borides ...) have metallic properties very similar to those of the pure transitional metals, well described in terms of a band picture with strong d-character (Friedel, 1965).

We shall therefore assume in what follows that a band picture with strong d-character is a reasonable starting point in the description of the transitional metals. We shall restrict ourselves here to the (uncorrelated) *Hartree* approximation, and relegate to later sections the study of *correlations*, of importance in the questions of lattice and magnetic structures.

8.1.2   The Hartree band picture; 'd-band' versus 'virtual d-levels'

There is no great difficulty of principle in computing the band structure of a transitional metal in the simplest Hartree approximation. But the complexity of the results and their sensitiveness to the choice of atomic potentials or to the approximations used are such that not too much store must be put on the details of these results (Lomer, 1967; Phillips, 1967 etc....).

One result, however, stands out fairly clearly from these computations: there is a rather narrow range of energies, about 5–10 eV typically, in which a number of moderately narrow bands with largely d-character are concentrated.

The classical viewpoint on this complex situation is that it can be considered as deduced, by a slight sd-mixing, from the superposition of an independent set of narrow d-bands and of a broad nearly electron free sp-band (Mott & Jones, 1936; Lomer, 1962; Mueller, 1967; Hodges *et al.* 1966 *b*). One can, for instance, first consider an element such as Zn, Cd or Hg, on the right of the corresponding transitional series, where X-ray emission and ultra-violet absorption spectra clearly show that the two sets of bands are separated in energy (Fig. 8.2a). One then treats a transitional metal with the same crystalline structure as obtained by introducing a small repulsive potential v on each atom, corresponding to the reduction in nuclear charge.

The matrix elements of v between two sp-states or between two d-states will shift the bands to higher energies, and the d-band more than

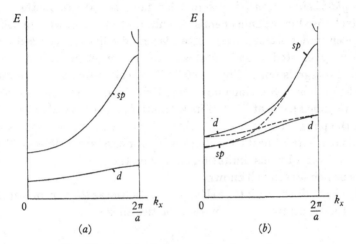

Fig. 8.2. sd-mixing between a d- and an sp-band.

the sp-band because the d-states are more localized than the sp ones (Fig. 8.1); they will also broaden the d-band, by increasing the dd overlap integrals. The matrix elements of v between d- and sp-states will mix the two sets of bands, as schematized Fig. 8.2b.

As long as v is small, i.e. at least for elements towards the end of the series, one can hope that the matrix elements of v are small compared with the initial width of the d-band. Then all the transitional metals with the same crystalline structure should have similar d-band structures, little affected by sd-mixing or by the place in the periodic table.

This classical point of view will be followed here, and we shall explain later the reasons why. Let us only point out now that if, on the contrary, the width of the d-band was small compared with the sd-matrix elements of v, it might be better to start from isolated atomic d-states, mix each of them with the sp-band into a 'virtual d state', and finally treat their interactions via the sp-band, and, eventually, through direct dd overlap (Blandin, private communication, 1965; Phillips, 1967; Anderson & McMillan, 1967). This procedure is valid for alloys with small concentrations of transitional elements (Caroli, 1967; Blandin, 1967; Alexander & Anderson, 1964); it is probably *not* valid for pure transitional metals.

8.1.3 d-band in tight binding

Computations of pure and narrow d-bands have mostly been done in the *tight-binding* approximation. This famous L.C.A.O. method, extensively used in organic chemistry and, up to the last war, in solids, has been under a cloud ever since Slater's doubts on the wisdom of neglecting excited atomic states with higher energies but larger overlap integrals (cf. Phillips, 1967). This 'configuration interaction' problem seems less serious now; this tight-binding method seems to have regained some at least of its attraction, owing to the successes it met, the physical insight such a simple picture allows and its possible extensions to treat problems of impurities or crystalline defects. There are, however, obvious limitations in the method.

Its principles are well known:

(1) Write down the lattice (Hartree) potential as the sum of 'atomic' potentials V_i centred on the various lattice sites i

$$V \simeq \sum_i V_i. \tag{8.1}$$

(2) Write down each electronic state in the solid as a linear combination of atomic (d) functions. For each site, i, there are 5 such atomic functions, denoted $|im\rangle$, where the orbital moment m goes from 1 to 5; they are eigenfunctions of V_i, with energy E_0; their overlap integrals over two sites are neglected

$$|\psi(E)\rangle \simeq \sum_{i,m} a_{im}|im\rangle, \tag{8.2}$$

$$(T + V_i)|\psi_{im}\rangle = E_0|\psi_{im}\rangle, \tag{8.3}$$

$$\langle im|jm'\rangle \simeq \delta_{ij}\delta_{mm'} \tag{8.4}$$

and

$$|a_{im}|^2 = 1. \tag{8.5}$$

(3) Of the matrix elements $\langle im|V_i|jm'\rangle$, only the *two*-centre integrals between first (or second) neighbours are retained.

The set of linear equations satisfied by the a_{im} coefficients is then of the type

$$(E_0 + \alpha_{im} - E)\,a_{im} + \sum_{j \neq i,\,m'} \beta_{im}^{jm'} a_{jm'} = 0 \tag{8.6}$$

with
$$\alpha_{im}\langle im|\sum_{j|\neq i}V_j|im\rangle \qquad (8.7)$$

and
$$\beta_{im}^{jm'} = \langle im|V_j|jm'\rangle. \qquad (8.8)$$

The α integrals merely *shift* the energy of the atomic levels $\psi_{im}(E_0)$, while the β integrals mix them into molecular states extending over the whole solid.

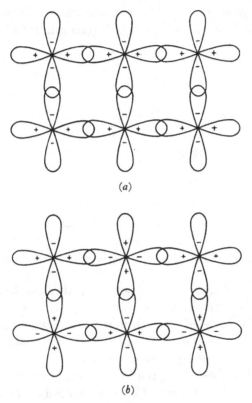

(a)

(b)

Fig. 8.3. a-'bonding'; and b-'antibonding' states in the d-band.

Writing
$$E = \frac{\langle\psi|H|\psi\rangle}{\langle\psi|\psi\rangle} \simeq \sum_{i,m}|a_{im}|^2\alpha_{im} + \sum_{\substack{i,m\\j\neq i,m'}}a_{im}^*a_{jm'}\beta_{im}^{jm'}, \qquad (8.9)$$

we see that the contribution of the β integrals to the energy E varies with the value of the coefficient a_{im}, from a minimum where most (or all) of the β terms are negative (Fig. 8.3a) to a maximum where most (or all) of these terms are positive (Fig. 8.3b). These two states, of energies E_b and E_a, correspond to the formation of *d*-'bonds' or 'antibonds' between most pairs of atoms.

In the bonding state of Fig. 8.3*a*, the electronic density is increased along the bonds, as compared with the atomic densities; in the antibonding state of Fig. 8.3*b*, the density is decreased.

The $5N$ atomic *d*-states $|im\rangle$ give rise to a band of $5N$ levels which are distributed quasi-continuously between these two extremes in energy: the β integrals give rise to the *width w* of the band; they are akin to the bonding molecular integrals (Coulson, 1953) (Fig. 8.4).

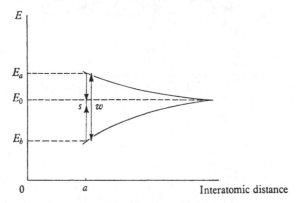

Fig. 8.4. Energy shift (s) and width (w) of a narrow band.

There is a progression from the bonding state E_a to the antibonding state E_b by the introduction of new surfaces S of antibonds across the volume V considered. At each step, the corresponding increase in energy is of the order of $S|\beta|/V \simeq N^{-\frac{1}{3}}|\beta|$; thus it tends to zero when the number N of atoms goes to infinity.

Again as in a dimolecule, the broadening in energy can be considered as due to a *resonance between the atomic d-levels*, which allows electrons to jump from one atom to another throughout the lattice.

The multiple overlaps of each atomic function with its neighbours lead to the various paths an electron can take through the metal, owing to the *multiple dd-bond* each transitional atom builds with its neighbours.

This is clear when computing *the change δT of kinetic energy* of an electron when going from a localized atomic state $|im\rangle$ to a travelling band state $|\psi\rangle$. Using (8.1) to (8.5) and assuming that

$$\langle im|V_i|im'\rangle = 0 \quad \text{if} \quad m' \neq m,$$

a relation true for atomic potentials V_i of spherical symmetry,† one easily shows that δT is just opposite to the broadening β terms in the energy (8.9):

$$\delta T = \langle \psi | T | \psi \rangle - \langle im | T | im \rangle$$

$$= - \sum_{\substack{im \\ j \neq i, m'}} a_{im}^* a_{jm'} \beta_{im}^{jm'}. \qquad (8.10)$$

Thus the bonding state of energy E_b, Fig. 8.4, has the higher (positive) interatomic kinetic energy, while the antibonding state, of energy E_a, Fig. 8.4, has a negative interatomic kinetic energy, i.e. less kinetic energy than in isolated atoms.

Finally, again as in a dimolecule, the energy width $w = h/\tau$ gives the average life time τ of an electron on each atom, and thence the mean *interatomic jump frequency* $1/\tau$.

This result, well known for crystalline solids, can be obtained directly by looking at the evolution of an electron placed at $t = 0$ on a definite atom ψ_0 of the metal. The solution of

$$i\hbar \frac{d\psi}{dt} = H\psi$$

with
$$|\psi(t)\rangle = \sum_{im} a_{im'}(t) | \psi_{im'} \rangle$$

and
$$a_{jm'}(0) = \delta_{ji} \delta_{mm'}$$

is, for small time t,

$$\left.\begin{aligned} a_{im} &\simeq \exp\left(-i\frac{E_0 + \alpha_{im}}{h}t\right), \\ a_{jm'} &\simeq \exp\left(-i\frac{E_0 + \alpha_{jm'}}{h}t\right)\frac{\beta_{jm}^{im'}t}{ih}, \end{aligned}\right\} \qquad (8.11)$$

for j neighbour to i, and zero otherwise.

8.1.4 Width of the *d*-band

The shape of the *d*-band thus obtained clearly depends on the values taken for the α and β integrals (8.7), (8.8). It must be owned that these are hard to compute accurately. The main difficulty arises from the definition of the atomic potentials V_i.

It is first clear that, in principle, the lattice potential $\sum_i V_i$ should be computed in a self-consistent way. Because the 'bonding' character of the occupied states in the *d*-band varies with the interatomic distances, the V_i should be taken as varying with the lattice parameter.

† Or with cubic symmetry in a cubic lattice.

In actual fact, the V_i are usually taken as those of the positive ions, instead of those of the neutral atoms required by the Hartree scheme. The argument is, as in the first Wigner–Seitz method, that the correlation hole in the valence band has roughly atomic dimensions.

This procedure has the obvious advantage of using the same energy E_0, potentials V_i and atomic orbitals $|im\rangle$ in the solid and in the gas. But it must be clear that it goes further than the simple Hartree scheme; this rough but reasonable assumption about correlation allows one to compare directly the energies in the solid and in the gas (cf. §8.2.4). The main drawback of this scheme is that in computing

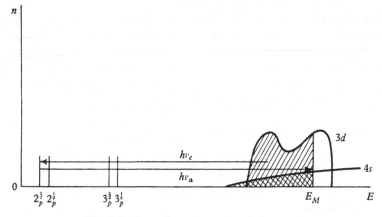

Fig. 8.5. X-ray absorption and emission spectra.

the α and β integrals, one actually uses the long range parts of the ionic potentials V_i. These are certainly screened out to some extent at that range by the d- and s-electrons. The values obtained for α and β are thus at best rough extimates, which must be checked by other computations or by some experiments.

Now all computations in metals point to d-bands *with a width w larger than their shift s*, as pictured in Fig. 8.4. Typical values are $w \simeq 5\text{–}10\,\text{eV}$; $s \simeq 1$ to $2\,\text{eV}$.

Such widths are consistent with experimental data from X-ray spectra (Bonnelle, 1964).

In the tight-binding scheme, the atomic selection rules should hold to a good approximation. They predict a strong transition probability for absorption from or emission to inner p-shells. For elements of the first transitional series, transitions with $2p$-states give rise to strong bands, with widths definitely larger than the Auger life time involved. Adding up the widths of the absorption and emission bands (Fig. 8.5),

AA

one obtains d-band widths of typically 5 eV. Similar but less convincing results can be deduced from far u.v. transitions involving the $3p$-states and, for noble metals where the d-band is full, from u.v. absorptions from the d-band to the Fermi level (Minor & Meier, 1940; Ehrenreich & Philipp, 1962; Mueller & Phillips, 1967).

It would be hard to accept that the uncertainties involved in the analyses of these data (width due to Auger effect, shifts due to energy dependent correlation corrections), introduce corrections larger than a fairly small fraction of the measured widths. This is therefore probably the most direct way of measuring w; it cannot easily be off by more than a factor 2.

8.1.5   Possible splitting of the d-band

These results can be used to show that, for reasonably compact packings, the d-band *is not expected to be split by an energy gap*. This splitting could originate in three different ways, each of which will be described.

8.1.5 a *Crystal field splitting.* If the α integrals had values differing by amounts large compared with the values of the β integrals, each type of α integral could give rise to a separate d sub-band. For instance, in a lattice where each atomic site has a cubic environment, one would then expect $e_g(x^2 - y^2, 2z^2 - x^2 - y^2)$ and $t_{2g}(xy, yz, zx)$ sub-bands. The smallness of the computed values of the α integrals point on the contrary to a situation where $|\alpha| \ll |\beta|$: hence *no* crystal-field splitting. Indeed tight-binding computations suggest that, in cubic crystals, the e_g and the t_{2g} parts of the density of states $n(E)$ are fairly uniformly distributed over most of the bandwidth; only the top and bottom of the band have definite (e_g or t_{2g}) character (Fletcher & Wohlfarth, 1951; Belding, 1959; Asdente & Friedel, 1961, 1962). The situation for such pure metals seems therefore to be quite the opposite from that sometimes suggested for some cubic transitional compounds, with metallic conductivity but large ionic character (oxides, sulphides, selenides, tellurides) (Goodenough, 1962).

8.1.5 b *Covalency splitting.* It is well known that, in covalent structures such as diamond, the atomic s- and p-states hybridize into two sets of bands separated by an energy gap: a bonding or valence band, and an antibonding or conduction band. *No* such splitting is expected for close-packed structures, if only one type of atomic state (here a

d-state) is involved (Friedel, 1956). The reason is essentially that each atomic function is involved in more than one interatomic bond, so that one goes progressively from a bonding state, as in Fig. 8.3 a, to an antibonding one, as in Fig. 8.3 b by introducing new surfaces of anti-bonds. As already pointed out, this process increases the energy by infinitesimal steps, without ever creating a finite jump in energy.

As a consequence of this absence of splitting, an incompletely filled d-band corresponds necessarily to a *metallic* state in the Hartree approximation.

8.1.5 c *Spin-orbit splitting.* If spin-orbit coupling is included, the spins are no longer exactly quantized. In the tight-binding approximation, it is reasonable to write down the spin-orbit coupling terms as a sum of atomic terms, each playing on one atomic function only. One thus neglects the interatomic 'spin-other-orbit' coupling (Asdente & Friedel, 1961, 1962; Slater & Koster, 1954; Lehman, 1959; Friedel *et al.* 1964; Abate & Asdente, 1965; Lenglart *et al.* 1966). The corresponding matrix elements between atomic states of spins α and β are thus of the form

$$\langle im, \alpha | \lambda ls | jm', \beta \rangle = \delta_{ij} \bar{\lambda} \langle im, \alpha | ls | im', \beta \rangle, \qquad (8.12)$$

where $\bar{\lambda}$ is the spin-orbit factor and α, β the two spin states. Table 8.1 gives values of

$$\bar{\lambda} = \left\langle im, \alpha \left| \frac{\lambda}{2m^2 c^2 r} \frac{dV_i}{dr} \right| im, \alpha \right\rangle$$

deduced from atomic spectra.

Table 8.1. *Spin-orbit parameter $\bar{\lambda}$ (in eV)*

	Ni	Pd	Pt
$\bar{\lambda}$	0·07	0·17	0·42

The use of e_g and t_{2g} wave-functions, for which $\langle im, \alpha | l_z | im, \alpha \rangle = 0$, shows that the trace of the spin-orbit matrix is zero. Therefore the mean value of the energy levels is not affected by the spin-orbit interaction. *This contributes to the width, but not to the shift of the d-band.* If $\bar{\lambda}$ was much larger than the width w of the d-band without spin-orbit interaction, one would then obtain a splitting of the d-band into two sub-bands, with nearly pure $J = L + S = \frac{5}{2}$ and $J = L - S = \frac{3}{2}$

characters. Table 8.1 shows, however, that, with $w \simeq 5\,\mathrm{eV}$, this is far from being the case. On the contrary, the spin-orbit coupling should not alter the width nor the form of the d-band very much, except near the ends of the d-band, over an energy range comparable with $\bar{\lambda}$.

8.1.6 Density of states of the d-band

The total width of the d-band is essentially a function of the average β-integrals, (8.8), i.e. the *average* number, distance and directions of atoms surrounding each atom. More detailed information requires a better knowledge of the *atomic structure* of the metal. This is true already for the *density of states* $n(E)$, i.e. the number of states per *unit energy* in the d-band. This will be counted *per atom and per spin direction* in what follows.

8.1.6a *A general relation with the correlation functions.* It is easy to show that the successive moments $M_p = \int n(E)\,(E-E_0)^p\,dE$ of the density of states are related to the successive correlation functions $g_p(R_1, R_2, ..., R_p)$ of the atoms in the metal, which give the probability that one atom is at R_1, one in R_2, ... one in R_p (Cyrot, 1967).

For one can write
$$M_p = \sum_{\mathbf{k}} \langle \mathbf{k}|(H-E_0)^k|\mathbf{k}\rangle.$$

Using the pseudo-orthonormal set of atomic d-functions to compute the trace, we obtain
$$M_p = \sum_i \langle im\,|(H-E_0)^p|im\rangle \tag{8.13}$$
$$= \sum_{i \neq j \neq ... l \neq i} \langle im|H-E_0|jm'\rangle\langle jm'|H-E_0|km''\rangle$$
$$... \langle lm^{p-1}|H-E_0|im\rangle.$$

If we then retain only the two-centre integrals, assume that all the atoms have the same (spherically symmetrical) atomic potential V_i, and write
$$\langle im|V_j|im\rangle = A_m(\mathbf{R}_i, \mathbf{R}_j),$$
$$\langle im|V_j|jm'\rangle = B_m^{m'}(\mathbf{R}_i, \mathbf{R}_j),$$
we have
$$M_1 = \int g_2(\mathbf{R}_1, \mathbf{R}_2)\,A_m(\mathbf{R}_1, \mathbf{R}_2)\,d_3\mathbf{R}_1 d_3\mathbf{R}_2,$$
$$M_2 = \int g_2(\mathbf{R}_1, \mathbf{R}_2)\,B_m^{m'}(\mathbf{R}_1, \mathbf{R}_2)^2\,d_3\mathbf{R}_1 d_3\mathbf{R}_2$$
$$+ \int g_3(\mathbf{R}_1, \mathbf{R}_2, \mathbf{R}_3)\,A_m(\mathbf{R}_1, \mathbf{R}_2)\,A_m(\mathbf{R}_1, \mathbf{R}_3)\,d_3\mathbf{R}_1 d_3\mathbf{R}_2 d_3\mathbf{R}_3,\ \text{etc.},\,...$$
$$\tag{8.14}$$

In general, if we neglect the smaller α integrals with respect to the larger β ones

$$M_p \simeq \sum_{i \,\neq\, j \,\neq\, k \,...\, \neq\, i} \langle im|V_i|jm'\rangle \langle jm'|V_j|km''\rangle ... \langle lm^{(p-1)}|V_l|im\rangle. \quad (8.15)$$

These equations show indeed that the shift M_1 of the d-band is given by the α integrals, that its width M_2 is given by the β ones (if larger than the α ones). But it indicates that more details of the shape of the band require a knowledge of more than two body correlation functions g_p $(p > 2)$. Indeed the knowledge of the pth moment M_p of the density of states requires a knowledge of the correlation functions g_q up to order p if one neglects the asymmetry produced by the α integrals, and up to order $p+1$ if one includes the α intregrals.

8.1.6 b *Crystalline metals.* In perfect crystalline structures, the correlation functions g_p are known exactly to all orders, and $n(E)$ can therefore be computed exactly within the tight-binding approximation, at least in principle. In actual fact, $n(E)$ turns out to have a fairly complex behaviour, with many peaks: it is therefore better computed directly, using the lattice symmetries.

$\psi(E)$ can then be taken as a *Bloch function.* Within the tight-binding approximation, this means that

$$a_{im} = a_{im}^0 e^{i\mathbf{k}\mathbf{R}_{i0}}; \quad (8.16)$$

$\mathbf{R}_{i0}$ is the centre of the lattice cell to which the ith atom belongs, and the coefficient a_{im}^0 only depends on the relative position of atom i in its cell. The set of $5N$ equations (8.6) reduces to a set of $5q$ equations in a_{im}^0, if there are q atoms per cell. These give rise to $5q$ sub-bands $E(\mathbf{k})$, each with N/q possible values of $\mathbf{k}$. And, at least if the potentials V_i are spin invariant, each of these states is doubly degenerate in spin.

If now spin-orbit coupling is included, the spin degeneracy is in principle lifted, and one has then a 10 by 10 determinant to solve for each $\mathbf{k}$-value. Owing to the invariance of the Hamiltonian under time reversal K, the $E(\mathbf{k})$ curves are symmetrical in $\mathbf{k}$-space, each $\mathbf{k}$-state corresponding to a $-\mathbf{k}$-state of equal energy and (mainly) opposite spin direction. This is the Kramers degeneracy (see Elliott, 1954). If each atom site is a centre of symmetry for the crystal, as in cubic structures, the product of the corresponding spatial inversion J by the time reversal K transforms each $\mathbf{k}$-state into a state with the same **k** and (mainly) opposite spin: the double degeneracy of each k-state is then preserved.

The density of states curve $n(E)$ can be deduced from the knowledge of the $E(\mathbf{k})$ curves, since

$$n(E) = \int_{S(E)} \frac{dS}{|\nabla_{\mathbf{k}} E|}, \qquad (8.17)$$

where $S(E)$ is the surface of energy E in the first Brillouin zone. This has been done in tight binding with some accuracy only in the cubic (B.C.C., F.C.C.) phases.

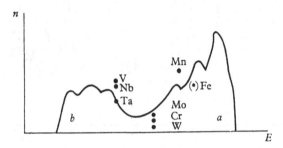

Fig. 8.6. Density of states $n(E)$ for the d-band in B.C.C. chromium: • • •: values of $n(E_M)$ deduced from the electronic specific heat of pure metals.

Results in the tight-binding approximation are pictured in Figs. 8.6 and 8.7. Because the bandwidth is large compared with spin-orbit corrections, these do *not* change the main features of $n(E)$ (Fig. 8.7 a, b,c).

Even in these cases, the results cannot be taken as very accurate, owing to uncertainties in the values of the α and β integrals and corrections for sd-mixing. Two features emerge, however, from these computations:

First of all *the band is subdivided in the* B.C.C. *structure* into a bonding b half and an antibonding a half connected by a region of low density of states (Friedel, 1955, 1958, 1962; Belding, 1959; Asdente & Friedel, 1961, 1962). This strong subdivision merely reflects the fact that, in such a structure, the neighbours of each atom are placed in a favourable position for each d-wave-function to build simultaneously a great number of bonds, as in Fig. 8.3. As already emphasized, the computed subdivision is not affected by spin-orbit corrections; it has nothing to do with crystal-field splitting; indeed, the component $n_{e_g}(E)$ and $n_{t_{2g}}(E)$ densities have nearly the same form as the total

$n(E)$ density, both showing the bonding and antibonding peaks connected by a minimum of density.

Large values of $n(E_M)$ at the Fermi level, separated by low ones

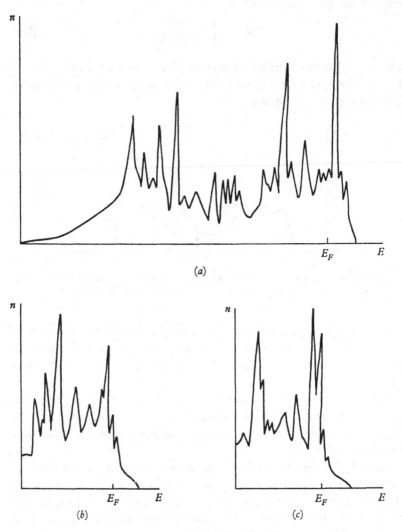

Fig. 8.7. Density of states $n(E)$ for the *d*-band in F.C.C. metals: *a*, Ni; *b*, Pd; *c*, Pt.

for elements in the middle of the series (Cr, Mo, M) are indeed a feature of B.C.C. metals and alloys, noticeable in measurements of $n(E_M)$ by electronic specific heat γ (Blanpain, 1957; Heiniger *et al.*

1966) (Fig. 8.6) or—in non-magnetic metals—by magnetic suscepti-bility χ (Kriessman & Callen, 1954; Vogt, 1966). No exact agreement must be expected between the computed and measured $n(E_M)$: fairly large corrections are expected from the values of γ and χ computed in the Hartree scheme, owing to Van Vleck and correlation corrections (cf. § 8.3) and, for γ, to coupling with phonons (Clogston, 1964; Krebs, 1963). The existence of the minimum of $n(E)$ in the middle of the band, and the order of magnitude of $n(E)$ in the bonding and antibonding peaks seem however well established; they agree again, for the band-width, with about 5 to 10 eV for the first series and perhaps a little more for the second and third series.

The second feature is the existence of a *sharp peak* in $n(E)$ near the top of the band in F.C.C. *metals* (Fig. 8.7). This peak does not correspond to a Van Hove singularity (Allan, 1967, cf. Van Hove, 1953) in $n(E)$. But it is not very sensitive to spin-orbit coupling. For a reasonable number of d-holes, it is expected to be not much below the Fermi level (para-magnetic) in Ni, Pd and Pt. There seems to be some evidence for the existence of such a peak at least in Pd and Pt (cf. § 8.3).

It would be of interest to have similar results for other phases, especially the H.C.P. one.

8.1.6c   *Liquid metals.* Little is known about the nature of their short-range order (Cyrot, 1967); it is therefore impossible to compute $n(E)$ accurately. It is of course clear that the picture of Fig. 8.4 should still apply, with energy limits E_a and E_b not much different from those of close-packed crystalline structures; one also expects the band not to show any strong splitting such as in the B.C.C. structure, because of the more random nature of the interatomic bonds.

Indeed, simple approximations for the pair correlation function g_2 show that the width as measured by the second moment M_2 of the d-band of a liquid should be smaller but of the same order of magnitude as that of a compact (F.C.C.) crystal structure with the same atomic volume. But it is clear from the earlier discussion that a more detailed computation of $n(E)$ would require a better knowledge of the many-body atomic correlations than at present measurable. Contrary-wise, a measurement of $n(E)$ would throw some light on these correlations. Practically no work has been done so far in this field.

8.1.7 Form of the Fermi surface in crystals

This is even less well known than the $n(E)$ curves, as it requires a detailed study of the $E(k)$ curves in crystals, and is more sensitive to corrections due to spin-orbit coupling or sd-mixing.

For face-centred cubic metals, the top of the d-band is computed in tight binding to be on the square faces of the Brillouin zone (Fletcher & Wohlfarth, 1951) (Fig. 8.8). One can then hope that, for elements such as Ni at the end of the series, the Fermi structure of the d-band does not mix too much with the s-band, which should be fairly spherical and concentrated more at the centre of the Brillouin zone.

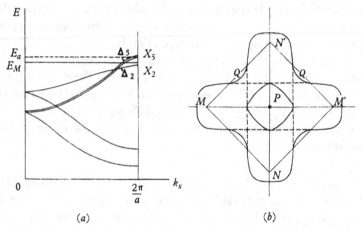

(a) (b)

Fig. 8.8. $E(\mathbf{K})$ for the d-band of (paramagnetic) Ni in the tight-binding approximation: (a) along $\{100\}$; (b) Fermi surface on the square faces.

The top E_a of the d-band is then made up of the diagonals of the square faces (MM', NN', Fig. 8.8b), when spin-orbit corrections are negligible, as in Ni. The surfaces of constant energy near to E_a are then elongated and intersecting hyperboloids; this two-dimensional geometry corresponds to a *finite* density of states $n(E_a)$ at the band edge (Gautier, 1962) (Fig. 8.7a). At the Fermi level E_M of Ni, fourth-order terms in $E_a - E_M$ should have removed the intersections of the hyperboloids, producing small pockets P of holes centred on the square faces, and more complex surfaces Q which possibly interact somewhat with the Fermi surface of the s-band (Friedel, 1955, 1958, 1962).

A somewhat similar situation should hold for Pd and Pt, except for the increasing spin-orbit correction (Lenglart, 1967; Allan, 1967), which removes the linear degeneracy at the top E_a of the d-band, thus producing a parabolic variation of $n(E)$ near E_a (Fig. 8.7b, c). In Pt the spin-orbit correction is larger than the distance from the Fermi energy E_M to the top E_a of the band for a likely number of d-holes. As a result, one expects the d-holes to be built up with atomic d-orbitals of nearly pure $J = L + S = \frac{5}{2}$ character (Friedel et al. 1964). This is indeed obvious in X-rays absorption spectra (Fig. 8.5), where only the L_{II} absorption from $2p$ ($J = \frac{3}{2}$) state gives rise to a strong absorption band. As expected from the numbers of Table 8.1, the L_{II} and L_{III} absorptions, from both $2p$-states ($J = \frac{3}{2}$ and $J = \frac{1}{2}$) give rise to equal absorption bands in Ni, while Pd is an intermediary case (Bearden & Snyder, 1941; Mott, 1949).

The Fermi surfaces of Pd and Pt should have forms somewhat similar to those pictured for Ni Fig. 8.8, but with the spin-orbit coupling increasing the distance between the pockets P and the sheets Q; as a result, in Pt, the pockets P might well vanish and the sheets Q should look more nearly like spheres centred on the centre of the square faces. The question whether small pockets arising from the X_2 branch, Fig. 8.8, exist, is also not completely settled.

Measurements of the Fermi surface of (ferromagnetic) Ni, Pd and Pt are still incomplete (Fawcett & Reed, 1962; Tsui & Stark, 1966; Vuillemin, 1966; Alekseevkii et al. 1964; Ketterson et al. 1966). However, something like a nearly spherical Fermi surface centred at the origin for the s-band and d-pockets near the centre of the square faces seem well-established features. The Fermi surfaces of Rh and Cr, Mo, W, have also been studied (Coleridge, 1965; Rayne et al. 1962, 1963, 1964; Fawcett & Walsh, 1962; Watts, 1964b; Sparlin & Marcus, 1966). They show as expected, a complex behaviour, that has been interpreted in terms of the band structure, with large sheets of Fermi surface caused by the d-band (Sawada & Fukuda, 1961; Iwamoto & Sawada, 1962; Tachiki & Nagamiya, 1963).

8.1.8 *sd*-mixing—the case of noble metals

We have assumed so far that the sd-mixing effects were small compared with the d-bandwidth (Fig. 8.2). There are good reasons for this. First of all the sd-mixing can be thought as due to the matrix elements of a small perturbation potential v that brings the d-band in the same energy range as the sp-conduction band. The situation is very similar to that of the widening of a d-shell into a virtual d-shell when a transitional impurity such as Mn is introduced into a matrix such as Cu. The same matrix elements are involved; they lead to widths of the order of a few eV, thus smaller than the d-bandwidth w. Secondly, the fair success achieved in explaining some features of the electronic structure using a pure d-band would be hard to understand if this was very strongly perturbed by sd-mixing. Thirdly, band computations by various methods (O.P.W., A.P.W.) can be fitted to a mixture of narrow d-bands described in tight binding and free or nearly free

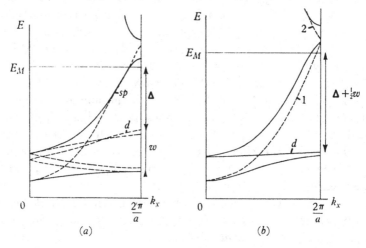

Fig. 8.9. sd-mixing in Cu: (a) real d- and sp-components; (b) schematic.

electron sp-bands, using an sd-mixing perturbation of the order of 1 eV. Finally, an interesting if extreme case is provided by the noble metals Cu, Ag, Au. These metals are known to have a full d-band, and indeed the strong optical absorption observed above an energy Δ given in Table 8.2 has been interpreted (Mott & Jones, 1936; Friedel, 1952) as due to transitions from the d-band to the Fermi level E_M (Fig. 8.9.)

As Δ is not much larger than the likely sd-mixing parameter, one expects the Fermi electrons to have an appreciable d-character; this should lead to specific effects, which would give a measure of the mixing strength. We shall detail some of them, to show that they point to a v_{sd} matrix element of the order of at most 1·5–2 eV.

Most of the Fermi surface is far enough from Brillouin zone limits for the sp-band states to be represented by one free electron or O.P.W. function ψ_1. Furthermore, the Fermi energy is far enough from the d-band for its width w to be neglected when studying the Fermi states, if its distance is taken as $\Delta + \frac{1}{2}w$ (Fig. 8.9b). Standard perturbation theory then shows that the sd-mixing shifts the energies of the sp-band by an amount

$$\delta E_1 \simeq \frac{v_{sd}^2}{E_1 - E_M + \Delta + \frac{1}{2}w}.$$

This amount decreases with increasing energy E_1; this can explain the large *effective masses* m^* measured by electronic specific heat or infra-red absorption (Rayne, 1956; Roberts, 1960):

$$\frac{m^*}{m} = \frac{N}{N_{\text{eff}}} = \frac{1}{1 + \left(\dfrac{d\,\delta E_1}{dE_1}\right)_{E_M}} \simeq \frac{1}{1 - \dfrac{v_{sd}^2}{(\Delta + \frac{1}{2}w)^2}}.$$

With $\Delta + \frac{1}{2}w \simeq 4\,\text{eV}$, observed increases of the order of 50% are explained if $|v_{sd}| \simeq 1\cdot 5\,\text{eV}$.

Table 8.2. *Energy difference* Δ *between the top of the d-band and the Fermi level, in noble metals* (*in* eV)

	Cu	Ag	Au
Δ	2	4	2·5

The Fermi surface of noble metals is known to touch the Brillouin zone boundaries near the centre of the (111) faces (Harrison, 1960). sd-mixing can again explain the order of magnitude of the corresponding large *energy gaps* on the Brillouin zone boundaries. We use again standard perturbation theory, with now two sp-bands $\psi_1(E_1)$ and $\psi_2(E_2)$ and the simplified model of Fig. 8.9b, where one neglects the d-bandwidth and the energy gap between the two sp-bands. sd-mixing then leads to an energy gap

$$G \simeq \frac{2v_{sd}^2}{\Delta + \frac{1}{2}w}.$$

This is of the required order of about $2\,\text{eV}$ if again $|v_{sd}| \simeq 1\cdot 5\,\text{eV}$ to $2\,\text{eV}$. Furthermore, it can be shown that the form of the Fermi surface near to the Brillouin zone boundary should be very similar to what one would compute by the standard two-o.p.w. procedure.

8.2 COHESION AND LATTICE STRUCTURES IN TRANSITIONAL METALS†

In this section the cohesion of the main condensed phases will be discussed. This will also include an account of their elastic properties, their phonon dispersion curves and finally their metastability.

The physical properties invoked so far, optical and X-rays spectra, electronic specific heat, Pauli paramagnetism, involved only the change of energy of a few electrons, for whom the Hartree approximation is reasonable. In problems involving cohesion, all the valence electrons are in principle involved. The question then arises how to take into account the electron–electron correlations, which certainly play a role. We shall first use a *Wigner–Seitz scheme*, which will then be discussed.

8.2.1 Cohesive energy of transitional metals, latent heats of phase changes

The regular variation of *cohesion* with the filling of the d-band, peaking to a maximum for the refractory metals in the middle of the series, is a well-known feature of the transitional metals (Smithells, 1949).

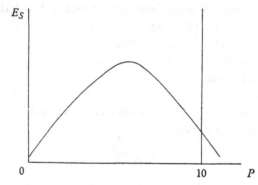

Fig. 8.10. Cohesive energy per atom E_S versus the atomic number in the transitional series (schematic, 2^d and 3^d series).

It is especially clear in the second and third series, where no magnetic complications occur in the solid state (Fig. 8.10). The magnitude of this cohesion peak, and its regularity, clearly shows that it must be related to d-band formation in the solid, and that factors such as the

† Cf. the general discussion of cohesion and structure in §1.6.

changes in Coulomb correlation, magnetism, etc., when one goes from
the free atoms to the solid are only details of secondary importance.

We shall therefore use in this field a simple-minded Hartree-like
scheme, where the total energy is obtained by summing the energies
of the occupied one-electron states. We shall discuss the validity of
this approximation in §8.4.2. We shall show that it is justified if the
shift of the band is small compared with its width and if the atomic
potentials V_i are taken as those of singly ionized atoms, as for free
atoms; this is indeed usually assumed in actual computations (cf.
§8.1).

The cohesive energy E_S per atom is then simply given by

$$E_S \simeq -\frac{2}{N} \sum_{k_{occ}} E_k + pE_0 = 2 \int^{E_M} (E_0 - E)\, n(E)\, dE \qquad (8.18)$$

if there are p d-electrons per atom.

It is clear from this equation and from Fig. 8.4, that, in an incompletely filled d-band, more 'bonding' than 'antibonding' states
are occupied. The effect should be a maximum when all the bonding
states are occupied, and no antibonding states, i.e. for a Fermi level
E_M equal to the atomic energy E_0. Fig. 8.10 shows that the peak of
cohesion occurs near to the middle of the transitional series, and that it is
fairly symmetrical, at least in the second and third series. According to
(8.18), this is consistent with *a d-band moderately symmetrical in energy
with respect to E_0*, hence with α integrals small compared with the
β ones; *the shift s of the band is small compared with its width w.*

With

$$p = 2 \int^{E_M} n(E)\, dE \qquad (8.19)$$

(8.18) gives indeed $dE_S/dp \gtrless 0$ for $E_M \lessgtr E_0$.

The experimental data can be roughly fitted to a shifted parabola

$$E_S(p) \simeq Ap(10-p) + Bp$$

if one assumes that p increases linearly from 0 to 10 through a series.
This would correspond to a rectangular d-band, $n = $ const, of width
$w = 20A$, shifted by $s = B$ towards low energies. Table 8.3 shows
that the values of the widths w thus deduced are of a right order of
magnitude, and that the shifts s are small.

It is easy to check (Friedel, 1964) that, for a given width, $E_S(p)$

is not very sensitive to the details of $n(E)$. This explains why E_S varies fairly smoothly through a series, though there are differences in lattice structure from one element to the next. It also explains why the latent heats of phase changes, melting included, are usually much smaller than the heat of sublimation: because of their numerous lobes, the atomic d-functions are not very sensitive to the number and direction of interatomic bonds they make, as long as their number is high and the overlaps similar.

Table 8.3. *Energy with w and shift s of the d-band, estimated from cohesion (in eV)*

	1st series	2nd series	3rd series
w	4	5	6
s	0·25	0·3	0·35

The *relative stability of various possible phases*, for a given metal, has not been studied in detail. Besides the study of metastability described below, one can only point out that the especially strong splitting of the B.C.C. phase, Fig. 8.6, must make it more stable than other phases for elements in the middle of the series; this might explain the B.C.C. structure observed for the refractory metals V, Nb, Ta; Cr, Mo, W. Similarly, the presence of a strong peak in the upper part of $n(E)$ in F.C.C. metals should tend to stabilize this structure for metals at the end of the series, when the Fermi level falls above or in that peak. This might possibly explain at least in part the F.C.C. structure observed in Ni, Co, Pd, Pt. Also the reduction of second moment of $n(E)$ by melting should lead to a positive latent heat of melting L_m. L_m is indeed roughly proportional to the energy of cohesion E_S, with a peak in the middle of the transitional series. The coefficient of proportionality points to a reasonable reduction of the bandwidth $M_2^{\frac{1}{2}}$ by melting, of the order of $\frac{1}{10}$.

Finally, it is clear in Fig. 8.10 that the d-band, although full, still plays a role in the *cohesion of the noble metals* (Friedel, 1952). This can be accounted for by the sd-mixing described in §8.1.8.

We have seen that the off-diagonal elements v_{sd} responsible for sd-mixing broaden the band structure without shifting its average energy. The energy increase δE_1 of the empty states above E_M thus corresponds to an equal lowering of energy of the occupied states.

Summing δE_1 over all empty states of the sp-band in the Brillouin zone gives the contribution to cohesion per atom of the sd-mixing:

$$\delta E_s = \frac{2V}{8\pi^3 N} \int_{\text{Fermi sphere}}^{\text{Brillouin zone limits}} \delta E_1 d_3 R_1 \simeq \gamma \frac{v_{sd}^2}{\Delta + \frac{1}{2}w},$$

where γ is a factor of the order of $\frac{1}{2}$.

A contribution of the order of 1 eV per atom is difficult to account for in Cu by any other means than sd-mixing. With $\Delta + \frac{1}{2}w \simeq 4$ eV, this leads to $|v_{sd}| \simeq 1\cdot5$ eV, in good agreement with previous estimates.

8.2.2 Energy of distortion

Equation (8.18) can be used to study the energy change produced by a small lattice distortion

$$\mathbf{u}_i = \mathbf{R}_i' - \mathbf{R}_i. \tag{8.20}$$

For simplicity, we consider a lattice with one atom per cell; we neglect the degeneracy of the atomic function $|i\rangle$, and the α integrals with respect to the β ones (8.6). We shall also assume that the distortion merely shifts the atomic potentials V_i and functions $|i\rangle$ to their new positions, *without change of form* (Lomer, 1966; Labbé, 1968).

Writing for a wave-function in the perturbed lattice

$$|\psi_{(E')}'\rangle = \sum_i a_i' |i'\rangle, \tag{8.21}$$

(8.6) gives

$$(E' - E_0) a_i' \simeq \sum_{j' \neq i'} \langle i' | V_{j'} | j' \rangle a_j'.$$

A Fourier analysis

$$a_i' = \frac{1}{\sqrt{N}} \sum_{\mathbf{k}} a_{\mathbf{k}}' e^{i\mathbf{k}\mathbf{R}_i}$$

and a development of the overlap integral with respect to the displacements easily give

$$(E' - E_k) \sum_{\mathbf{K}_{RR}} a_{\mathbf{k}+\mathbf{K}_{RR}}',$$
$$= \sum_{\mathbf{k}} a_{\mathbf{k}}' \sum_j e^{i(\mathbf{k}'-\mathbf{k})\mathbf{R}_j} \sum_{i \neq j} (\mathbf{u}_j - \mathbf{u}_i)(\nabla_{ji} V_{ij}) e^{i\mathbf{k}(\mathbf{R}_j - \mathbf{R}_i)} + O_2(\mathbf{u}); \tag{8.22}$$

$\mathbf{K}_{RR}$ are the periods of the reciprocal lattice, and $E_{\mathbf{k}}$ the energy of the Bloch state of vector $\mathbf{k}$.

For a *plane distortion wave*

$$\sum_i \mathbf{u}_i = \epsilon \mathbf{u} \Sigma_i e^{(i\mathbf{K}\mathbf{R}_i)}, \tag{8.23}$$

where **u** is a unit vector, (8.22) reduces to

$$(E' - E_\mathbf{k}) \sum_{\mathbf{K}_{RR}} a'_{\mathbf{k}+\mathbf{K}_{RR}} = \epsilon [\mathbf{u} \sum_{i \neq 0} (1 - e^{-i\mathbf{K}\mathbf{R}_i} e^{i\mathbf{k}\mathbf{R}_i} \nabla_{i0} V_{0i}] \sum_{\mathbf{K}_{RR}} a'_{\mathbf{k}+\mathbf{K}_{RR}-\mathbf{K}}. \quad (8.24)$$

Using the fact that, when $\mathbf{u} \to 0$, $a'_k \to a_k \delta_{kk'}$, the total change of energy by distortion is easily obtained to be

$$\delta E_\mathbf{k} = 2 \sum_{\mathbf{k}_{occ}} (E'_\mathbf{k} - E_\mathbf{k}) = 2\epsilon^2 \sum_{\mathbf{K}_{RR}} \sum_{\mathbf{k}_{occ}} \left\{ A_\mathbf{k} + B_\mathbf{k}^2 \frac{1}{E_\mathbf{k} - E_{\mathbf{k}+\mathbf{K}_{RR}-\mathbf{K}}} \right\} + O_4(\epsilon)$$

$$(8.25)$$

with

$$A_\mathbf{k} = \sum_{i \neq 0} V_{0i} g(\mathbf{u}, \mathbf{R}_i) \cos \mathbf{k} \mathbf{R}_i \sin^2 \frac{\mathbf{K}\mathbf{R}_i}{2}, \quad (8.26)$$

$$\left. \begin{array}{c} B_\mathbf{k} = q \sum_{i \neq 0} V_{0i} \frac{\mathbf{u}\mathbf{R}_i}{R_i} \cos \frac{(\mathbf{K} - 2\mathbf{k})\mathbf{R}_i}{2} \sin \frac{\mathbf{K}\mathbf{R}_i}{2}, \\[2ex] g(\mathbf{u}, \mathbf{R}_i) = q^2 \left(\frac{\mathbf{u}\mathbf{R}_i}{R_i} \right)^2 + q \frac{(\mathbf{u}\mathbf{R}_i)^2 - R_i^2}{R_i^3} \end{array} \right\} \quad (8.27)$$

and use has been made of the exponential decrease of the overlap integral $V_{0i} \propto \exp(-qR_i)$.

Before applying these equations, it is worth noting the following points: (*a*) They are based on (8.18), and are thus better suited to study *shear* waves than longitudinal waves, where the local changes of atomic volume might lead to sizeable contributions from changes in correlation energies. This is especially true for the uniform modes. (*b*) They neglect possible *sd*-charge transfer owing to the distortion, which might lower the energy. (*c*) They neglect the changes in the self-consistent V_i due to distortion. (*d*) Finally, and more important, a contribution due to the *valence s-electrons* must be added. This must be of the same order of magnitude as in normal (non-transitional) metals.

8.2.3 Elastic constants. Phonon dispersion curves. Metastability of lattice structures

As long as the distortions $\mathbf{u}_i$ are small, each of the Fourier component (8.23) can be treated separately. The new period introduced by such a plane wave of strain produces energy gaps at the new Brillouin zone boundaries (Fig. 8.11). As the mixing of the two bands is symmetrical in energy, this will necessarily decrease the average energy of the occupied states, whatever the position of the Fermi surface. This lowering of energy will vary most rapidly with the position of the Fermi surface when this nearly touches the new Brillouin zone limit over a large region.

BB

These qualitative predictions can be made somewhat more accurate with the help of (8.25).

It should first be pointed out that, although each contribution $E'_k - E_k$ diverges when E_k approaches the new Brillouin zone limit, the sum (8.25) converges and gives the correct answer to second order in perturbation, whatever the position of the Fermi level. This result holds for Fermi surfaces in two or three dimensions (Pick & Blandin, 1964).

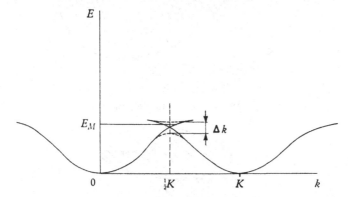

Fig. 8.11. Energy gap produced at new Brillouin zone boundary by a plane wave of strain of wave-vector **K**.

Some general properties of $\delta E_{\mathbf{K}}$ are also useful to note:

1. The summation extends for **k** over the volume V within the Fermi surface, and for $\mathbf{k}' = \mathbf{k} + \mathbf{K}_{RR} - \mathbf{K}$ over the same volume V', shifted by $\mathbf{K} - \mathbf{K}_{RR}$ (Fig. 8.12).

2. In V', only the regions outside V contribute to the term in B_k^2, because in the common volume V'' contributions where **k** and **k**′ interchange cancel each other. As $B_{\mathbf{k}}$ is real, the second term in $\delta E_{\mathbf{K}}$ is necessarily *negative*.

3. An important contribution to the second term in $\delta E_{\mathbf{K}}$ comes obviously from $E_{\mathbf{k}} \simeq E_{\mathbf{k}'}$. From the previous remarks, this comes from the regions in V and V' near to the intersection line I of the two Fermi surfaces S and S'.

4. For a fixed direction of $\mathbf{K} - \mathbf{K}_{RR}$, this contribution varies most rapidly with the *length* of $|\mathbf{K} - \mathbf{K}_{RR}|$ when the two Fermi surfaces S and S' just touch each other (Fig. 8.13a, b).

5. This rapid variation will be especially important for special *directions* of $\mathbf{K} - \mathbf{K}_{RR}$ when the parts of S and S' in contact have nearly equal *curvatures* (Fig. 8.14a, b), or possibly several points of contact (Fig. 8.13b). It is clear from these examples that the condition for a

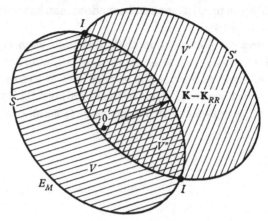

Fig. 8.12. Summation for $\delta E_{\mathbf{K}}$ and $\chi_{\mathbf{K}}$.

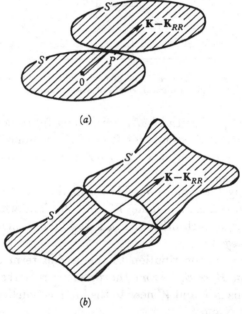

(a)

(b)

Fig. 8.13. Conditions for strong variation of $\delta E_{\mathbf{K}}$ or $\chi_{\mathbf{K}}$
with $|\mathbf{K} - \mathbf{K}_{RR}|$.

Brillouin zone limit associated with **K** to be tangent to the Fermi surface only holds for simple cases such as Fig. 8.14a, or for cubic crystals.

6. When, as in Fig. 8.13 or 8.14a, the surfaces S and S' are convex with well defined curvatures near their point of contact P, a develop-

ment of $E(\mathbf{k})$ near P shows that $\delta E_{\mathbf{K}}$ has a logarithmic anomaly with a vertical tangent for the condition of contact

$$\delta E_{\mathbf{K}} \simeq M^2(v)\,(m_1 m_2)^{\frac{1}{2}}\,v\ln|v| + N(v), \qquad (8.28)$$

M and N are slowly varying functions of v, the difference between $|\mathbf{K} - \mathbf{K}_{RR}|$ and its value for contact; m_1 and m_2 are the two effective masses associated with the two curvatures of the Fermi surface at point P. Rather similar anomalies obtain for more complex geometry.

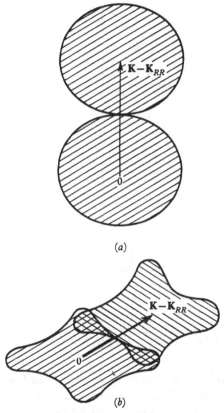

(a)

(b)

Fig. 8.14. Examples of conditions for maximum variation of $\delta E_{\mathbf{K}}$ or $\chi_{\mathbf{K}}$ with $|\mathbf{K} - \mathbf{K}_{RR}|$.

7. Finally, for $K \to 0$, (8.25) shows that $\delta E_K \propto \epsilon^2 K^2$, as expected.

From these general remarks, we can conclude that (a) The d-electrons contribute the term (8.25) to the energy $\delta E_{\mathbf{K}} \propto \epsilon^2 \omega^2$ of plane distortion waves, and especially shear waves. This is the sum of a first term which varies continuously with K, and a second negative term with a logarithmic anomaly. (b) As a result, logarithmic anomalies should occur on the phonon dispersion curve for the conditions of Figs. 8.13 and 8.14, as shown schematically in Fig. 8.15 (Kohn anomalies, see Kohn 1959a). (c) It might be that, for some metals, the total contribution of the d-electrons to an elastic constant is negative and larger than the normal positive contribution from the s-electrons.

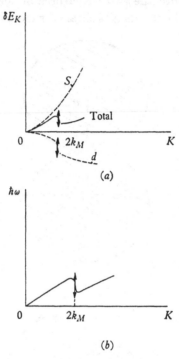

Fig. 8.15. Phonon dispersion curve for a transition metal (schematic):
(a) energy $\delta E_K \propto \hbar^2 \omega^2$; ($b$) frequency $\hbar\omega$.

In such a case, $\omega(K)$ is imaginary at small K, and the structure considered should be unstable with respect to the corresponding *uniform* mode of strain (Fig. 8.16a). This is a straight extension to band structures of the (classical) Jahn–Teller effect for impurity states. It might also be that this occurs near a Kohn anomaly. The structure is then unstable with respect to a periodic distortion (Fig. 8.16b). The unstable modes are *near to* but not exactly at the conditions of Figs. 8.13 or 8.14. The most unstable modes occur near to conditions of contact over a large area (Fig. 8.14). If neither of these situations occurs, the structure considered should be at least *metastable*. It could still be unstable with respect to strongly different lattice structures. Also, if conditions of instability prevail (Fig. 8.16 a, b), the previous analysis, only valid for small distortions, cannot predict what is the stable structure.

Some anomalies in the phonon dispersion curves of W have been attributed to Kohn anomalies (W. M. Lomer, private communication); a low temperature phase change observed in V has been explained as

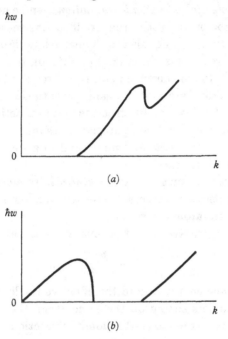

(a)

(b)

Fig. 8.16. Conditions of complete instability of a lattice structure:
(a) versus a uniform mode; (b) versus a periodic mode.

a Jahn–Teller effect. However, it must be admitted that quantitative comparison between theory and experiment is lacking in this field except in the very special case of transitional compounds, with βW structure (V_3Si, Nb_3Sn, ...) (Labbé & Friedel, 1966; Barisic & Labbé, 1968). Detailed study of the metals at the end of the transitional series would be of especial interest.

8.2.4　Electron–electron correlations

Finally, it is of interest to discuss the validity of the scheme which has been used in this section. First the *pure Hartree scheme* used in §8.1, assuming that the d-electrons are completely uncorrelated, clearly allows too large charge fluctuations: it allows electrons to be too near to each other within an atom so that *intra-atomic correlations* ought to be considered, and it allows too many electrons to jump at the same time onto the same atom so that *kinetic interatomic correlations* ought to be considered.

By analogy with the simpler cases of free electrons and atoms, it is clear on the other hand that the correlation effects are rather short range, so that *potential interatomic correlations* can be neglected.

The *Virial theorem* can justify the modified Hartree scheme used in §8.2: the cohesive energy equals the change in kinetic energy between the metal and free atoms; from (8.9) and (8.10), this is obtained by summing $E_0 - E_k$ for the occupied states if three conditions are met: one-electron functions are used; the same potentials V_i are used in both cases; $|\alpha| \ll |\beta|$. The last condition seems met in practice. The second condition, fulfilled in most computations, assumes, reasonably, the same *interatomic correlations* in a metal and in a free atom with the same number p of carriers.

One-electron wave-functions neglect *kinetic intra-atomic correlations*. The large fluctuations of carriers due to their randomness thus affects their average intra-atomic correlations.

The atomic concentration of atoms with n carriers would be (Friedel, 1952)

$$C_n = \frac{p^n}{n! \, e^p}. \tag{8.29}$$

If u is the average energy due to the effective Coulomb interactions between two carriers sitting on the same atom, the contribution of the atoms with n carriers to the Coulomb interaction would be

$$\epsilon_n = \frac{n(n-1)}{2} C_n u \tag{8.30}$$

and the total contribution per atom

$$\epsilon = \sum_n \epsilon_n = \tfrac{1}{2} p^2 u, \tag{8.31}$$

as one would expect from the fact that there are on the average $\tfrac{1}{2} p^2$ pairs of carriers per atom. Taking energy ϵ_n to be much the same in the condensed phase as in free atoms or ions, the corresponding term for the free atoms will be $\tfrac{1}{2}\{p(p-1)\} u$. Thus, at constant number p of d-electrons per atom, the charge fluctuations allowed by complete neglect of the kinetic interatomic correlations increase the energy by $\delta V = \tfrac{1}{2} p u$ per atom. For complete disorder of carriers to be reasonable, this should be small compared with the cohesive energy, thus u must be small compared with the bandwidth.

If on the contrary the kinetic interatomic correlations were dominant, they would reduce the charge fluctuations on each atom to at most unity: only the two ionic charges nearest to the average number p would be allowed. Most of the bonding interatomic kinetic energy would be suppressed, and what remained would not vary much with the progressive filling of the d-band. It would indeed vanish for every integral value of p.

The real situation is clearly intermediary between these two extremes; on the one hand, the general behaviour of the cohesive energy, Fig. 8.10, and its agreement with (8.18) show the interatomic kinetic energy of the d-electrons to dominate cohesion, and to be not far from that given by the scheme of §8.2, neglecting the kinetic interatomic correlations; on the other hand, the Coulomb correlation energy u between d-electrons is probably too large compared with the bandwidth w to be treated as a very small perturbation.

With u of the order of a sizeable fraction of w, one expects the large charge fluctuations, which produce most of the Coulomb term $\delta V = \frac{1}{2}pu$, to be suppressed without the average kinetic energy of the d-electrons to be much reduced, in agreement with the scheme used in this chapter. Table 8.4 shows for instance that, for $p = 5$, most of the energy δV comes, in the Hartree scheme, from the charge fluctuations producing ions with 9, 10 or more electrons. Thus, with

$$\delta V_n = \left[\frac{n(n-1)}{2} - \frac{p(p-1)}{2} \right] C_n u,$$

we see that, in the Hartree scheme, the small charge fluctuations $2 \times 5 \to 4 + 6$, $3 + 7$ or $2 + 8$ only account for 12% of δV in this case.

In conclusion, the success of the scheme of §8.2 in explaining the cohesion of transitional metals suggests that only the large charge fluctuations are suppressed by correlations. This is consistent with a Coulomb interaction energy u between d-electrons somewhat less than the bandwidth w.

No mention has been made so far of *exchange*. It is clear experimentally that magnetic phenomena, both in the solid and in the free atom, are of an order of magnitude less than the cohesive energies discussed here. This is not surprising, as the antisymmetrization of the wave-functions has but a small influence on the correlations. It forbids two d-electrons of parallel spins to sit at the same time on the same d-orbital. This only touches, on the average $\frac{1}{10}$ of all the possible configurations of an electron pair, and can therefore only play a minor role. It prevents for instance configurations with more than 10 electrons per atom; but these are anyway of little importance in the Hartree scheme (cf. Table 8.4). It also introduces only moderately small exchange energy terms between two d-electrons of parallel spins which are on different d-orbitals.

Table 8.4. Probability C_n, in the Hartree scheme, for having an ion with n electrons. Proportion $\delta V_n/\delta V$ of the Coulomb contribution to cohesion. Case of $p = 5$

n	C_n	$\delta V_n/\delta V$
0	0·007	—0·03
1	0·034	−0·14
2	0·084	−0·30
3	0·141	−0·39
4	0·261	−0·41
5	0·176	0
6	0·147	0·29
7	0·105	0·46
8	0·065	0·47
9	0·036	0·38
10	0·018	0·26
11	0·008	0·15
...	...	...

One can therefore neglect such complications in discussing the absolute magnitude of cohesion. Exchange terms should only play an important role in specific magnetic effects such as magnetostriction.

8.3 MAGNETISM OF TRANSITIONAL METALS

The study of magnetism in metals requires to take some account of the spin dependence of the correlation corrections to the Hartree scheme. We wish only to stress some points in this well trodden field (Mott, 1964; Herring, 1966; Van Vleck, 1953; Wohlfarth, 1953; Beeby, 1967; Mattis, 1965). We then review the nature of the magnetic interactions and the magnetic properties of metals at 0 °K and without spin-orbit coupling. We finally review the possible influence of temperature and spin-orbit coupling.

8.3.1 Spin dependence of correlation effects in narrow bands

This requires a more detailed analysis than cohesion does (cf. §8.2.4). The usual way to discuss this problem is first to introduce exchange corrections, then discuss in a semiquantitative way the corrections due to Coulomb correlations. For magnetic problems, the two types of correction largely cancel each other; it might therefore be better to invert the order of corrections.

8.3.1 a *Exchange.* This is introduced by going from the Hartree to the Hartree–Fock scheme. The determinants built up for electrons with parallel spins introduce exchange terms in the interaction energy of two Bloch functions, which reduce in tight binding to linear combinations of terms of the form

$$-J_{m\,m'}^{ij} = -\left\langle \psi_{im}(1)\,\psi_{jm'}(2)\left|\frac{1}{r_{12}}\right|\psi_{jm'}(1)\,\psi_{im}(2)\right\rangle. \qquad (8.32)$$

First it is obvious that the *interatomic terms* ($i \neq j$) are much smaller than the *intra-atomic ones* ($i = j$), because of the long-range exponential decrease of the atomic functions $\psi_{im} \to e^{-qr}$. It is also clear that the interatomic terms should be small compared with the β integrals responsible for the d-bandwidth; this is because the J terms are proportional to e^{-2qb}, if b is the interatomic distance, while the β terms should be proportional to e^{-qb}. Thus, we expect, as in free atoms or insulators, that the stability of the atomic magnetic moments will depend on intra-atomic terms J^{ii}; and it is clear that the interatomic magnetic couplings must come from the one electron kinetic energy terms β, and not from interatomic exchange terms J^{ij} (Slater, 1930). We can be sure therefore that the classical foundation of a Heisenberg picture is *not* valid. In what follows, we shall neglect the interatomic exchange terms altogether.

The intra-atomic exchange terms can themselves be separated into what is classically known as the 'self-energy' terms $U = J_{mm}^{ii}$ and the '*exchange*' terms $J_{mm'}^{ii}$ ($m' \neq m$). The self-energy terms are *subtracted* from the energy of electrons of parallel spin if, as assumed here, the Hartree Hamiltonian and wave-functions correspond to (neutral) atomic potentials.† This merely reflects the fact that the Pauli exclusion principle applies fully, in the tight-binding approximation, to electrons sitting on the same atom; the Hartree–Fock determinants automatically prevent two electrons of parallel spin from sitting on the same atom with the same orbital moment. As a result, the Coulomb interaction U corresponding to such an occurrence is just removed

† In the recent literature on magnetism, one usually takes a Hartree Hamiltonian of bare ions; the self-energy term must then be added to the energy of electrons with antiparallel spins. We pointed out earlier that practical Hartree band structures are usually computed with an intermediary scheme, corresponding to potentials of monovalent ions.

from the energy of the system. Estimated orders of magnitude are computed to be

$$U \simeq 20\,\mathrm{eV} \quad \text{and} \quad J^{ii}_{m+m'} \simeq J \simeq 1\,\mathrm{eV}.$$

We can define an *average* exchange energy $\overline{U}$ *per pair of electrons* (or holes) *per atom*. This is the average energy gained to switch from antiparallel spins to parallel spins. As the 5 quantum numbers m are roughly equally represented at each energy (cf. §8.1.6) this crude Hartree–Fock scheme would give

$$\overline{U}_{HF} \simeq \frac{U + 4J}{5} \simeq 5\,\mathrm{eV}.$$

8.3.1*b*   *Coulomb correlations.* Their influence leads obviously to a screening out of the exchange interactions, hence to a strong reduction in $\overline{U}$. And the purpose of this section is to discuss by what reduced effective values one must replace U, J and $\overline{U}$ if one wants to treat their effect as if they were small perturbations acting on uncorrelated Bloch functions.

The essential role of the dd Coulomb correlations is to create a small 'Coulomb-hole' in the correlation function of electrons with antiparallel spins. This stabilizes somewhat a pair of such electrons, by reducing the effect of their short-range repulsive potential. It thus reduces its difference of energy $\overline{U}$ compared with a pair of electrons with parallel spin.

As emphasized in §8.2.1, one can distinguish two effects in the tight-binding approximation; intra-atomic Coulomb correlations and kinetic interatomic correlations.

Intra-atomic Coulomb correlations: the way two d-electrons (or d-holes) of opposite spin tend to avoid each other when sitting on the same atom can be taken as similar to what it is in a free atom or ion. To take this effect into account, one can therefore keep the Hartree–Fock scheme, but replace the exchange terms J by *modified* values J' deduced from optical spectra of free atoms or ions. Thus, for $U = J^{ii}_{mm}$ values for positive holes in nickel can be deduced from the following transitions

$$U' = 2(3d)^9 \rightarrow 3d^{10} + 3d^8$$

$$= (3d^9 \rightarrow 3d^8) - (3d^{10} \rightarrow 3d^9)$$

where, for $3d^8$, one must take the two holes of opposite spins in the same d-orbital $|m\rangle$. This gives

$$U' \simeq 7{\cdot}5\,\mathrm{eV}, \quad \text{or} \quad 13\,\mathrm{eV},$$

depending on whether one assumes a supplementary $4s$-electron to be present or not. Thus, in metallic Ni with 0·5 s-electrons per atom, U' is likely to be of the order of 10 eV.

This is certainly an *upper limit* to the value of U'. Smaller estimates are likely in *other transitional metals*, because their smaller nuclear charges lead to larger s-orbits, hence smaller Coulomb interactions. These estimates should be further reduced by the *ds correlations*, which should to a certain degree screen out the dd interactions. Indeed, Herring, assuming that such a screening should leave each atom neutral, deduced values of $U' \simeq 2$ eV from atomic reactions such as $2 \times 3d^9 4s \rightarrow 3d^8 4s^2 + 3d^{10}$ for Ni (Herring, 1966).

Such an extreme assumption is not very reasonable and one expects on the contrary the dd correlations to dominate the ds ones, owing to the larger number and smaller kinetic energy of the d-carriers; and indeed d-electrons seem more effectively than s-electrons to screen out excess nuclear charges introduced by impurity atoms (Mott & Jones, 1936; Friedel, 1958). The reduction of U' by ds correlation, although probably appreciable, is therefore certainly less than Herring's estimate.

In conclusion, average values

$$U' \leqslant 5 \, \text{eV}, \tag{8.33}$$

which are a sizeable fraction of the bandwidth w, seem reasonable. They are consistent with the previous discussion on cohesion (cf. §8.2.4).

For $J^{ii}_{m+m'}$, values deduced from the optical spectra of free atoms and ions depend somewhat on m' and m and on the degree of ionization. One can, however, define likely average values J', which decrease slowly through a transitional series, and from one series to another. As a rough order of magnitude, we can take (Slater, 1930; Friedel, 1958)

$$J' \simeq 0·6 \, \text{eV}. \tag{8.34}$$

Kinetic interatomic correlations: except perhaps in some extreme cases with few d-carriers (Ni, Pd, Pt), the perturbing potential associated with the Coulomb hole can be taken as at most of atomic dimensions. There is then no direct potential interatomic correlation. But the way two d-electrons (or holes) of opposite spin tend to avoid jumping onto the same atom at the same time is obviously a matter of compensating the kinetic energy lost by the potential energy gained. This will reduce still further the average exchange energy $\overline{U}$.

First of all it is possible to write the following equation for the potential energy spent in bringing two carriers on the same atom together.

$$U_{mm'}^{ii} = \left\langle m(1)\, m'(2) \left| \frac{1}{r_{12}} \right| m(1)\, m'(2) \right\rangle \qquad (8.35)$$

and to a degree this depends on $m - m'$; but its average value deduced from atomic optical spectra is not much less than $U_{mm}^{ii} = U' \leqslant 5\,\text{eV}$ (8.33). The kinetic energy spent on preventing two carriers from jumping onto the same atom is on the other hand of the order of $w \geqslant 5\,\text{eV}$ (cf. 8.10). The kinetic interatomic correlations will therefore only prevent *large* charge fluctuations from occurring. As pointed out in §8.2.4, these are rather rare anyway. A first approximation is therefore to *neglect the kinetic interatomic correlations altogether*. In this scheme intra-atomic correlations reduce the average difference of energy between two electrons of parallel and non-parallel spins to

$$\bar{U}' = \tfrac{1}{5}(U' + 4J') \leqslant 1\cdot5\,\text{eV}. \qquad (8.36)$$

In this simplified approximation, we see that the self-energy term U' contributes nearly $\frac{2}{3}$ of $\bar{U}'$ and the exchange terms J' responsible for magnetism in free atoms and ions (Hund's rule) play a minor role only.

To improve on the approximation, a complete study is still lacking (Hubbard, 1963 a, b, 1964; Kanamori, 1963; Gutzwiller, 1964, 1965), although it is fairly easy to imagine at least a simple variational exploration of it.

The only case treated completely so far is the limit of a *very small number* of d-carriers (Kanamori, 1963). It is then clear that one can study the correlation of two d-carriers, neglecting all others; also, when studying one of these carriers one can assume the other to be fixed on one atom. The problem is then reduced to a carrier moving in a narrow band and scattered by a large atomic repulsive potential u. Standard scattering theory shows (De Witt, 1956; Toulouse, 1966) that the self-energy of the scattered carrier is changed by an amount proportional to its phase shift $\eta(E)$. For small kinetic energies E and repulsive potentials u

$$\eta(E) \simeq tg\,\eta(E) = -\frac{\pi\tfrac{1}{5}n(E)\,u}{1 - \tfrac{1}{5}F(E)\,u}, \qquad (8.37)$$

where n is the density of states per atom in the band and

$$F(E) = \int \frac{n(E')\,dE'}{E - E'}. \qquad (8.38)$$

We have assumed for simplicity that the various d-orbitals $|m\rangle$ have the same density of states (cf. §8.1.6); and the factor $\frac{1}{5}$ takes into account the degeneracy of the d-band.

If u were small, the Born approximation $\eta = -\pi\frac{1}{5}n\,u$ would hold. For large u's one can still consider that the Born approximation holds, but with an *effective* potential

$$U_{\text{eff}} = \frac{u}{1 - \frac{1}{5}F(E)\,u}. \tag{8.39}$$

The value of $F(E)$ for a nearly empty band depends on the form $n(E)$ of the band. But in the cases of interest here, we can take

$$\tfrac{1}{5}F(E) = -\frac{\alpha}{w} \tag{8.40}$$

with† $\alpha \simeq 2$ to 3.

The physical meaning of the reduction (8.39) is clear; for large repulsive potentials u, the carriers prefer to lose the kinetic energy $|\beta| \simeq \frac{1}{2}w$ involved in interatomic jumps, in order to avoid one another.

For two carriers in orbitals $m \neq m'$, we have, with the previous notations,

$$u = U'_{mm'} - J'_{mm'} \quad \text{for the triplet state,}$$

$$u = U'_{mm'} + J'_{mm'} \quad \text{for the singlet state.}$$

If we take, as usual, the state of non-parallel spins as the average of the singlet and one triplet state, we can write

$$U^{\uparrow\downarrow}_{mm'} = \frac{1}{2}\left\{ \frac{U'_{mm'} - J'_{mm'}}{1 - \frac{1}{5}F(E)\,(U'_{mm'} - J'_{mm'})} + \frac{U'_{mm'} + J'_{mm'}}{1 - \frac{1}{5}F(E)\,(U'_{mm'} + J'_{mm'})} \right\} \tag{8.41}$$

for non-parallel spins and

$$U^{\uparrow\uparrow}_{mm'} = \frac{U'_{mm'} - J'_{mm'}}{1 - \frac{1}{5}F(E)\,(U'_{mm'} - J'_{mm'})} \tag{8.42}$$

for parallel spins.

The differences of these quantities give the quantities which replace U' (for $m' = m$) and J' (for $m' \neq m$) in (8.36). As $U'_{mm} = J'_{mm}$ we have

$$\overline{U}_{\text{eff}} = \tfrac{1}{5}(U_{\text{eff}} + 4J_{\text{eff}}) \tag{8.43}$$

† For a rectangular band where $n = \text{const}$ for $E_0 - \dfrac{w}{2} < E < E_0 + \dfrac{w}{2}$,

$$\tfrac{1}{5}F(E) = \frac{1}{w}\ln\left|\frac{w - (E - E_0)}{E - E_0}\right|.$$

For paramagnetic Ni, $p = 0\cdot5$ d-holes per atom and $\alpha \simeq 3$; for paramagnetic Co, $p \simeq 1\cdot6$ d-holes per atom and $\alpha \simeq 1\cdot7$.

with

$$U_{\text{eff}} = \frac{1}{2} \frac{2U'}{1 - \frac{2}{5}U'F(E)} \simeq \frac{U'}{1 + 4(U'/w)}, \tag{8.44}$$

$$J_{\text{eff}} \simeq \frac{J'}{\{1 + (2/w)\,(U'_{mm'} + J')\}\{1 + (2/w)\,(U'_{mm'} - J')\}}. \tag{8.45}$$

With $U' \simeq U'_{mm'} \simeq \frac{1}{2}w \gg J'$, we see that the *kinetic interatomic correlations reduce both the self-energy term U' and the exchange terms*† J'. Reasonable values of w, U', J' and $U'_{mm'}$ lead to values of $\bar{U}_{\text{eff}}$ *of the order of* 0·5 to 1 eV.

It must be stressed that such an estimate is only valid for nearly empty (or full) bands. For larger numbers of carriers, the intra-atomic Coulomb correlations are expected to be more effective in reducing the self-energy terms U while the interatomic Coulomb correlations are expected to be less effective. It is probable that the total reduction is still fairly large, leading to values of $\bar{U}_{\text{eff}}$ not much in excess of 1 eV.

8.3.2 Stability and couplings of small magnetic moments at 0 °K

The positive sign of $\bar{U}_{\text{eff}}$ favours the creation of magnetic moments. The kinetic energy associated with the bandwidth works in the opposite direction. If the total balance is in favour of the creation of magnetic moments, the kinetic energy tends to transfer magnetism from one atom to its neighbours, and is thus responsible for the magnetic couplings.

We restrict ourselves first to cases of small magnetic moments at 0 °K. This allows a study of the relative stability of the paramagnetic state of no permanent moment with respect to states with small moments and various ferro- or antiferromagnetic couplings. The possibility of large magnetic moments and the influence of temperature will be treated next (Lomer, 1962; Sawada & Fukuda, 1961; Iwamoto & Sawada, 1962; Tachiki & Nagamiya, 1963; Izuyama *et al.* 1963; Overhauser, 1962; Mattis, 1965; Mitsudo *et al.* 1965; Blandin & Lederer, 1965, 1966; Lederer, 1966, 1967; Penn, 1965; Alexander & Horwitz, 1966, 1968).

8.3.2a *Magnetic energy.* Let us start with a metal with no magnetic moment, and create a distribution of small atomic moments

$$\mu_i = 2S_i\mu_B; \tag{8.46}$$

† Some of the $U_{mm'}$ are definitely smaller than U', thus probably much smaller than w. For such electron pairs, $J_{\text{eff}} \simeq J'$.

S_i is the average spin moment on atom i, μ_B the Bohr magneton. We can write for the distribution

$$\mu = \sum_i \mu_i = \sum_{\mathbf{K}} \mu_{\mathbf{K}} e^{-i\mathbf{K}\mathbf{R}_i}. \tag{8.47}$$

We shall consider the change of energy ΔE associated with this process.

In the spirit of the extended *Hartree–Fock* scheme developed in §8.3.1, we can associate to each atomic moment μ_i an effective atomic exchange potential

$$V_{i\,\mathrm{exch}} = \mp \overline{U}_{\mathrm{eff}} S_i, \tag{8.48}$$

where the − and + signs refer to electrons of spin respectively parallel and antiparallel to S_i. This potential is to be added to a spin-independent potential with the period of the non-magnetic lattice. The V_i's can be considered as producing a small internal magnetic field H_i, different for each atom, parallel to S_i and such that

$$\mu_B H_i = V_{i\,\mathrm{exch}}. \tag{8.49}$$

We can, furthermore, treat each Fourier component of the moment separately (Fig. 8.17), and write for the magnetic energy stored per atom and per Fourier component

$$\Delta E_{\mathbf{K}} = \tfrac{1}{2}\chi_{\mathbf{K}} H_{i\mathbf{K}}^2 = \tfrac{1}{2}\mu_{\mathbf{K}}^2/\chi_{\mathbf{K}}. \tag{8.50}$$

$\chi_{\mathbf{K}}$ is here the *susceptibility per atom of the metal under a small magnetic field of amplitude $H_{\mathbf{K}}$ and wave-vector $\mathbf{K}$*. If $\chi_{\mathbf{K}}^0$ is the Hartree susceptibility, neglecting correlation corrections, the relation

$$\mu_{\mathbf{K}} = \chi_{\mathbf{K}} H_{\mathbf{K}} = \chi_{\mathbf{K}}^0 (H_{\mathbf{K}} + H_{i\mathbf{K}}) \tag{8.51}$$

gives finally

$$\frac{1}{\chi_{\mathbf{K}}} = \frac{1}{\chi_{\mathbf{K}}^0} - \frac{\overline{U}_{\mathrm{eff}}}{2\mu B^2} \tag{8.52}$$

and

$$\Delta E = \tfrac{1}{2} \sum_{\mathbf{K}} \mu_{\mathbf{K}}^2 \left(\frac{1}{\chi_{\mathbf{K}}^0} - \frac{\overline{U}_{\mathrm{eff}}}{2\mu B^2} \right). \tag{8.53}$$

This can also be rewritten in terms of the atomic moments μ_i:

$$\Delta E = \frac{1}{2N} \Phi(0) \sum_i \mu_i^2 + \frac{1}{2N} \sum_{i \neq j} \Phi(\mathbf{R}_{ij}) \mu_i \mu_j, \tag{8.54}$$

where

$$\Phi(\mathbf{R} \neq 0) = \frac{1}{N} \sum_{\mathbf{K}} \frac{e^{i\mathbf{K}\mathbf{R}}}{\chi_{\mathbf{K}}} = \frac{1}{N} \sum_{\mathbf{K}} \frac{e^{i\mathbf{K}\mathbf{R}}}{\chi_{\mathbf{K}}^0} \tag{8.55}$$

and

$$\Phi(0) = \frac{1}{N} \sum_{\mathbf{K}} \frac{1}{\chi_{\mathbf{K}}} = \frac{1}{N} \sum_{\mathbf{K}} \frac{1}{\chi_{\mathbf{K}}^0} - \frac{\overline{U}_{\mathrm{eff}}}{2\mu B^2}. \tag{8.56}$$

Within the second-order approximation valid for small moments, ΔE can therefore be thought of as the sum of a *self-energy term* $\frac{1}{2}\Phi(0)\mu_i^2$ and a *coupling-energy term* $\Phi(\mathbf{R}_{ij})\mu_i\mu_j$ between different atomic moments.

Let us recall that basic to this approximation are the assumptions of small moments and a Hartree–Fock average of correlation interactions, independent of time and momentum.

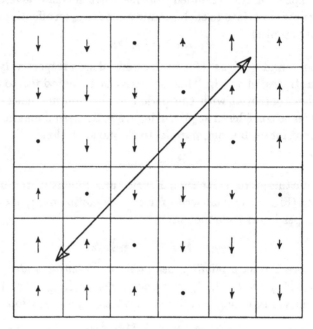

Fig. 8.17. Fourier component $K = 2\pi/\lambda$ of the magnetization.

8.3.2*b Magnetic susceptibility*. The discussion of (8.52)–(8.56) rests on that of $\chi_{\mathbf{K}}^0$. In the tight-binding limit, we can neglect all inter-atomic matrix elements of the perturbation with respect to intra-atomic ones. Straight second-order perturbations lead then to the formula

$$\chi_{\mathbf{K}}^0 = -2\mu_B^2 \sum_{\mathbf{K}_{RR}} \sum_{\mathbf{k}_{occ}} \frac{1}{E_{\mathbf{k}} - E_{\mathbf{k}+\mathbf{K}_{RR}-\mathbf{K}}}; \qquad (8.57)$$

$\mathbf{K}_{RR}$ are the reciprocal lattice vectors and summation over $\mathbf{k}$ is within the Fermi surface.

The variation of $\chi_{\mathbf{K}}^0$ with $\mathbf{K}$ gives rise to the same type of discussion as for the change of energy $\delta E_{\mathbf{K}}$ due to a lattice distortion, §8.2.2.

CC

382   J. FRIEDEL

$\chi_{\mathbf{K}}^0$ is necessarily positive. For a given direction of $\mathbf{K} - \mathbf{K}_{RR}$, the contribution of $\mathbf{K}_{RR}$ to $\chi_{\mathbf{K}}^0$ has a logarithmic anomaly in $|\mathbf{K} - \mathbf{K}_{RR}|$ when conditions such as Fig. 8.13 are fulfilled. This anomaly is especially large when conditions such as Fig. 8.14 are fulfilled. For $K = 0$, $\chi_{\mathbf{K}}^0$ reduces to the Pauli susceptibility

$$\chi_{\mathbf{K}}^0 = 2\mu_B^2 n(E_M). \qquad (8.58)$$

For nearly empty or nearly full bands in a cubic crystal, one can sometimes assume the Fermi surface to be a sphere of radius k_M. $\chi_{\mathbf{K}}^0$ is then the sum of well-known free-electron functions

$$\chi(\mathbf{k}) = -\frac{2\mu_B^2 \pi m}{8\pi^3 \hbar^2}\left[k_M + \frac{4k_M^2 - k^2}{4k} \ln\left|\frac{2k_M + k}{2k_M - k}\right|\right] \qquad (8.59)$$

centred on each node of the reciprocal lattice ($\mathbf{k} = \mathbf{K} \pm \mathbf{K}_{RR}$). It is then clear that, for k_M much smaller than the period of the reciprocal lattice, $\chi_{\mathbf{K}}^0$ has its maximum value for $K = 0$ (Fig. 8.18a). For larger k_M, the umklapp processes, i.e. the contributions from the various reciprocal periods $\mathbf{K}_{RR}$, interfere so as to make $\chi_{\mathbf{K}}^0$ a maximum for $\mathbf{K} \neq 0$ (Fig. 8.18b). It is then clear that the maximum of $\chi_{\mathbf{K}}^0$ occurs for K *near to* $2k_M$ or $|K_{RR} - 2k_M|$, i.e. near the condition of contact previously defined, but not exactly for it. It is easy to show that, for small K's,

$$\chi_{\mathbf{K}}^0 \simeq \chi_0^0\left[1 - \frac{K^2}{12k_M^2} + \ldots + \tfrac{4}{3}k_M^2 \sum_{\mathbf{K}_{RR} \neq 0} \left\{ \frac{1}{K_{RR}^2} + 4\frac{(KK_{RR})^2}{K_{RR}^6} - \frac{K^2}{K_{RR}^4} + \ldots \right\}\right]. \qquad (8.60)$$

For cubic lattices, one thus finds that the change of curvature of $\chi_{\mathbf{K}}^0$ associated with the change from case a to case b, Fig. 8.18, occurs for $2k_M$ of the order of half the smallest period K_{RR}. Similar results hold for more complex geometries (ellipsoidal Fermi surfaces; other lattice structures...).

8.3.2c *Paramagnetism.* One expects small magnetic moments to be unstable, and thus the paramagnetic phase to be (meta-) stable, for $\Delta E > 0$, if the condition

$$\frac{1}{\chi_{\mathbf{K}}} = \frac{1}{\chi_{\mathbf{K}}^0} - \frac{\overline{U}_{\text{eff}}}{2\mu_B^2} > 0 \qquad (8.61)$$

holds for all values of $\mathbf{K}$. As $\chi_{\mathbf{K}}^0$ is finite and positive for all values of $\mathbf{K}$, this condition is always fulfilled if $\overline{U}_{\text{eff}}$ is small enough. The uniform susceptibility χ_0 is then given from (8.52) and (8.58) by

$$\chi_0 = \frac{2\mu_B^2 n(E_M)}{1 - \overline{U}_{\text{eff}} n(E_M)}.$$

Such a paramagnetism is indeed observed in all transitional metals except those at the end of the first series (Kriessman & Callen, 1954).

Their magnetic susceptibility χ_0 is roughly proportional to the density of states $n(E_M)$, as deduced from the electronic specific heat, but is larger than the value χ_0^0 predicted in the Hartree scheme.

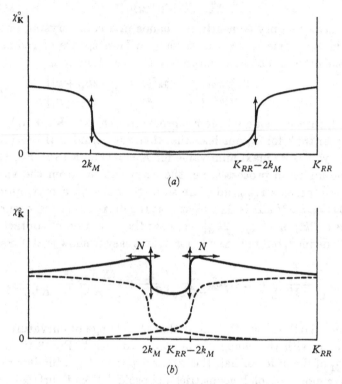

Fig. 8.18. Schematic behaviour of $\chi_\mathbf{k}^0$ for a spherical Fermi surface: (a) small k_M; (b) larger k_M.

The *Stoner enhancement factor* (Stoner, 1946)

$$\chi_0/\chi_0^0 = [1 - \overline{U}_{\text{eff}} n(E_M)]^{-1}$$

thus obtained is fairly large, of the order of 2 to 4, in rough agreement with a value of $\overline{U}_{\text{eff}}$ of the order of 1 eV. A detailed comparison would be worthless, because fairly large correction terms are present both in the specific heat and in the magnetic susceptibility (electron–phonon coupling, Van Vleck term, spin-wave term).

We shall only discuss here the existence of a *Van Vleck term*, which can be large, owing to the small width of the *d*-band (Van Vleck,

1932; Kubo & Obata, 1956; Denbigh & Lomer, 1963). This term is due to the angular momentum part of the magnetic Hamiltonian, $-\mu_B \mathbf{l}H$, where H is the applied magnetic field.

Second-order perturbations lead to a change of energy per atom of

$$\Delta E' = \frac{2}{N} \sum_{\substack{\mathbf{k'} \text{inocc} \\ \text{kocc}}} \frac{\langle \mathbf{k}| - \mu_B \mathbf{l}H|\mathbf{k'}\rangle \langle \mathbf{k'}| - \mu_B \mathbf{l}H|\mathbf{k}\rangle}{E_\mathbf{k} - E_\mathbf{k'}}. \tag{8.62}$$

This is not zero, because the non-diagonal elements $\langle \mathbf{k}|\mathbf{l}|\mathbf{k'}\rangle$ are not necessarily zero for Bloch functions $|\mathbf{k}\rangle$. Indeed it leads to a term in the energy which is negative and proportional to H^2, hence to a contribution to the paramagnetic susceptibility.

In the tight-binding approximation, we can keep again only the intra-atomic orbital terms in the matrix elements involved. Using known forms of $\langle im|lz|im'\rangle$, the whole numerator is clearly of the order of $N\mu_B^2 H^2$. Thus

$$\chi_{VV} = -\frac{1}{H}\frac{d\Delta E'}{dH} \tag{8.63}$$

is small for a nearly empty or nearly full band; it goes through a maximum of the order of $20\mu_B^2/w$ for half-filled d-bands. It should therefore be comparable or even larger than the Pauli term for B.C.C. metals near the middle of the transitional series, with small densities of states $n(E_M)$. But it can probably be neglected otherwise, especially in metals like Pd, Pt with large $n(E_M)$ and nearly full d-bands.

8.3.2d   *Magnetic phases.* Equations (8.53), (8.54) are still valid when $1/\chi_\mathbf{K} \leqslant 0$ for some values of $\mathbf{K}$. The corresponding periodic magnetic structures with wave-number $\mathbf{K}$ are more stable than the paramagnetic phase; and the corresponding negative values of $\Delta E_\mathbf{K}$ measures their stability relative to that phase.

Such magnetic phases (with small moments) become more stable than the paramagnetic phase when $\chi_\mathbf{K} < 0$, i.e. when $\mu_B^2/2\overline{U}_{\text{eff}}$ is smaller than the maximum value of $\chi_\mathbf{K}^0$ (8.52). For a given small length of atomic moments μ_i, the most stable arrangement corresponds to the most negative $\Delta E_\mathbf{K}$, hence the largest value of $\chi_\mathbf{K}^0$ (8.53). From the discussion above, this should be *ferromagnetic* for nearly empty or nearly full d-bands and *antiferromagnetic* with a period nearly related to the Fermi wavelength according to the constructions of Fig. 8.14 for intermediary cases of half-filled bands. The stability of such magnetic phases might arise partly from the stability of the magnetic moments μ_i and partly from their magnetic couplings (8.54).

Indeed, for not very large values of $\Delta E_{\mathbf{K}}$, isolated atomic moments μ_i should be unstable; they are stabilized by their mutual couplings.

It is fairly easy to show and nearly obvious that the coupling energy $\Phi(\mathbf{R}_{ij})\,\mu_i\mu_j$ is the same as for two isolated atomic moments μ_i, μ_j at a distance $\mathbf{R}_{ij}$ in an otherwise non-magnetic metal (8.55) (Blandin & Lederer, 1965, 1966; Lederer, 1966, 1967). General results

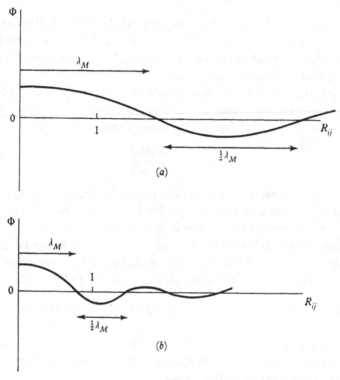

Fig. 8.19. Schematic variation with distance R_{ij} of the coupling energy $\Phi(R_{ij})$. (a) Ferromagnetic coupling; (b) antiferromagnetic coupling.

of scattering theory (Gautier, 1962; Friedel, 1958; De Witt, 1956; Toulouse, 1966) show that $\Phi(\mathbf{R}_{ij})$ is an oscillating function of the distance R_{ij}, with an amplitude decreasing as a high power of R_{ij}, a wavelength related to the Fermi wavelength $\lambda_M = 2\mu/k_M$ and a whole geometry related at large distances to that of the Fermi surface (Fig. 8.19). It is then clear (Friedel *et al.* 1961) that the ferromagnetic coupling arises for small numbers of carriers, when the Fermi wavelength is large compared with interatomic distances, and when Φ is still moderately positive at nearest neighbour distances (Fig. 8.19a).

Antiferromagnetic couplings arise on the other hand for larger numbers of carriers, when the coupling energy Φ oscillates with a wavelength comparable with or smaller than the interatomic distance (Fig. 8.19b).

The nature of the magnetic couplings is therefore intimately related to the form of the Fermi surface. The situation is very similar to that of the lattice structures discussed in §8.2. In both cases, it must be stressed that an exact knowledge of the Fermi surface is not, in principle, completely sufficient to discuss the magnetic (or lattice) structures: the most stable phases, associated with the maxima in $\chi_{\mathbf{K}}^0$, correspond to wave-vectors $\mathbf{K}$ which are only near the logarithmic anomalies associated with the Fermi surface (Fig. 8.18); this is also clear in real space, where the Fermi surface only regulates the asymptotic oscillations of the magnetic couplings $\Phi(\mathbf{R}_{ij})$ (Fig. 8.19).

The general features of magnetic phases in transitional metals are in agreement with these predictions, although clearly some corrections should be made for the large moments observed (cf. §8.3.3). Magnetic phases are observed only at the *end of the first transitional series*, where $\overline{U}_{eff}$ is maximum and the d-band narrowest, because the d-shells have their smallest radii†. *Ferromagnetism* is observed at the end of the series (αFe, Co, Ni), while *antiferromagnetic* phases are observed in the middle of the series (Cr, Mn).

In antiferromagnetic (B.C.C.) Cr, the major component $\mathbf{K}$ in the oscillations of the magnetic moments is not commensurable with the lattice periods; it varies continuously with temperature from about $\frac{1}{21}$ to about $\frac{1}{27}$ of a reciprocal lattice vector, between $0\,^\circ$K and T_N (Wilkinson *et al.* 1962; Shirane & Takei, 1962). Its order of magnitude has been related to the form of the Fermi surface (Lomer, 1962; Sawada & Fukuda, 1961; Iwamoto & Sawada, 1962; Tachiki & Nagamiya, 1963; Izuyama *et al.* 1963). This seems to present a fairly extended and nearly flat area which should give rise to a large logarithmic anomaly for translations $\mathbf{K}$ such as in Fig. 8.14. One expects the corresponding elements Mo, W of the other series to have fairly similar Fermi surfaces. A pronounced Kohn anomaly with a similar wave-vector $\mathbf{K}$ has indeed been observed in the phonon dispersion curve of W.

---

† Ferromagnetism is observed for nearly empty d-bands only in compounds such as ZrZn$_2$ where the d-band is very narrow, hence the density of s-states high, owing to the large distances between transitional atoms (Matthias & Bozorth, 1958; Shirane *et al.* 1964).

8.3.3   Large magnetic moments at 0 °K

The second-order perturbation method developed so far has some obvious limitations. It cannot predict the equilibrium length of the magnetic moments in a magnetic phase. If this length is large, i.e. if the third-order terms in $\Delta E(\mu_i)$ are small or negative, the whole perturbation method breaks down. Direct couplings between more than two atomic moments have to be considered; thus coupling between Fourier components may arise, leading to ferrimagnetic arrangements or various kinds of non-sinusoidal antiferromagnetism. Because of

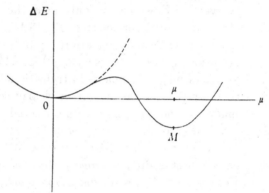

Fig. 8.20. Possibility of a stable magnetic phase when the paramagnetic phase is metastable.

these co-operative interactions, magnetic phases with finite moments might be more stable than a metastable paramagnetic phase (Fig. 8.20). Condition (8.61) is thus *not* a sufficient condition for the paramagnetic phase to be stable.

A general discussion of such effects is at present clearly impossible. Only a few remarks concerning the various type of couplings will be made.

8.3.3*a*   *Strong and weak ferromagnetism.* In the simple case of ferromagnetism, $\Delta E(\mu)$ can be directly computed for large atomic moments μ. Fig. 8.21 shows that

$$\Delta E(\mu) = \int_{E_M}^{E_1} n(E)\,E\,dE - \int_{E_1}^{E_M} n(E)\,E\,dE - \frac{1}{2}\left(p_1^2 + p_2^2 - \frac{p^2}{2}\right)\overline{U}_{\text{eff}} \quad (8.64)$$

with

$$p_i = \int^{E_i} n(E)\,dE \quad (8.65)$$

and

$$p_1 + p_2 = p. \quad (8.66)$$

Besides Stoner's criterion for metastability of the paramagnetic phase

$$\left(\frac{d^2\Delta E}{d\mu^2}\right)_0 > 0 \quad \text{hence} \quad n(E_M) > \frac{1}{\overline{\overline{U}}_{\text{eff}}}, \qquad (8.67)$$

one obtains easily for the condition of stability of the ferromagnetic phase (point M; Fig. 8.20):

$$\frac{d\Delta E}{d\mu} \leqslant 0$$

if
$$\overline{n}_{12} = \frac{p_1 - p_2}{E_1 - E_2} = \frac{1}{E_1 - E_2}\int_{E_2}^{E_1} n(E)\,dE \geqslant \frac{1}{\overline{\overline{U}}_{\text{eff}}}. \qquad (8.68)$$

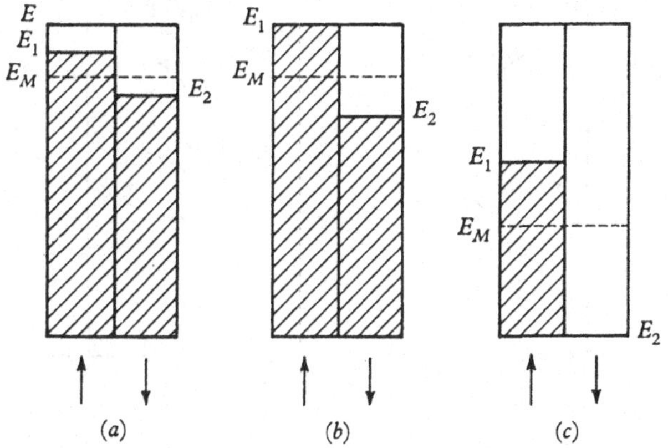

Fig. 8.21. Ferromagnetic d-band: (a) weak ferromagnetism; (b) strong ferromagnetism ($p \leqslant 5$); (c) strong ferromagnetism ($p \geqslant 5$).

For a given number p of electrons, the possible magnetic moments $\mu = (p_1 - p_2)\mu_B$ have a maximum value, equal to $p\mu_B$ (Fig. 8.21 b for less than half-filled bands) or to $(10 - p)\mu_B$ (Fig. 8.21 c, for more than half-filled bands). As a result, two cases might arise.

Either ΔE has a minimum value for some length μ smaller than this maximum value (Fig. 8.21 a). Condition (8.68) is then fulfilled with the equality sign. Both halves of the d-band with opposite spin directions are partially filled. This requires the two half d-bands with opposite spin directions to have the same Fermi level when the energy of each electron is corrected by $\pm \frac{1}{2}(\mu/\mu_B)\overline{U}_{\text{eff}}$ for the correlations (Fig. 8.22 a). This case of *weak ferromagnetism* can be treated by the second-order perturbation method of §8.3.2 if the equilibrium value of μ is small enough.

Alternatively, ΔE has no minimum value for μ less than its maximum length. This maximum length $\mu/\mu_B = p$ (Fig. 8.21 b) or $10 - p$

(Fig. 8.21 c) is then the equilibrium value. Condition (8.68) then applies with usually the inequality sign. As a result, the Fermi level of the partially filled half d-band falls above the top or below the bottom of the other half when the energy E is corrected by the correlation term $\frac{1}{2}(\mu/\mu_B)\overline{U}_{\text{eff}}$ (Fig. 8.22 b). This case of *strong ferromagnetism* can never be treated correctly by the second-order perturbation method.

The choice between these two types of solutions obviously depends on the form $n(E)$ of the d-band, its filling p and the strength $\overline{U}_{\text{eff}}$ of correlation effects.

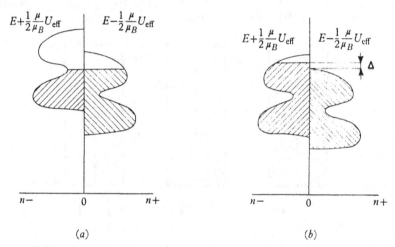

Fig. 8.22. Relative position of the two half d-bands with opposite spin directions: (a) weak ferromagnetism (Fe); (b) strong ferromagnetism (Ni, Co).

For *nearly full or nearly empty bands*, where usually $dn/dE \neq 0$ and $d^2n/dE^2 \leqslant 0$, an increase in $\overline{U}_{\text{eff}}$ leads progressively from the paramagnetic state through a weak to a strong ferromagnetic one. It is fairly obvious and easily checked in simple cases that weak ferromagnetism only occurs for a limited range of values of $\overline{U}_{\text{eff}}$. Thus for a trapezoidal band $n = A + BE$ (Fig. 8.23 a), conditions (8·67) and (8·68) lead, for $E_2 = 0$, to

$$\overline{U}_{\text{eff}}^{-1} = \sqrt{(A^2 + Bp)} \quad \text{and} \quad \overline{U}_{\text{eff}}^{-1} = \tfrac{1}{2}\{A + \sqrt{(A^2 + 2Bp)}\}$$

respectively (Fig. 8.23 b). Analogous results would hold for a parabolic band $n \propto E^{\frac{1}{2}}$. Thus one understands why metals like Ni or Co, where Stoner's condition (8.67) is certainly amply fulfilled, are strong ferromagnetics.

For *increasing numbers of carriers*, there might come a moment when strong ferromagnetism is no longer possible and is replaced by weak ferromagnetism. This is clear from the fact that

$$E_1 - E_2 = (p_1 - p_2)\,\overline{U}_{\text{eff}} \simeq (p_1 - p_2)\,\text{eV} \leqslant 5\,\text{eV},$$

while we have seen that the width w of the d-band is at least $5\,\text{eV}$.

In fact, the metals now discussed have usually the B.C.C. structure, with a d-band fairly well divided into two sub-bands (Fig. 8.6). One can then expect strong ferromagnetism to apply to all the $2 \cdot 5$ carriers of the upper sub-band but weak ferromagnerism to start for larger numbers of carriers (Fig. 8.22 a, b). The two sub-bands are indeed

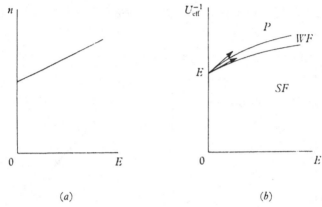

(a)   (b)

Fig. 8.23. Simple example of ferromagnetism: (a) density of states curve; (b) stability of various phases: P paramagnetism; WF weak ferromagnetism; SF strong ferromagnetism.

so well separated that one can expect from condition (8.68) the upper sub-band to empty completely before the lower one begins to empty appreciably. This explains quantitatively the Pauling–Slater curve (Slater, 1937; Pauling, 1938) giving the average magnetic moment $\bar{\mu}$ of ferromagnetic metals and alloys versus their average atomic number $\overline{N}$, i.e. the number of d-electrons. The peak in $\bar{\mu}(\overline{N})$ separates the strong from the weak ferromagnetics (Fig. 8.24), and its value is indeed near to $2 \cdot 5$ Bohr magnetons.

The fact that the $2 \cdot 5$ carriers of one sub-band can align ferromagnetically, but not the 5 carriers of a full band sets limits to the ratio of $\overline{U}_{\text{eff}}/w$. Taking into account that the upper sub-band of Fig. 8.6 has a peak with a width of about $\frac{1}{3}w$, criterion (8.68) gives

$$\frac{1}{3 \times 2,5} < \frac{\overline{U}_{\text{eff}}}{w} < \frac{1}{5}.$$

A width $w \simeq 7\,\text{eV}$ leads to $0,8 < \overline{U}_{\text{eff}} < 1 \cdot 4\,\text{eV}$, in good agreement with the estimates above.

Various aspects of the magnetic and electrical properties of impurities in these metals and alloys are fully consistent with this distinction between strong and weak ferromagnetism. We can also note that it is the antibonding part of the *d*-band that becomes empty (Figs. 8.6 and 8.7): as a result, the unpaired magnetic electrons should be more concentrated on the atoms than the *d*-electrons of free atoms. Neutron and X-ray data† seem on the whole to agree with this (Mook & Shull, 1966; Hodges *et al.* 1966*b*; Shull & Yamada, 1961).

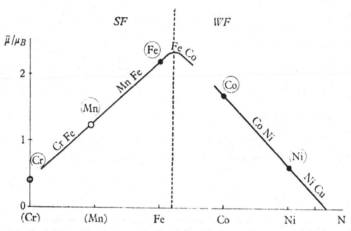

Fig. 8.24. Pauling–Slater curve giving the magnetic moments $\bar{\mu}$ of ferromagnetic metals versus their atomic numbers.

One can finally note that the moments for pure Mn and Cr fall on the Pauling–Slater curve (Nathans & Pickart, 1963). However, their couplings are not ferromagnetic and will be discussed below. Also the emptying of the lower *d*-sub-band does not lead to any magnetism in the early elements of the transitional series. This is possibly due to the increase in size and overlap of the *d*-shells, which lead to a broadening of the *d*-band and lowers $n(E)$. Also the stronger mixing with the

† If we neglected the overlap corresponding to the bonding or antibonding effect, the anisotropy γ of the magnetic form factor should be given by the ratio of the e_g character of the unpaired electrons to its value 0·4 for an isotropic distribution of *d*-characters. The value of γ measured in Ni is smaller than unity, as expected from the t_{2g} character of the top of the *d*-band in the F.C.C. structure (Fig. 8.8*a*). Its value 0·2/0·4 ≃ 0·5 is of the order of magnitude expected from the computed ratio of e_g character for the *d*-holes (Allan, 1967; Mook & Shull, 1966; Hodges *et al.* 1966; Shull & Yamada, 1961). It does not seem therefore necessary to invoke the overlap corresponding to antibonding, to explain the anisotropy factor observed. This overlap should increase the e_g-like character of the electronic distribution, thus increasing γ.

s-band should reduce $\bar{U}_{\text{eff}}$ (Friedel, 1955, 1958, 1962). Both effects work against criteria (8.67) and (8.68) for ferromagnetism. Indeed the only known cases of ferromagnetic alloys with nearly empty d-bands refer to compounds with well-separated transitional atoms, hence very narrow d-bands.

8.3.3b Antiferromagnetism. Ferrimagnetism. Another case that can in principle be treated exactly is that of a *sinusoidal* antiferromagnetic arrangement $\mu_{\mathbf{K}}e^{i\mathbf{KR}_i}$ with a large amplitude $\mu_{\mathbf{K}}$ (Fig. 8.17). The new energy gaps thus produced (Fig. 8.11) can then be analysed by the usual two beams procedure, which keep two Bloch functions of the paramagnetic lattice as important in the development of the wave-functions of the magnetic state. This 'dynamical' approximation is usually a sufficient improvement on the one beam, second-order perturbation or 'kinematic' approximation developed so far (Slater, 1951b; Lidiard, 1954; Rajagopal & Brooks, 1967).

One could, in the same spirit, analyse the possible mixing of Fourier components of different wave-vectors $\mathbf{K}$ by admixing in a self-consistent way Bloch functions of corresponding $\mathbf{K}$ vectors.

Our knowledge of the basic structure of the non-magnetic metals is probably too rough for such analysis to have much interest, except to show the possibility in principle of explaining the complex magnetic arrangements observed in a metal like Cr.

8.3.4   Magnetism at finite temperatures

Besides the usual independent electron excitations and collective Coulomb excitations (plasmas) common to all solids, some at least of the transitional metals show fairly strong collective spin-dependent excitations (spin-waves). These various excitations can be thermally excited. Therefore the first and last types can play a role in the temperature variation of magnetism, as well as alter through their zero point motion the magnetic properties at 0 °K.

8.3.4a Independent electron excitations. Within the Hartree–Fock type of scheme developed here, these excitations alter the *Pauli* susceptibility of a *paramagnetic metal* by their effect on χ^0. Thus (8.52) is still valid with (Mott & Jones, 1936; Seitz, 1940)

$$\chi^0(T) = \chi^0(0)\left[1 + \frac{\pi^2}{6}k_B^2\left\{\frac{d^2n}{ndE_M^2} - \left(\frac{dn}{ndE_M}\right)^2\right\}T^2 + \ldots\right]. \quad (8.69)$$

The first correction term comes from the change in the average effective density of states at fixed Fermi level, due to the broadening $k_B T$ of the Fermi distribution. The second term comes from the shift in the Fermi level necessary to keep a constant number of *d*-carriers. This second term neglects a small correction term due to possible *sd* transfers. Other small corrections can come from the change in $n(E)$ due to lattice expansion and from a small contribution from the temperature variation of the Van Vleck susceptibility.

This low temperature development is actually valid up to the melting point, at least as long as the Fermi level E_M is not too near a strong peak of $n(E)$. Applied to the computed form of $n(E)$ it seems to explain reasonably at least the general behaviour, sign and order of magnitude of $\chi(T)$ in B.C.C. metals (Friedel, 1955, 1958, 1962). The anomalous and strong temperature (Kriessman, 1956) variation of

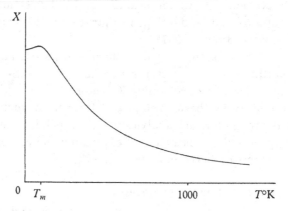

Fig. 8.25. Temperature variation of the magnetic susceptibility in Pd.

$\chi(T)$ in Pd (Fig. 8.25) can be similarly explained by the temperature variation of $\chi(T)$ due to a strong peak of $n(E)$ near to E_M (Fig. 8.7*b*): the maximum in $\chi(T)$ would arise when $k_B T_m$ is of the order of the energy distance from the peak to the Fermi level. The emptying of this initially full peak would also produce a small peak in the electronic specific heat around T_m, which is observed (Shimizu *et al.* 1962, 1963, 1964; Allan, 1967). The development (8.69) would then hold only at temperatures well below T_m.

Similar one-electron excitations are possible in *magnetic phases*. In most cases (antiferromagnetism, weak ferromagnetism), both spin directions are present at the same surface of the Fermi sea. In the weak ferromagnetic case, the densities of states of both spin directions now have different slopes dn/dE_M (Fig. 8.22*a*). The necessity to keep a constant Fermi level then leads to a linear shift of magnetization with temperature. This is obtained by equating the number of spin flips to the difference in shift of the Fermi levels at constant-spin populations

$$\frac{\Delta\mu}{\mu_B} = \tfrac{1}{2}n(E_M)(\Delta E_- - \Delta E_+) \simeq \left[\left(\frac{dn}{dE_M}\right)_- - \left(\frac{dn}{dE_M}\right)_+\right]k_B T. \quad (8.70)$$

In B.C.C. Fe, the difference of slopes should be positive, leading to a small increase of magnetization with temperature. A correction term should again be introduced for the effect of lattice expansion.

No such linear term is expected in antiferromagnetics, where the density of states is the same for both spin directions. Indeed in antiferromagnetics, one expects the one-electron excitations to affect magnetization appreciably only when $k_B T$ is of the order of the energy gaps created at $0\ ^\circ$K (Fig. 8.11). Thermal spin flips should then reduce magnetization rapidly. This process therefore sets an *upper limit* to the Néel temperature

$$k_B T_N \leqslant |\Delta_K| = \bar{U}_{eff}\left|\frac{\mu_K}{\mu_B}\right|. \quad (8.71)$$

In *strong ferromagnetics*, one expects again no appreciable reduction of the one-electron excitations on magnetization until a temperature comparable with the energy gap $\Delta = \bar{U}_{eff}(\mu/\mu_B) - E_c$ (Fig. 8.22b), where E_c is the kinetic energy of the magnetic carriers (electrons if $p \leqslant 5$; positive holes if $p \geqslant 5$). It is clear that, for all ferromagnetics, magnetization will have disappeared at temperatures larger than the shift $(\mu/\mu_B)\bar{U}_{eff}$ between the two half-bands. This sets again an upper limit to the Curie temperature

$$k_B T_c \leqslant \frac{\mu}{\mu_B}\bar{U}_{eff}. \quad (8.72)$$

More accurate and lower values can be deduced from a detailed knowledge of the density of states. This limit has little physical interest as collective spin-fluctuations should often dominate at T_c or T_N.

8.3.4b *Small spin fluctuations in magnetic metals.* Collective excitations are associated with the correlation effects discussed in §§8.2.1 and 8.3.1. Coulomb correlations produce *plasmon* excitations. Because of the large number of easily excitable d-electrons, the plasmon frequencies are fairly large. They play no role, except in the definition of $\bar{U}_{eff}$ by their zero-point motion. Spin-dependent correlations produce *spin-waves* or *spin fluctuations*.

We meet here with the same difficulty as for the 0 °K magnetization; small fluctuations, as found in paramagnetic metals or at low temperatures in magnetic ones, are easily treated by perturbation methods and large fluctuations, as found possibly near Curie or Néel points, have not been treated successfully so far.

We shall restrict ourselves to fluctuations of *small amplitude*. They can be treated by using the concept of a *dynamical magnetic susceptibility* $\chi(\mathbf{K}, \omega)$ (cf. Sawada & Fukuda, 1961; Iwamoto & Sawada, 1962; Tachiki & Nagamiya, 1963; Izuyama *et al.* 1963; Blandin & Lederer, 1965, 1966; Lederer, 1966, 1967; Doniach, 1967), relating an external applied frequency-dependent magnetic field

$$\mathbf{H} = \sum_i \mathbf{H}(\mathbf{K}, \omega) \exp\{i(\mathbf{K}\mathbf{R}_i - \omega t)\}\, \delta(\mathbf{r} - \mathbf{R}_i) \qquad (8.73)$$

to the magnetization produced

$$\Delta\boldsymbol{\mu} = \sum_i \Delta\boldsymbol{\mu}(\mathbf{K}, \omega) \exp\{i(\mathbf{K}\mathbf{R}_i - t)\}\, \delta(\mathbf{r} - \mathbf{R}_i). \qquad (8.74)$$

Thus

$$\Delta\boldsymbol{\mu}(\mathbf{K}, \omega) = \chi(\mathbf{K}, \omega)\mathbf{H}(\mathbf{K}, \omega). \qquad (8.75)$$

The first assumption inherent in such a concept is that $\mathbf{H}$ is *small* enough for first-order perturbation to apply; thus each Fourier component $(\mathbf{K}, \omega)$ can be treated separately. Also, in the Hartree–Fock type of scheme developed here, the internal field $\mathbf{H}_i$ (8.49) can be treated as following *instantaneously* any change in magnetization. This approximation is certainly valid for the spin-waves of interest, with frequencies much smaller than the Fermi frequencies, which measure the frequencies of fluctuations of $\mathbf{H}_i$. This allows us to assume $\Delta\boldsymbol{\mu}$ to oscillate with the same frequency as $\mathbf{H}$. Equation (8.52) is then still valid, and

$$\frac{1}{\chi(\mathbf{K}, \omega)} = \frac{1}{\chi^0(\mathbf{K}, \omega)} - \frac{\overline{U}_{\text{eff}}}{2\mu_B^2}, \qquad (8.76)$$

where χ^0 is the susceptibility computed for non-interacting electrons.

If we write for the magnetization

$$\Delta\boldsymbol{\mu}(\mathbf{R}) = \psi^*(\mathbf{R})\, \sigma\psi(\mathbf{R}), \qquad (8.77)$$

where σ is the Pauli spin matrix, and

$$\psi(\mathbf{R}) = \begin{bmatrix} \psi_\uparrow(\mathbf{R}) \\ \psi_\downarrow(\mathbf{R}) \end{bmatrix}, \quad \psi^*(\mathbf{R}) = [\psi_\uparrow(\mathbf{R}), \quad \psi_\downarrow(\mathbf{R})], \qquad (8.78)$$

standard second-order perturbations give respectively for the susceptibility measured *perpendicular* and *parallel* to the field.

$$\chi_\perp^0(\mathbf{K}, \omega) = -2\mu_B^2 \sum_{\mathbf{k}} \sum_{\mathbf{K}_{RR}} \frac{n_\mathbf{k}^\downarrow - n_{\mathbf{k}+\mathbf{K}_{RR}-\mathbf{K}}^\uparrow}{E_\mathbf{k}^{\prime\downarrow} - E_{\mathbf{k}+\mathbf{K}_{RR}-\mathbf{K}}^{\prime\uparrow} - \hbar\omega + i\epsilon}, \qquad (8.79)$$

$$\chi_\parallel^0(\mathbf{K}, \omega) = \tfrac{1}{2}[\chi_\uparrow^0(\mathbf{K}, \omega) + \chi_\downarrow^0(\mathbf{K}, \omega)]. \qquad (8.80)$$

In these expressions, $n_\mathbf{k} = 1$ if the k-state is occupied, and zero
otherwise ; $\chi_\uparrow^0$ is deduced from formula (8.79) by taking for all n and
E' quantities referring to spin-up states. Finally, the $E_\mathbf{k}'$ are *renormalized* energies, taking into account the possible effects of a *constant*
internal magnetic field H_{i0}:

$$\left.\begin{array}{l} E_\mathbf{k}^{\prime\uparrow} = E_\mathbf{k}^\uparrow + \mu_B H_{i0}, \\ E_\mathbf{k}^{\prime\downarrow} = E_\mathbf{k}^\downarrow - \mu_B H_{i0}. \end{array}\right\} \qquad (8.81)$$

In a paramagnetic metal, $H_{i0} = 0$; it is then clear that

$$\chi_\perp^0(\mathbf{K}, \omega) = \chi_\parallel^0(\mathbf{K}, \omega) = \chi^0(\mathbf{K}, \omega),$$

and these expressions reduce to $\chi^0(\mathbf{K})$ as given by (8.57) when $\omega \to 0$.
In a ferromagnetic metal,

$$\mathbf{H}_{i0} = -\frac{\overline{U}_{eff}\mu_i}{2\mu_B^2},$$

where μ_i is the atomic moment.

The generalized magnetic susceptibility given by (8.76), (8.79),
(8.80) is a straight extension of the concept of dielectric constant.
It is given by the same type of formula, and is also a *complex* quantity,
showing that, in general, the fluctuation $\Delta\mu$ in magnetization is not
exactly in phase with the applied perturbing field. It is however easy
to check that $\chi^0(\mathbf{K}, 0)$ is real.

In a *ferromagnetic* metal, *spin-waves* or *magnons* are small transverse
fluctuations of magnetization, with positive energies of excitation
(Herring & Kittel, 1951). In the limit of small temperatures and hence
small amplitudes, the preceeding equations are valid. Each Fourier
component $\mu_T(\mathbf{K}, \omega)$ of the fluctuation can be treated separately. Its
energy is given by a low-frequency pole of the corresponding susceptibility $\chi(\mathbf{K}, \omega)$, and thus, according to (8.76) (Mills & Lederer, 1966;
Blandin & Lederer, 1965, 1966; Lederer, 1966, 1967):

$$\chi_T^0(\mathbf{K}, \omega) = \frac{2\mu_B^2}{\overline{U}_{eff}}. \qquad (8.82)$$

An approximate solution of (8.82) is possible for small frequencies ω, where the imaginary part of χ_T^0 vanishes. Developing χ_T^0 *for small* ω and K then gives

$$\omega(\mathbf{K}) \simeq -\left[\frac{\mu(\mathbf{K}\nabla_{\mathbf{K}})^2 \chi_T^0}{\chi_T^{02}}\right]_{0,0}. \tag{8.83}$$

Use has been made of the fact that, if μ is the ferromagnetic magnetization,

$$\chi_T^0(0, \omega) = \frac{\mu_B \mu}{\overline{U}_{\text{eff}}(\mu/\mu_B) - \hbar\omega} \tag{8.84}$$

and

$$\nabla_{\mathbf{K}} \chi_T^0(0, 0) = 0. \tag{8.85}$$

As expected, the curvature of the parabolic variation of $\omega(\mathbf{K})$ is related to the energy $\Delta E_T(\mathbf{K})$ necessary to produce a transverse fluctuation of frequency ω and small amplitude $\Delta\mu_T(\mathbf{K}, \omega)$. Equation (8.53) applies

$$\Delta E_T(\mathbf{K}, \omega) = \frac{\Delta\mu_T^2}{2\chi_T(\mathbf{K}, \omega)} = \frac{\Delta\mu_T^2}{2}\left(\frac{1}{\chi_T^0(\mathbf{K}, \omega)} - \frac{\overline{U}_{\text{eff}}}{2\mu_B^2}\right). \tag{8.86}$$

Again with the help of (8.84) and (8.85), this gives for *small* $\mathbf{K}$ and ω

$$\Delta E_T(\mathbf{K}, \omega) \simeq -\frac{\overline{U}_{\text{eff}}}{8\mu_B^2}[(\mathbf{K}\nabla_{\mathbf{K}})^2 \chi_T^0]_{0,0} \Delta\mu_T^2(\mathbf{K}). \tag{8.87}$$

By analogy with the paramagnetic case, one can also write the energy ΔE_T of a magnetic fluctuation in terms of the transverse magnetic fluctuations $\Delta\mu_{Ti}$ on each atom. Formulae such as (8.54–8.56) are obtained. However, it is only the transverse fluctuations that are coupled by a formula of the type (8.54). Also the self-energy and coupling energy term involve $\chi_T^0(\mathbf{K}, \omega)$ which itself depends on the longitudinal magnetization. Indeed a detailed study of the coupling energy shows, as one would expect, that its spatial variation differs somewhat from that obtained in the paramagnetic state. Indeed, in a strong ferromagnetic, the coupling energy should die out exponentially at large distances with a rate related to the energy gap Δ between the two spin directions (Fig. 8.22b), while in a weak ferromagnetic the long-range oscillations of the coupling energy are connected with wave-vectors $\mathbf{K}$ which bring into mutual contact the two Fermi surfaces of different forms corresponding to the two spin directions (Blandin & Lederer, 1965, 1966; Lederer, 1966, 1967).

These expressions can be extended to spin-waves in *antiferromagnetics*. The development of magnetic energy then occurs near the wave-vector defining the antiferromagnetic ordering.

DD

It is clear that, so far, one's knowledge of the d-bands is too poor to make any quantitative estimates of the coefficients involved in (8.83) and (8.86).

8.3.4c Small spin fluctuations in paramagnetic metals. In a paramagnetic metal, the symmetry between up and down spins in (8.79), (8.80) shows that the real part of χ is an even function of ω, while its imaginary part is an odd function of ω.

As, in a paramagnetic metal, $\chi^{0}(\mathbf{K}, 0) > 2\mu_{B}^{2}/\overline{U}_{\text{eff}}$ by hypothesis (cf. §8.3.2), we see that the real part of $\chi^{0}(\mathbf{K}, \omega)$ could satisfy condition (8.82) only for finite pulsations ω, for which the imaginary part of $\chi^{0}(\mathbf{K}, \omega)$ would be large: *no real low frequency spin-waves exist in such paramagnetic metals.*

If however the magnetic correction in $\overline{U}_{\text{eff}}$ is large enough for the paramagnetic state to be nearly unstable with respect to a magnetic state with wave-vector $\mathbf{K}_{0}$, the real part of $1/\chi(\mathbf{K}, \omega)$ will be small for $\mathbf{K} \to \mathbf{K}_{0}$ and $\omega \to 0$. As χ is real for $\omega = 0$, the whole complex susceptibility $\chi(\mathbf{K}, \omega)$ should be large at small frequencies ω and for $\mathbf{K}$ near to $\mathbf{K}_{0}$. One can say that, in this range of frequencies, 'virtual' spin-waves can be excited; their finite life time is due to their coupling with one-electron excitations; one can also say that each one-electron excitation is 'clothed' by a collective magnetic reaction of the metal. Such virtual spin-waves have been called *paramagnons.* (Berk, 1966; Berk & Schrieffer, 1966; Doniach & Engelsberg, 1966; Mills & Lederer, 1966.) It must be understood that this rather vague concept merely describes the easier low frequency magnetic fluctuations associated with a strong exchange correction $\overline{U}_{\text{eff}}$.

Writing, for low frequencies ω,

$$\chi^{0}(\mathbf{K}, \omega) = \chi_{\mathbf{K}}^{0} + iA_{\mathbf{K}}\omega + B_{\mathbf{K}}\omega^{2} + O_{3}(\omega),$$

where $A_{\mathbf{K}}$ and $B_{\mathbf{K}}$ are two real coefficients, we have

$$\chi(\mathbf{K}, \omega) = \frac{\chi_{\mathbf{K}}^{0} + iA_{\mathbf{K}}\omega + B_{\mathbf{K}}\omega^{2}}{1 - (\overline{U}_{\text{eff}}/2\mu_{B}^{2})\,[\chi_{\mathbf{K}}^{0} + iA_{\mathbf{K}}\omega + B_{\mathbf{K}}\omega^{2}]} + O_{3}(\omega).$$

Thus

$$Im\chi(\mathbf{K}, \omega) = \frac{A_{\mathbf{K}}\omega}{C_{\mathbf{K}}^{2} + D_{\mathbf{K}}\omega^{2}} + O_{3}(\omega),$$

where

$$C_{\mathbf{K}}^{-1} = \left[1 - \frac{\overline{U}_{\text{eff}}}{2\mu_{B}^{2}}\chi_{\mathbf{K}}^{0}\right]^{-1}$$

is the Stoner reinforcement factor for wave-number $\mathbf{K}$ and

$$D_{\mathbf{K}} = \frac{\overline{U}_{\mathrm{eff}}^2}{4\mu_B^2} A_{\mathbf{K}}^2 + \left(1 - \frac{\overline{U}_{\mathrm{eff}}}{2\mu_B^2}\right) \frac{\overline{U}_{\mathrm{eff}}}{\mu_B^2} B_{\mathbf{K}} \chi_{\mathbf{K}}^0.$$

The imaginary part of $\chi(\mathbf{K}, \omega)$ thus presents a maximum value which becomes sharp and takes place at low frequency if $C_{\mathbf{K}}$ is small, i.e. near the magnetic instability condition. This can be taken as the frequency of the paramagnons. In the usual case where $D_{\mathbf{K}}$ is positive, this maximum is finite and occurs at

$$D_{\mathbf{K}}^{\frac{1}{2}} \omega = C_{\mathbf{K}}.$$

$D_{\mathbf{K}}$ has only been computed so far in the simple case of nearly ferromagnetic instability, i.e. $C_{\mathbf{K}}$ small for $\mathbf{K} = 0$, and for a parabolic band $E_{\mathbf{k}}$. It is then easy to show that the term in $A_{\mathbf{K}}^2$ predominates in $D_{\mathbf{K}}$ and gives

$$D_{\mathbf{K}}^{\frac{1}{2}} \omega \simeq \frac{\overline{U}_{\mathrm{eff}}}{2\mu_B^2} A_{\mathbf{K}} \simeq \left(\frac{\pi}{4} \frac{\hbar\omega}{E_F} \frac{k_F}{K}\right) \frac{\overline{U}_{\mathrm{eff}}}{2\mu_B^2} \chi_0^0 \qquad (8.88)$$

for

$$\frac{\hbar\omega}{E_F} < \frac{2K}{k_F} \ll 1,$$

and zero otherwise.

Here E_F and k_F are the Fermi energy and Fermi wave-number. In this simple case, the paramagnons have therefore, at small wave-vectors K and small energies ω, the following dispersion curve

$$\hbar\omega = \frac{4}{\pi} \frac{1-I}{I} E_F \frac{K}{k_F}, \qquad (8.89)$$

where

$$C_{\mathbf{K}} = 1 - I = 1 - \frac{\overline{U}_{\mathrm{eff}}}{2\mu_B^2} \chi_0^0$$

is the inverse of the Stoner reinforcement factor. It is clear that such collective spin excitation phenomena can be important only if the Stoner enhancement factor is very large. Indeed when $1 - I \to 0$, these fluctuations became very large; the preceding analysis, using first-order perturbation, is no longer sufficient, as emphasized in (8.3.4d).

Various possible consequences of such paramagnetic excitations have been stressed recently.

The coupling of the low energy one-electron excitations with these paramagnons should lead to an increase in their effective mass or in their density of states as measured by electronic specific heat. The effect is similar to that due to the coupling with phonons. It can be studied in the same spirit, adding to each one-electron energy a local field correction due to the spin polarization of the metal by the electron. The correction to the effective mass is large only if the Stoner factor $[1 - I]^{-1}$ is very large, as possibly for metals such as Pd or Pt.

Again as with phonons, one-electron excitations of frequencies much larger than that of the paramagnons should lose their magnetic clothing. This should produce a high temperature reduction in the Pauli magnetic susceptibility; this however, seems too small to explain the behaviour of χ in Pd (Fig. 8.25).

The clothing of one-electron excitations with paramagnons corresponds to a tendency for coupling electrons of parallel spins. This works against the tendency to form the Cooper pairs of opposite spins responsible for supraconductivity. This factor might particularly explain the absence of supraconductivity in paramagnetic metals such as Pd, where the usual B.C.S. theory would lead one to expect the large density of states at the Fermi level producing a high critical supraconductive temperature.

8.3.4d   *The problem of large spin fluctuations. Bloch walls; behaviour near a critical point.* The usual difficulties in applying perturbation methods arise when spin fluctuations of large amplitude are involved. Indeed, in only two cases are extensions fairly straightforward: for *Bloch walls* and *high temperature behaviour of small moments.*

Bloch walls are fluctuations of large amplitude but wavelength very large compared with interatomic distances (Blandin & Lederer, 1965, 1966 and Lederer, 1966, 1967). The changes in direction of magnetization are therefore small on an atomic scale, and one can reasonably expect the magnetic energy involved in this relative torsion of the atomic magnetic moments to be comparable to that in a small amplitude fluctuation with the same wavelength. As a result, (8.86), (8.87) are expected to be valid, with $\omega = 0$ and $\Delta\mu_T$ replaced by the full magnetic moment.

It is an inherent difficulty in this procedure, that the length of the atomic moments is not necessarily the same in a Bloch wall as in a magnetic domain. Since the creation of the Bloch wall necessitates the spin reversal of some one-electron wave-functions, the moments in the wall are expected to be somewhat smaller than in the domains. However, for usual widths of walls, the reduction should be barely measurable.

Comparing (8.83) and (8.87), we then see that there are obvious relations between the magnetic energies involved in spin-waves and Bloch walls. As with spin-waves, one can therefore set up an effective

Heisenberg interaction between atomic moments, but with the same restrictions on its meaning

High temperature behaviour of small moments (Friedel *et al.* 1961; Blandin & Lederer, 1964, 1966; Lederer, 1966, 1967): strong disorder of the atomic moments can only be treated by perturbation methods if the moments themselves are small. The energy of a spin fluctuation $\mu(\mathbf{K}, \omega)$ can then be written as

$$\Delta E(\mathbf{K}, \omega) = \frac{\mu^2(\mathbf{K}, \omega)}{2\chi(\mathbf{K}, \omega)} = \frac{\mu^2(\mathbf{K}, \omega)}{2}\left[\frac{1}{\chi^0(\mathbf{K}, \omega)} - \frac{\bar{U}_{\text{eff}}}{2\mu_B^2}\right], \quad (8.90)$$

where χ and χ^0 refer to the metal in its *paramagnetic* state (cf. (8.79) and (8.80)).

Equations (8.54) to (8.56) are therefore still valid, where $\chi_{\mathbf{K}}^0$ is replaced by $\chi^0(\mathbf{K}, \omega)$ for a fluctuation of frequency ω.

Furthermore, at least for small frequencies ω and not too large wave-vectors $\mathbf{K}$, the real part of $\chi(\mathbf{K}, \omega)$ does not depend much on ω, and its imaginary part is small (cf. 8.88).

Low-frequency spin fluctuations, (i.e. $\hbar\omega \ll \mu H_i$) corresponding to a fairly strong disorder of the atomic moments changing slowly with time, should therefore be fairly well-defined excited states of the magnetic metal. They should be described with the zero frequency self-energy $\Phi(0)$ and coupling energies $\Phi(R)$ by (8.55) and (8.56). Their importance should depend on the relative values of $\Phi(0)$ and $\Phi(R)$.

In one extreme case, the stability of the magnetic phase at $0\,°\mathrm{K}$ might come essentially from the self-energy term, the coupling energy playing a role only in the type of magnetic order chosen at $0\,°\mathrm{K}$. Thus $\Phi(0) \ll 0$ and $|\Phi(0)| \gg |\Phi(R)|$. One then expects excitations of low energy to arise from disordering the atomic moments without suppressing them. The *critical temperature* T_c or T_N and the *local order* observed at higher temperatures would then be obtained mainly from balancing the internal energy spent, involving the $\Phi(R)$, with the magnetic entropy gained by the magnetic disorder of slowly fluctuating atomic moments.

The stability of the magnetic phase at $0\,°\mathrm{K}$ might also arise mainly from the coupling energy. Thus $\Phi(0) > 0$ and $|\Phi(0)| \gtrsim |\Phi(R)|$. The magnetically disordered solution has now a high energy; it is not expected to be so well defined and has anyway an internal energy higher than the non-magnetic state. The *critical temperature* T_c or T_N is then expected to be derived from balancing the total internal energy spent by *suppressing* the moments with the magnetic disorder due to

26

one-electron excitations. Magnetic fluctuations are then expected to play only a reduced role above T_c or T_N.

These very qualitative predictions have not been developed so far. The main difficulty is that, as at $0\,^\circ$K, the second-order perturbation method leading to (8.90) is not sufficient to give the equilibrium length of the atomic moments.

One expects the preceding discussion to apply only qualitatively to the *high temperature behaviour of large moments*. This is especially clear in strong ferromagnetics. We have already pointed out that the magnetic susceptibilities and coupling energies involved should vary with the total magnetization. As a result, no single effective Heisenberg model can be set up to describe the magnetic behaviour of a (strong) ferromagnetic both at low and high temperatures. This means in particular that no exact relation is expected between the effective exchange constants derived from spin-waves or Bloch walls and from the Curie temperature. Also no relation is expected between the spatial variation of the coupling energy $\Phi(R)$ deduced from the spin-wave spectrum and from the high temperature local order.

However the main conclusions of the preceding discussion should subsist. Large moments should correspond to values of $\overline{U}_{\mathrm{eff}}$ much larger than its minimum value for magnetism to appear at $0\,^\circ$K. Thus in such cases slowly fluctuating and disordered atomic moments should subsist above the critical temperature.

The physical properties of Ni, Co, Fe and Mn are indeed consistent with such a prediction:

At the critical temperature, there exists a fairly narrow peak in the specific heat, compatible with the entropy of disordering of the magnetic moments observed at low temperatures. Near the critical temperature, critical neutron scattering phenomena are observed that can be analysed in terms of the degree of disorder of atomic moments (Jacrot *et al.* 1959, 1960, 1962; Kouvel & Fisher, 1964). Above the critical temperature, there is a magnetic scattering of neutrons (Jacrot *et al.* 1959, 1960, 1962; Kouvel & Fisher, 1964) and a scattering of the conduction electrons (White & Woods, 1958; Coles, 1958) that is well explained by the presence of disordered magnetic moments. Finally, above the critical temperature, there is a large temperature-dependent paramagnetism consistent with a Curie–Weiss law and a moment comparable with (although usually larger than) its low temperature value (Rhodes & Wohlfarth, 1963; Ubelacker, 1965).

None of these characteristic properties are observed in Cr near and above its Néel point (Wilkinson *et al.* 1962; Shirane & Takei, 1962). This leads to the assumption that Cr is the only transitional metal where the thermal disappearance of magnetism might be due mostly to one-electron excitations.

8.3.5 Effect of spin-orbit coupling on magnetism

As emphasized in §8.1.5, the energy shifts $\overline{\lambda}$ due to spin-orbit coupling are small compared with the width w of the d-band. The physical effects due to spin-orbit coupling can therefore be studied using *perturbation methods*, except in a range of energies of the order of $\overline{\lambda}$ near the band edges. Practically, this means that perturbation methods are acceptable except perhaps for Ni, Pd and Pt.

Within this scheme, we can write formally for the perturbed Bloch functions and energy shifts

$$|\mathbf{k}\alpha\beta\rangle = |\mathbf{k}\alpha\rangle + \overline{\lambda}\left\{\sum_{\mathbf{k}'}|\mathbf{k}'\alpha\rangle\frac{\langle\mathbf{k}'\alpha|\,\mathbf{ls}\,|\mathbf{k}\alpha\rangle}{E_{\mathbf{k}}-E_{\mathbf{k}'}} + \sum_{\mathbf{k}''}|\mathbf{k}''\beta\rangle\frac{\langle\mathbf{k}''\beta|\,\mathbf{ls}\,|\mathbf{k}\alpha\rangle}{E_{\mathbf{k}}-E_{\mathbf{k}''}}\right\} + \cdots,$$

$$(8.91)$$

$$\Delta E_{\mathbf{k}\alpha\beta} = -\lambda\langle\mathbf{k}\alpha|\,\mathbf{ls}\,|\mathbf{k}\alpha\rangle + \overline{\lambda}^2\left\{\sum_{\mathbf{k}'}\frac{|\langle\mathbf{k}\alpha|\,\mathbf{ls}\,|\mathbf{k}'\alpha\rangle|^2}{E_{\mathbf{k}}-E_{\mathbf{k}'}}\right.$$
$$\left.+ \sum_{\mathbf{k}''}\frac{|\langle\mathbf{k}\alpha|\,\mathbf{ls}\,|\mathbf{k}''\beta\rangle|^2}{E_{\mathbf{k}}-E_{\mathbf{k}''}}\right\} + \cdots;\quad(8.92)$$

α and β refer to the two spin directions.

The physical consequences of such perturbations have only been explored for ferro- and paramagnetic metals.

8.3.5a Ferromagnetic metals. Magnetic anisotropy; g-factors. The total change of energy due to spin-orbit coupling is obtained by summing (8.92):

$$\Delta E_{\text{total}} = \sum_{\text{occ}} (\Delta E_{\mathbf{k}\alpha\beta} + \Delta E_{\mathbf{k}\beta\alpha}).\qquad(8.93)$$

Using the fact that, for each $\mathbf{k}$ vector the spin-orbit coupling produces a broadening without energy shift (§8.1.6), this can also be written

$$\Delta E_{\text{total}} = -\sum_{\text{inocc}} (\Delta E_{\mathbf{k}\alpha\beta} + \Delta E_{\mathbf{k}\beta\alpha}).\qquad(8.94)$$

This form is useful in the actual ferromagnetic metals, which have more than half-full bands (Brooks, 1940; Fletcher, 1954; Asdente, 1968).

The summation has only been carried out for *cubic* crystals. Developing ΔE in cubic harmonics then gives

$$\Delta E_{\text{total}} = K_0(\alpha_1^2 + \alpha_2^2 + \alpha_3^2) + K_1(\alpha_1^2\alpha_2^2 + \alpha_2^2\alpha_3^2 + \alpha_3^2\alpha_1^2) + K_2(\alpha_1^2\alpha_2^2\alpha_3^2) + ...,$$
$$(8.95)$$

where α_1, α_2, α_3 are the direction cosines of the magnetic moment with respect to the cubic axes. The successive coefficients K_n correspond to successive terms in $\bar{\lambda}(\bar{\lambda}/w)^{1+2n}$; in the development (8.92), they decrease rapidly therefore for increasing n. The other terms disappear by symmetry. As the term $n = 0$ has full spherical symmetry, the energy of *magnetic anisotropy* is defined in practice by the two coefficients K_1 and K_2.

In the first series, $\bar{\lambda} \simeq 0.07$ eV and $w \simeq 7$ eV, then, as an *order of magnitude*, one expects $|K_1| \simeq \bar{\lambda}^4/w^3 \simeq 7 \cdot 10^{-8}$ eV/atom $\simeq 10^4$ ergs/cm³ and $|K_2| \simeq \bar{\lambda}^6/w^5 \simeq 7 \cdot 10^{-12}$ eV/atom $\simeq 1$ erg/cm³. Table 8.5 shows values of those orders at least for K_1. More detailed computations, using the explicit band structure, would be required to obtain the *signs* of the coefficients. It must be owned that those made so far have not been very successful numerically (Table 8.5); this is hardly surprising, as one is dealing with high-order perturbation terms on a not so well-known band structure.

Table 8.5. Experimental and computed values of the coefficients of magnetic anisotropy (in ergs cm⁻³)

| | K_1 | | K_2 |
	Measured	Computed	Measured
Ni	-5×10^4	-5×10^7	$\simeq 0$
Fe	4×10^5	5×10^3	1.5×10^5

In a non-cubic metal one expects the term in second order in the α_i's to contribute to the anisotropy. This should be of the order of $\bar{\lambda}^2/w \simeq 10^8$ ergs/cm³, i.e. larger than the coefficients in cubic crystals. The anisotropy observed in H.C.P. metals has indeed the required symmetry, and a coefficient of the order of 4×10^6 ergs/cm³, larger than those observed in Ni and Fe. If the band structure was known, one could hope that this coefficient would be computed with more accuracy than in cubic crystals.

The *gyromagnetic factors* g and g' have been similarly computed in terms of average components l_z and s_z of the orbital and spin moments along an applied magnetic field (see Kittel, 1956)

$$g' = \frac{\overline{l_z + 2s_z}}{\overline{l_z + s_z}}, \tag{8.96}$$

$$g = 2\,\overline{l_z + 2s_z}. \tag{8.97}$$

The average is made again over all occupied states; and, as is well known, it is equivalent to take the average over the unoccupied states, if we reverse the sign of $\overline{\lambda}$ (positive holes). It is then clear that, for the actual ferromagnetic metals with their more than half-filled d-bands, $g' < 2$ and $g > 2$. The leading term in the development (8.91) then gives

$$2 - g' \simeq g - 2 \simeq O(\overline{\lambda}/w) \simeq 10^{-2}. \tag{8.98}$$

This leading term is isotropic in cubic crystals, and again the development must be rapidly convergent.

More detailed computations have only been made for Ni. Here, the deviations from the value 2 are larger than as predicted by (8.98), because the Bloch states Δ_5 of the few positive holes mostly mix between themselves and with a neighbouring branch of states Δ_2. The energy differences involved are much smaller than w (Fig. 8.8a). The spin-orbit coupling factor $\overline{\lambda}$ is, however, still small enough for the perturbation scheme to be rapidly convergent (Table 8.6). As a result, no measurable anisotropy of the g-factor is expected. The slightly too large values computed are probably due to the too small bandwidth used in these computations.

Table 8.6. *Gyromagnetic factors in ferromagnetic nickel*

	Measured	Computed	
		to 1st order	to 2nd order
g	2·21 to 2·4	2·10	2·14
g'	1·85	1·85	1·84

It would again be of interest to study the g-factors and their possible anisotropy in ferromagnetic cobalt.

8.3.5b *Paramagnetic metals. g-factors; exchange enhancement factor.* Under an applied magnetic field H, the magnetic energy of a d-electron is (Lehman, 1959; Lenglart, 1967)

$$\epsilon = -[\mu_B(l_z + 2s_z) \pm 2S_z \overline{U}_{\text{eff}} s_z]. \tag{8.99}$$

Here, l and s refer to the electron considered, S to the total atomic spin moment created by the application of H. The second term is the internal field correction which only acts on s_z (cf. (8.48), (8.49)).

We can write for the total spin and orbital moments per atom

$$S_z = 2\overline{\epsilon s_z}\, n(E_M),$$

$$L_z = 2\overline{\epsilon l_z}\, n(E_M),$$

where the averages are made on the Fermi surface.

The magnetic susceptibility is then, taking into account both spin-orbit and internal field corrections,

$$\chi_{s0} = -\frac{\mu_B(L_z + 2S_z)}{H}$$

with $\qquad S_z = 2n(E_M)\,[2S_z\,\overline{U}_{\mathrm{eff}}\,\overline{s_z^{-2}} - \mu_B H(\overline{l_z s_z} + 2\overline{s_z^{-2}})].$

Finally $\qquad S_z = -\dfrac{2\mu_B H n(E_M)\,[\overline{l_z s_z} + 2\overline{s_z^{-2}}]}{1 - 4\overline{U}_{\mathrm{eff}}\,n(E_M)\overline{s_z^{-2}}},$ (8.100)

$$L_z = -2\mu_B H n(E_M)\left\{\overline{l_z^{-2}} + 2\overline{l_z s_z} + \frac{4\overline{U}_{\mathrm{eff}}\,\overline{l_z s_z}(\overline{l_z s_z} + 2\overline{s_z^{-2}})}{1 - 4\overline{U}_{\mathrm{eff}}\,n(E_M)\,\overline{s_z^{-2}}}\right\}$$ (8.101)

and

$$\chi_{s0} = 2\mu_B^2 n(E_M)\,\frac{(\overline{l_z + 2s_z})^2 + 4\overline{U}_{\mathrm{eff}}\,n(E_M)\,(\overline{l_z s_z^{-2}} - \overline{l_z^{-2} s_z^{-2}})}{1 - 4\overline{U}_{\mathrm{eff}}\,n(E_M)\,\overline{s_{\bar{z}}^{-2}}}.$$ (8.102)

Except in the cases of Pd and Pt considered below, the *perturbation scheme* of (8.91), (8.92) is expected to apply reasonably well even in the second and third series. Then

$$l_z = O\left(\frac{\overline{\lambda}}{w}\right), \quad 2s_z - 1 = O\left(\frac{\overline{\lambda}}{w}\,l_z\right) = O_2\left(\frac{\overline{\lambda}}{w}\right),$$

hence *the internal field factor is only altered to second order in* $\overline{\lambda}/w$, by a factor usually negligible. The condition of instability of the non-magnetic phase

$$4\overline{U}_{\mathrm{eff}}\,n(E_M)\,\overline{s_z^2} \geqslant 1$$ (8.103)

should *not* be significantly altered by spin-orbit coupling, except in Pd and Pt. Thus the absence of magnetic phases in the second and third series of transitional metals is probably related more to a general decrease in $\overline{U}_{\mathrm{eff}}$ and in $n(E_M)$ due to larger atomic orbits and stronger overlaps.

To first order in $\bar{\lambda}/w$, we can write

$$\chi_{s0} \simeq (\tfrac{1}{2}g)^2 \chi_0,$$

where χ_0 is the uniform susceptibility computed without spin-orbit coupling (8.52) and g is the Landé factor

$$g = 2[\overline{2s_z} + \overline{l_z}] = 2 + O(\bar{\lambda}/w).$$

The correction to be expected from spin-orbit coupling to the susceptibility is therefore *negligible* compared with other terms such as the exchange enhancement or the Van Vleck term.

8.3.5c *The cases of* Pd *and* Pt. As pointed out before, the spin-orbit coupling term $\bar{\lambda}$ is here equal to or larger than the Fermi energy of the positive holes. The perturbation method developed so far is expected to fail.

Indeed, in platinum, $\bar{\lambda}$ is definitely larger than the Fermi energy $E_a - E_M$, and also larger than the energy difference $X_5 - X_2$ between the top branches of the d band (Fig. 8.8a). Near the centre of a square face of the Brillouin zone along {100}, approximate Bloch functions can be taken as

$$\left. \begin{aligned} |k\alpha\beta\rangle &= \frac{1}{\sqrt{3}}[\,|k\phi_3\alpha\rangle + i|k\phi_2\alpha\rangle - |k\phi_4\beta\rangle\,], \\ |k\beta\alpha\rangle &= \frac{1}{\sqrt{3}}[\,|k\phi_3\beta\rangle - i|k\phi_2\beta\rangle + |k\phi_4\alpha\rangle\,], \end{aligned} \right\} \tag{8.104}$$

where $k\phi_2 = f(r)\,xy$ and $k\phi_3 = f(r)\,xz$ are the d-orbitals of the X_5 branch, and $k\phi_4 = \tfrac{1}{2}f(r)\,(x^2 - y^2)$ is the d-orbital of the X_2 branch. Such a solution would strictly make the Hamiltonian diagonal in the limit $|X_5 - X_2|/\bar{\lambda} \to 0$, and is expected to hold reasonably well in Pt.

It is very easy to show that the spin and orbital moments of the Kramers doublet (8.104) are *isotropic*, with reduced values

$$s_z = \tfrac{1}{6},$$

$$l_z = \tfrac{2}{3}.$$

The orbital moment of the positive holes is free to align itself parallel to the spin moment; therefore

$$\tfrac{1}{2}g = l_z + 2s_z = 1$$

and

$$\chi = \frac{2\mu_B^2\,n(E_M)}{1 - \dfrac{1}{g}\,\bar{U}_{\text{eff}}\,n(E_M)}.$$

This formula is the same as when the spin-orbit coupling is absent, except for a reduction by $\frac{1}{9}$ of the effective internal field energy $\overline{U}_{\text{eff}}$. Therefore for Pt the spin-orbit coupling should strongly reduce the Stoner enhancement factor and stabilize the non-magnetic state.

The situation of Pd is certainly more complex. A similar but weaker reduction of the Stoner factor is expected.

In conclusion of this study, it would be normal to analyse magnetostrictive effects, in an attempt to connect the studies of the last two sections. However, to do so would be difficult and even the simplest application, that of volume (hydrostatic) effects would be fairly complex (Patrick, 1954; Galperine et al. 1953; Kondorsky & Sedov, 1959, 1960, 1962; Bloch, 1966; Stoelinga et al. 1965; Fawcett & White, 1967).

AUTHOR INDEX AND BIBLIOGRAPHY

Abarenkov, I. V. & Heine, V.
(1965), *Phil. Mag.* **12**, 529   10, 12, 87
— *see* Heine, V.
Abate, E. & Asdente, M. (1965),
Phys. Rev. **140**, A 1303   351
Abrikosov, A. A., Gorkov, L. P.
& Dzyaloshinski, I. E. (1963),
*Methods of Quantum Field
Theory in Statistical Physics*
(English ed. London: Prentice-
Hall)   250
Aigrain, P. (1960), *Proc. Int. Conf.
Semiconductor Phys.* (Prague,
Czech Acad. Sci.), p. 224   246
Aldea, A., *see* Banyai, L.
Alekseevski, N. E., Korteus, G. E.
& Mozhaev, V. V. (1964), *Sov.
Phys. J.E.T.P.* **19**, 1333   358
Alexander, S. & Anderson, P. W.
(1964), *Phys. Rev.* **133**, A 1594   345
— & Horwitz, G. (1966), *Sol. State
Comm.* **4**, 513; *Phys. Rev.*   379
— (1968), *Phys. Rev.* (to appear)   379
Algie, S. H. *see* Hall, E. O.
Allan, G. (1967), Thèse de 3 ème
Cycle (Orsay)   356, 358, 391, 393
Alperin, H. A., Brown, P. J. &
Nathans, R. (1963), *J. Appl.
Phys.* **34**, 1201   324, 335
Altmann, S. L. (1958), *Proc. Roy.
Soc. A*, **244**, 141, 153   11
Amar, H., *see* Johnson, K. H.
Amundsen, T. (1966), *Proc. Phys.
Soc.* **88**, 757   248
Anderson, J. R. & Gold, A. V.
(1965), *Phys. Rev.* **139**, A 1459
1, 103
— , O'Sullivan, W. J. & Schirber,
J. E. (1967), *Phys. Rev.* **153**, 721   104
Anderson, P. W., *see* Alexander, S.
Anderson, P. W. & McMillan,
W. L. (1967), *Teoria del mag-
retisino nei metalli di transi-
zione.* Varenna Summer School
(New York), XXXVII   345
Animalu, A. O. E. (1965), *Phil.
Mag.* **11**, 379   15, 19, 299
— (1966), *Phil. Mag.* **13**, 53   22
— Bonsignori, F. & Bortolani, V.
(1966), *Nuovo Cimen to*, **44**B,
159   39
— & Heine, V. (1965), *Phil. Mag.*
12, 1249   13, 16, 24, 37, 42, 44

— *see* Vasvari, B.
Antoncik, E. (1959), *J. Phys.
Chem. Solids*, **10**, 314   6
Arase, T., *see* Brockhouse, B. N.   25
Armstrong, J. E. (1935), *Phys.
Rev.* **47**, 391   311
Asdente, M. (1968), to appear   403
— *see* Abate, E.
Asdente, F. & Friedel, J. (1961),
Phys. Rev. **124**, 384   350, 351, 354
— (1962), *Phys. Rev.* **126**, 2262
350, 351, 354
Ashcroft, N. W. (1963), *Phil. Mag.*
8, 2055   17, 18, 97, 99
— (1965), *Phys. Rev.* **140**, A 935   84
— (1966a), *Phys. Lett.* **23**, 48   21
— (1966b), *Phys. Lett.* **23**, 529   302
— & Guild, L. J. (1965), *Phys.
Lett.* **14**, 23   35, 299
— & Langreth, D. C. (1967), *Phys.
Rev.* **155**, 682   46
— & Lekner, J. (1966), *Phys.
Rev.* **145**, 83   299, 308
— & Wilkins, J. W. (1965), *Phys.
Lett.* **14**, 285   36, 60, 89
Auch, K., *see* Justi, E.
Austin, B. J. & Heine, V. (1966),
J. Chem. Phys. **45**, 928   25
— , Heine, V. & Sham, L. J. (1962),
Phys. Rev. **127**, 276   6
Azbel', M. Ya. (1954), *Doklady
Akad. Nauk.* **98**, 519   190
— (1960), *Sov. Phys. J.E.T.P.* **12**,
283   232, 235, 236
— (1963a), *Sov. Phys. J.E.T.P.*
17, 667   190
— (1963b), *Sov. Phys. J.E.T.P.*
17, 851   190, 191
— (1964), *Sov. Phys. J.E.T.P.*
19, 634   132
— & Gurzhi, R. N. (1962), *Sov.
Phys. J.E.T.P.* **15**, 1133   188
— & Kaner, E. A. (1955), *Sov.
Phys. J.E.T.P.* **2**, 749   206
— (1957), *Sov. Phys. J.E.T.P.* **5**,
730   209, 219, 221
— (1958), *J. Phys. Chem. Solids*,
6, 113   209, 218, 221, 226, 244
— & Peschanskii, V. G. (1965),
Sov. Phys. J.E.T.P. **22**, 399   191

Bacon, G. E. (1961), *Acta Cryst.*
14, 823   326

Bailyn, M. (1960a), *Phil. Mag.* **5**, 1059   277
— (1960b), *Phys. Rev.* **120**, 381   272
— (1961), *Phys. Rev.* **121**, 1336   253
Ballentine, L. E. (1966), *Can. J. Phys.* **44**, 2533   316
Baluffi, R. W., *see* Simmons, R. O.
Banyai, L. & Aldea, A. (1966), *Phys. Rev.* **143**, 652   311
Bardeen, J. (1937), *Phys. Rev.* **52**, 688   256, 257
— (1956), *Handb. Physik.* **15**, 274   151, 152
Bardos, D. I., *see* Shrinivasan, T. M.
Barisic, S. & Labbé, J. (1968), *J. Phys. Chem. Solids* (to appear)   370
Barron, T. H. K., *see* MacDonald, D. K. C.
Basinski, Z. S., Dugdale, J. S. & Howie, A. (1963), *Phil. Mag.* **8**, 1989   268
Bate, R. T. Martin, B. & Hille, P. F. (1963), *Phys. Rev.* **131**, 1482   187
Batterman, B. W. (1959), *Phys. Rev.* **115**, 81   332
— , Chipman, D. R. & De Marco, J. J. (1961), *Phys. Rev.* **122**, 68   332
Baym, G. (1964), *Phys. Rev.* **135**, A1691   253, 255, 290
Bearden, J. A. & Snyder, T. M. (1941), *Phys. Rev.* **59**, 162   358
Beck, P. A., *see* Shrinivasan, T. M.
Beeby, J. L. (1967), *Teoria del magnetismo nei metalli di transizione.* Varenna Summer School (New York), XXXVII   373
— & Edwards, S. F. (1963), *Proc. Roy. Soc. A*, **274**, 395   316
Belding, E. F. (1959), *Phil. Mag.* **4**, 1145   350, 354
Bennemann, K. H. (1964), *Phys. Rev.* **133**, A1045   7, 40
— (1965), *Phys. Rev.* **137**, A1497; **139**, A482   7, 40
Bennett, A. J. & Falicov, L. M. (1964), *Phys. Rev.* **136**, A998   128
Bennett, H. E. & Bennett, J. M. (1966), *Optical Properties and Electronic Structure of Metals and Alloys*, ed. F. Abeles (Amsterdam and New York; North-Holland and Wiley), p. 175.   184
Bergstresser, T. K., *see* Cohen, M. L.
Berk, N. F. (1966), Thesis   398

— & Schrieffer, J. R. (1966), *Phys. Rev. Lett.* **17**, 433   398
Berko, S. (1962), *Phys. Rev.* **128**, 2166,   76
Bernal, J. D. (1964), *Proc. Roy. Soc. A*, **280**, 299   286
Bhargava, R. N. (1967), *Phys. Rev.* **156**, 785   110
Bhatia, A. B. & Krishnan, K. S. (1948), *Proc. Roy. Soc. A*, **194**, 185   290
Black, P. J. (1956), *Acta Met.* **4**, 172   319
— & Cundall, J. A. (1966), *Acta Cryst.* **20**, 417   321
Blandin, A. (1963), *Metallic Solid Solutions*, eds J. Friedel & A. Guinier (New York: G. Benjamin), 50   37
— (1966), *Phase Stability in Metals and Alloys*, ed. P. S. Rudman (New York: McGraw Hill)   51
— (1967), *Teoria del magnetismo nei metalli di transizione*, Varenna Summer School (New York), XXXVII   345
— , Friedel, J. & Saada, G. (1967), *J. Phys. Rad.* (to appear)   40
— & Lederer, P. (1965), *Proc. Int. Conf. Mag., Nottingham*   379, 385, 395, 396, 397, 400, 401
— (1966), *Phil. Mag.* **14**, 363   379, 385, 395, 396, 397, 400, 401
— *see* Pick, R.
Blanpain, R. (1957), *Bull. Roy. Sci. Liege*, **4**, 165, 182   356
Blatt, F. J. (1957), *Phys. Rev.* **108** 285   261
— (1963), *Phys, Lett.* **8**, 235   279
— & Satz, H. G. (1960), *Helv. Phys. Acta*, **33** 1007   188
— *see* Le Page, J
Bloch, D. (1966), *Ann. Phys.* **1**, 93   408
Bloch, F. (1928), *Z. Phys.* **59**, 208   252
Bloom, F. A., *see* Joseph, A. S.
Bloomfield, P. (1966a), *Physica*, **32**, 1189   218, 219
— (1966b), *Bull. Am. Phys. Soc.* **11**, 170   194, 197
Blount, E. I. (1962), *Phys. Rev.* **126**, 1636   114, 129
Bohm, H. V. & Easterling, V. J. (1962), *Phys. Rev.* **128**, 1021   92
Bonnelle, C. (1964), *Thèse (Paris)*   349
Bonsignori, F., *see* Animalu, A. O. E.
Bornemann, K. & von Rauschenplat, G. (1912), *Metallurgie*, **16**, 505   309

Eckstein, Y. (1966), *Phys. Lett.*
20, 142   74
— *see* Priestley, M. G.
Edington, J. W. & Smallman,
R. E. (1965), *Phil. Mag.* 11, 1109   39
Edwards, S. F. (1962), *Proc. Roy.*
Soc. A, 267, 518   316
— *see* Beeby, J. L.
Egelstaff, P. A., *see* Enderby, J. E.
Ehrenreich, H. & Phillipp, H. R.
(1962), *Phys. Rev.* 128. 1622   350
— *see* Hodges, L.
Elliott, R. J. (1954), *Phys. Rev.*
96, 280   354
El Naby, M. H., *see* Mayer, H.
Enderby, J. E., *see* Howe, R. A.
— , North, D. M. & Egelstaff, P. A.
(1966), *Phil. Mag.* 14, 961   306
Endo, H. (1963), *Phil. Mag.* 8, 1403   304
Engelsberg, S., *see* Doniach, S.
Evans, W. A. B. (1966), *Proc.*
Phys. Soc. 88, 723   311

Faber, T. E. (1966), *Adv. Phys.*
15, 547   291, 292, 295, 313, 314
— (1967a), *Phil. Mag.* 15, 1   284
— (1967b), *Adv. Phys.* 16, 637 309, 310
— & Ziman, J. M. (1965), *Phil.*
Mag. 11, 153   261, 263, 306
Fadikov, I., *see* Kikoin, I.
Falicov, L. M. (1962), *Phil. Trans.*
A, 255, 55   109
— & Lin, P. J. (1966), *Phys. Rev.*
141, 562   2, 36
— & Penn, D. (1967), *Phys. Rev.*
158, 476   53
— , Pippard, A. B. & Sievert, P. R.
(1966), *Phys. Rev.* 151, 498
163, 171
— & Sievert, P. R. (1964), *Phys.*
Rev. Lett. 12, 550   169
— & Stachowiak, H. (1966), *Phys.*
Rev. 147, 505   139, 142
— & Stark, R. W. (1967), *Prog.*
L. T. Phys. 5, 235   129, 173
— *see* Bennett, A. J.
— *see* Cohen, M. H.
— *see* Cohen, M. H. & Golin, S.
— *see* Lee, M. J. G.
— *see* Lin, P. J.
Fal'ko, V. L., *see* Kaner, E. A.
235, 240
Fawcett, E. (1964), *Adv. Phys.*
13, 139   163
— & Reed, W. A. (1962), *Phys.*
Rev. Lett. 9, 336   358
— & Walsh, N. H. (1962), *Phys.*
Rev. Lett. 8, 476   358

— & White, G. K. (1967), *Int.*
Conf. Mag. (Boston)   408
Fert, A., *see* Campbell, I. A.
Fielder, M. L. (1967), *Adv. Phys.*
16, 681   311
— *see* Cusack, N. E.
Fisher, L. M., *see* Sharvin, Yu. V.
Fisher, M. E., *see* Kouvel, J. S.
Fletcher, G. C. (1954), *Proc. Phys.*
Soc. 67 A, 505   403
— & Wohlfarth, E. P. (1951), *Phil.*
Mag. 42, 106   350, 357
Forsyth, J. B. & Brown, P. J.
(1964), *Proc. Conf. Magnet.*
Nottingham (London: Inst. of
Phys. and Phys. Soc.), p. 524   325
— , Pickart, S. J. & Brown, P. J.
(1966), *Proc. Phys. Soc.* 88,
333   326, 327
— *see* Brown, P. J.
Fournet, G. (1957), *Handb. Physik.*
(Berlin: Springer), 33, 238   289
Foxon, C. T. B., *see* Rider, J. G.
Franck, E. U. & Hensel, F. (1966),
Phys. Rev. 147, 109   305
Freeman, A. J., *see* Hodges, L.
Friedel, J. (1952), *Proc. Phys. Soc.*
65 b, 769   359, 363, 371
— (1954), *Adv. Phys.* 3, 446   263, 310
— (1955), *J. Phys.* 16, 829
342, 354, 357, 392, 393
— (1956), *J. Phys. Chem. Solids*,
1, 175   350
— (1958), *Nuovo Cim. Suppl.* 7,
287 342, 354, 357, 376, 385, 392, 393
— (1962), *J. Phys.* 23, 501
342, 354, 357, 392, 393
— (1964), *Trans. AIME*, 230, 616   362
— (1965), *Bull. Sté. Chim. de*
France, 80, 1186   343
— , Leman, G. & Olszewski, S.
(1961), *J. Appl. Phys.* 32, 3255
385, 401
— , Lenglart, P. & Leman, G.
(1964), *J. Phys. Chem. Solids*,
25, 781   351, 358
— *see* Asdente, F.
— *see* Blandin, A.
— *see* Labbé, J.
Fuchs, K. (1935), *Proc. Roy. Soc.*
A, 151, 585   37
— (1938), *Proc. Camb. Phil. Soc.*
34, 100   176, 183, 186
Fujiwara, K. & Sueoka, O. (1966),
J. Phys. Soc. Japan, 21, 1947   93
— , Sueoka, O. & Imura, T. (1966),
J. Phys. Soc. Japan, 21, 2738   93
Fukuda, N., *see* Sawada, K.

Galperine, F. *et al.* (1953), *Dok.*
Akad. Nauk. SSSR, **89**, 429   408
Gantmakher, V. F. (1963), *Soviet*
Phys. J.E.T.P. **15**, 982; **16**,
247   70, 231
— (1967), *Prog. L.T. Phys. V*
(Amsterdam: North-Holland),
181   231
— & Kaner, E. A. (1963), *Soviet*
Phys. J.E.T.P. **18**, 988
                   232, 233, 235, 241
— (1965), *Soviet Phys. J.E.T.P.*
21, 1053   249
— & Krylov, I. P. (1965), *Soviet*
Phys. J.E.T.P. **22**, 734   237
— & Sharvin, Yu. V. (1965),
Soviet Phys. J.E.T.P. **21**, 720   233
Garber, M., *see* LePage, J.
Gautier, F. (1962), *J. Phys.* **23**,
738   357, 385
Gavenda, J. D., *see* Deaton, B. C.
Geller, S. (1956), *Acta Cryst.* **9**,
885   318
— (1957), *Acta Cryst.* **10**, 380   318
Gersdorf, R., *see* Stoelinga, J. H. M.
Gerstenkorn, H. (1952), *Ann.*
Phys. **10**, 49   290
Gertner, E., *see* Joseph, A. S.
Gingrich, N. S., *see* Wilkinson,
M. K.
Gold, A. V. (1958), *Phil. Trans.*
A, **251**, 85   97
— *see* Anderson, J. R.
Golin, S. (1968), *Phys. Rev.* **166**,
643   21
— *see* Cohen, M. H.
Goodenough, J. B. (1962), *Mag-*
netism and the Chemical Bond
(Interscience: New York)   350
Gordon, W. L., *see* Joseph, A. S.
— *see* Larson, C. O.
— *see* Plummer, R. D.
Gorkov, L. P., *see* Abrikosov,
A. A.
Gorter, F. W., *see* Klemens, P. G.
Graham, G. M. (1958), *Proc. Roy.*
Soc. A, **248**, 522   186
Greene, M. P. & Kohn, W. (1965),
Phys. Rev. **137**, A 513
                             255, 259, 272
Greene, R. F. (1966), *Phys. Rev.*
141, 687   183
Greenfield, A. J. (1966), *Phys.*
Rev. Lett. **16**, 6   302
Greenwood, D. A. (1958), *Proc.*
Phys. Soc. **71**, 585   250
Grimes, C. C. & Kip, A. F. (1963),
Phys. Rev. **132**, 1991   225, 228

— , Kip, A. F., Spong, F., Strad-
ling, R. A. & Pincus, P. (1963),
Phys. Rev. Lett. **11**, 455   228, 229
Gugan, D., *see* Dugdale, J. S.
Guild, L. J., *see* Ashcroft, N. W.
Güntherodt, H.-J., *see* Busch, G.
Gurevich, V. L. (1958), *Sov. Phys.*
J.E.T.P. **8**, 464   191, 197
Gurney, R. W., *see* Mott, N. F.
Gurzhi, R. N., *see* Azbel', M. Ya.
Gützwiller, M. C. (1964), *Phys.*
Rev. **134**, A 923   377
— (1965), *Phys. Rev.* **137**, A 1726   377

Häcker, W., *see* Bross, H.
Halder, N. C., *see* Wagner, C. N. J.
Hall, E. O. & Algie, S. H. (1966),
Metallur. Rev. **11**, 61   321
Halse, M. R. (1967), Ph.D. Thesis
(Cambridge University) and to
be published   91, 92, 93, 94
Ham, F. S. (1962), *Phys. Rev.* **128**,
82, 2524   32, 87
Hargreaves, M. F., *see* Clare-
brough, L. M.
Harper, P. G. (1955), *Proc. Phys.*
Soc. A, **68**, 879   172
Harrison, M. J., *see* Cohen, M. H.
Harrison, W. A. (1956), *Phys.*
Rev. **104**, 1281   264
— (1958), *J. Phys. Chem. Solids*,
5, 44   261, 266
— (1959), *Phys. Rev.* **116**, 555   17
— (1960), *Phys. Rev.* **118**, 1190
                         97, 98, 192, 360
— (1963), *Phys. Rev.* **129**, 2503,
2512   15, 37, 260
— (1966), *Pseudopotentials in the*
Theory of Metals (New York:
Benjamin)
    16, 34, 37, 38, 40, 49, 51, 256, 261, 263
— *see* Cohen, M. H.
Hartmann, L. E. (1966), Thesis
Columbia Univ., New York
(unpublished)   221
— & Luttinger, J. M. (1966),
Phys. Rev. **151**, 430; **156**, 1038   221
Hasegawa, A. (1964), *J. Phys.*
Soc. Japan, **19**, 504   272, 273
Hashitsume, N., *see* Kubo, R.
Häussler, P. & Welles, S. J. (1966),
Phys. Rev. **152**, 675   224, 226
Hebborn, J. E., Luttinger, J. M.,
Sondheimer, E. H. & Stiles,
P. J. (1964), *J. Phys. Chem.*
Solids, **25**, 741   152
Heine, V. (1957), *Phys. Rev.* **107**,
431   210

Heine, V. (1962), *Phil. Mag.* **7**, 775 58
— (1965*a*), *Low Temperature Physics*, ed. J. G. Daunt (New York: Plenum Press), p. 698 17
— (1965*b*), *Phil. Mag.* **12**, 53 281
— (1966), *Phase Statility in Metals and Alloys*, ed. P. S. Rudman (New York: McGraw-Hill) 25
— (1967), *Phys. Rev.* **153**, 673 28, 31, 33
— (1968), *J. Phys. C.* **1**, 222 50
— & Abarenkov, I. V. (1964), *Phil. Mag.* **9**, 451 86, 299
— , Nozieres, P. & Wilkins, J. W. (1966), *Phil. Mag.* **13**, 741 61
— & Weaire, D. (1966), *Phys. Rev.* **152**, 603 37, 48, 49
— *see* Abarenkov, I. V.
— *see* Animalu, A. O. E.
— *see* Austin, B. J.
— *see* Austin, B. J. & Sham, L. J.
— *see* Cohen, M. H.
— *see* Vasvari, B.
Heiniger, F., Bucher, E. & Muller, J. (1966), *J. Phys. Cond. Matter*, **5**, 243 356
Henry, W. G., *see* Crisp, R. S.
Hensel, F., *see* Franck, E. U.
Herman, F. (1964), *Physics of Semiconductors*. Proceedings of 7th International Conference (Paris: Dunod). 14
Herpin, A. & Meriel, P. (1961), *J. Phys. Rad.* **22**, 337 326
Herring, C. (1966), *Exchange Interactions among Itinerant Electrons, Magnetism IV* (New York) 342, 373, 376
— & Kittel, C. (1951), *Phys. Rev.* **81**, 5, 869 396
Hille, P. F., *see* Bate, R. T.
Hodges, C. H. (1967), *Phil. Mag.* **15**, 371 39, 40
Hodges, L., Ehrenreich, H. & Lang, N. D. (1966*a*), *Phys. Rev.* **152**, 505 29
— , Lang, N. D., Ehrenreich, H. & Freeman, A. J. (1966*b*), *J. Appl. Phys.* **37**, 1449 344, 391
Hodgson, J. N. (1959), *Phil. Mag.* **4**, 183 313
— (1960), *Phil. Mag.* **5**, 272 313
— (1961), *Phil. Mag.* **6**, 509 313
— (1962), *Phil. Mag.* **7**, 229 313
— (1963), *Phil. Mag.* **8**, 735 282
Holz, A., *see* Bross, H.

Horwitz, G., *see* Alexander, S.
Howe, R. A. & Enderby, J. E. (1967), *Phil. Mag.* **16**, 467 311
Howie, A. (1960), *Phil. Mag.* **5** 251 264, 265
— (1961), *Phil. Mag.* **6**, 1191 265
— *see* Basinski, Z. S.
Hubbard, J. (1963*a*), *Proc. Roy. Soc. A*, **276**, 238 377
— (1963*b*), *Proc. Roy. Soc. A*, **277**, 237 377
— (1964), *Proc. Roy. Soc. A*, **281**, 401 377
Huebner, R. P. (1966), *Phys. Rev.* **146**, 490 278
Hunter, S. C. & Nabarro, F. R. N. (1953), *Proc. Roy. Soc. A*, **220**, 542 266
Hurd, C. M. (1965), *Phil. Mag.* **12**, 47 281
Hutchinson, P., *see* Johnson, M. D.

Imura, T., *see* Fujiwara, K.
Iwamoto, F. & Sawada, K. (1962), *Phys. Rev.* **126**, 887 358, 379, 386, 395
Izuyama, T., Kim, D. J. & Kubo, R. (1963), *J. Phys. Soc. Japan*, **18**, 1025 379, 386, 395

Jacobs, R. L. (1968), *J. Phys. C.* **1**, 1296 and 1307 29, 33
Jacrot, B. *et al.* (1959), *J. Phys. Rad.* **20**, 178 402
— (1960), *J. Phys. Chem. Solids*, **13**, 235 402
— (1962), *J. Phys. Soc. Japan*, **17**, 35 402
Jennings, L. D., Chipman, D. R. & De Marco, J. J. (1964), *Phys. Rev.* **135**, A1612 332
— *see* Chipman, D. R.
Johnson, K. H. (1966), *Phys. Rev.* **150**, 429 12
— & Amar, H. (1965), *Phys. Rev.* **139**, A760 11
Johnson, M. D., Hutchinson, P. & March, N. H. (1964), *Proc. Roy. Soc. A*, **282**, 283 40
Jones, B. K., *see* Chambers, R. G.
Jones, H. (1937), *Proc. Phys. Soc.* **49**, 250 51
— *see* Mott, N. F.
Joseph, A. S. & Gordon, W. L. (1962), *Phys. Rev.* **126**, 489 65
— & Thorsen, A. C. (1965), *Phys. Rev.* **138**, A1159 158

Joseph, A. S., Thorsen, A. C. &
  Bloom, F. A. (1965), *Phys. Rev.*
  140, A2046   158
—, — Gertner, E. & Valby, L. E.
  (1966), *Phys. Rev.* **148**, 569   91
— *see* Thorsen, A. C.
Joshi, S. K., *see* Sharma, K. C.
Justi, E. & Auch, K. (1963), *Z.
  Naturforsch.* **18**, 767   167

Kadanoff, L. P., *see* Prange, R. E.
Kanamori, J. (1963), *Progr. Theor.
  Phys.* **30**, 275   377
Kanematsu, K. (1962), *J. Phys.
  Soc. Japan*, **17**, 85   325
Kaner, E. A. (1958), *Sov. Phys.
  Doklady*, **3**, 314   232
— (1963), *Sov. Phys. J.E.T.P.* **17**,
  700   235
— (1967), *Physics*, **3**, 285   241
— & Fal'ko, V. L. (1966), *Sov.
  Phys. J.E.T.P.* **24**, 392   235, 240
— & Skobov, V. G. (1963), *Sov.
  Phys. J.E.T.P.* **18**, 419   245
— (1964), *Sov. Phys. Solid State*,
  6, 851   244
— (1966), *Sov. Phys. Uspekhi*, **9**,
  480   244, 246
— *see* Azbel', M. Ya.
— *see* Gantmakher, V. F.
Kao, Y. H. (1963), *Phys. Rev.*
  129, 1122,   110
— (1965), *Phys. Rev.* **138**, A1412   190
Kaplan, M., *see* Kleppa, O. J.
Kasper, J. S. & Kouvel, J. S. (1959),
  J. Phys. Chem. Solids, **11**, 231   326
— *see* Kouvel, J. S.
Keeton, S. C. & Loucks, T. L.
  (1967), *Phys. Rev.* **152**, 548   54
Kendall, P., *see* Cusack, N. E.
Ketterson, J. B., Priestley, M. G.
  & Vuillemin, J. J. (1966), *Phys.
  Lett.* **20**, 452,   358
— & Stark, R. W. (1967), *Phys.
  Rev.* **156**, 748   71, 105, 106
— *see* Priestley, M. G.
Khaikin, M. S. (1960), *Sov. Phys.
  J.E.T.P.* **12**, 152   222
— (1962*a*), *Sov. Phys. J.E.T.P.*
  14, 1260   69, 231
— (1962*b*), *Sov. Phys. J.E.T.P.*
  15, 18   228
— (1966), *J.E.T.P. Lett.* **4**, 113   222
Kikoin, I. & Fadikov, I. (1935),
  Phys. Z. Sowjet. **7**, 507   311
Kim, D. J., *see* Izuyama, T.
King-Smith, P. E. (1965), *Phil.
  Mag.* **12**, 1123   67

Kip, A. F., Langenberg, D. N. &
  Moore, T. W. (1961), *Phys.
  Rev.* **124**, 359   68
— *see* Grimes, C. C.
— *see* Koch, J. F.
— *see* Spong, F. W.
Kittel, C. (1963), *Phys. Rev. Lett.*
  10, 339   155
— (1956), *Introduction to Solid
  State Physics* (New York)   405
— *see* Herring, C.
Kjeldaas, T. (1959), *Phys. Rev.*
  113, 1473   73
Kleinman, L. (1963), *Phys. Rev.*
  130, 2283   36
— (1966), *Phys. Rev.* **146**, 472   38, 43
— & Phillips, J. C. (1962) *Phys.
  Rev.* **125**, 819   49
— *see* Phillips, J. C.
Klemens, P. G., van Baarle, C. &
  Gorter, F. W. (1964), *Physica*,
  30, 1470   277
Kleppa, O. J., Kaplan, M. &
  Thalmayer, C. E. (1961), *J.
  Phys. Chem.* **65**, 843   307
Koch, J. F. (1968), *Proc. 1967
  Simon Fraser Summer School,
  Electrons in Metals* (New York:
  Gordon and Breach)   222
— & Kuo, C. C. (1966), *Phys. Rev.*
  143, 470   222
—, Stradling, R. A. & Kip, A. F.
  (1964), *Phys. Rev.* **133**, A240
                                            93, 231
— & Wagner, T. K. (1966), *Phys.
  Rev.* **151**, 467   241
Koehler, W. C., Cable, J. W., Child,
  H. R., Moon, R. M. & Wollan,
  E. O. (1966), *I.U.Cr. Moscow
  Congr. Abstr.* **7**, 19   330
Koenigsberg, E. (1953), *Phys.
  Rev.* **91**, 8   190
Kohler, M. (1948), *Z. Phys.* **124**,
  772   269
Kohn, W. (1959*a*), *Phys. Rev. Lett.*
  2, 393   77, 369
— (1959*b*), *Phys. Rev.* **115**, 1460   114
— (1961), *Phys. Rev.* **123**, 1242   154
— *see* Greene, M. P.
— & Luttinger, J. M. (1957), *Phys.
  Rev.* **108**, 590   250
— & Rostoker, N. (1954), *Phys.
  Rev.* **94**, 1111   32
Kondorsky, E. I. & Sedov, V. L.
  (1959), *J. Phys. Rad.* **20**, 185   408
— (1960), *J. Appl. Phys.* **31**, 331   408
— (1962), *Sov. Phys. J.E.T.P.* **2**,
  561   408

Morgan, G. J. (1966), *Proc. Phys. Soc.* **89**, 365   11, 12
Morin, F. J. (1959), *Phys. Rev. Lett.* **3**, 34   343
Morton, V. M. (1960), Ph.D. Thesis (Cambridge University)   96
Motizuki, K., *see* Mitsudo, R.
Mott, N. F. (1934*a*), *Proc. Phys. Soc.* **46**, 680   273
— (1934*b*), *Proc. Roy. Soc.* A **146**, 465   283
— (1936), *Proc. Camb. Phil. Soc.* **32**, 281   262, 263
— (1949), *Proc. Phys. Soc.* **62**, 416   342, 358
— (1956), *Nature, Lond.* **178**, 1205   56
— (1964), *Adv. Phys.* **13**, 325   342, 373
— (1966), *Phil. Mag.* **13**, 989   305, 307, 313
— (1967), *Adv. Phys.* **16**, 49   305
— & Gurney, R. W. (1939), *Trans. Farad. Soc.* **35**, 364   286
— & Jones, H. (1936), *The Theory of the Properties of Metals and Alloys* (Oxford: Clarendon Press)   4, 26, 50, 51, 152, 257, 262, 276, 342, 343, 359, 376, 392
— & Stevens, K. W. H. (1957), *Phil. Mag.* **2**, 1364   342
Mozhaev, V. V., *see* Alekseevski, N. E.
Mueller, F. M. (1967), *Phys. Rev.* **153**, 659   29, 30, 343
— & Phillips, J. C. (1967), *Phys. Rev.* **157**, 600   350
— *see* Phillips, J. C.
Muller, J., *see* Heiniger, F.
Munarin, J. A. & Marcus, J. A. (1965), *Proc. Ninth Int. Conf. L.T.Phys.* (New York: Plenum)   197
— (1966), *Bull. Am. Phys. Soc.* **11**, 170   197
Murray, A. M. *see* March, N. H.
Myers, A. & Bosnell, J. R. (1965), *Phys. Lett.* **17**, 9   128

Nabarro, F. R. N., *see* Hunter, S. C.
— & Ziman, J. M. (1961), *Proc. Phys. Soc.* **78**, 1512   266
Nagamiya, T., *see* Mitsudo, R.
— *see* Tachiki, M.
Nakajima, S. (1955), *Suppl. Phil. Mag.* **4**, 363   151
Nathans, R. & Pickart, S. J. (1963), *Magnetism III* (New York)   391
— *see* Alperin, H. A.
— *see* Brown, P. J.
— *see* Pickart, S. J.

Nee, T. W., *see* Prange, R. E.
Nevitt, M. V. (1966), *Phase Stability in Metals and Alloys*, ed. P. S. Rudman (New York: McGraw-Hill)   46
— (1963), *Electronic Structure and Alloy Chemistry of the Transition Elements*, ed. P. A. Beck (Wiley), p. 101   319, 321
Nilsson, G., *see* Stedman, R.
Noguchi, S., *see* Takeuchi, T.
North, D. M., *see* Enderby, J. E.
Nozieres, P. (1964), *Theory of Interacting Fermi Systems* (New York: Benjamin)   56
— *see* Heine, V.
— & Pines, D. (1966), *The Theory of Quantum Liquids* (New York: Benjamin)   56, 61

Obata, Y., *see* Kubo, R.
Okazaki, M., *see* Onodera, V.
Okumura, K. & Templeton, I. M. (1965), *Proc. Roy. Soc.* A, **287**, 89   84, 86
— *see* Dugdale, J. S.
Olsen, J. L. (1958), *Helv. Phys. Acta*, **31**, 713   187, 188
— (1962), *Electron Transport in Metals* (New York and London: Interscience)   187
Olszewski, S., *see* Friedel, J.
Onodera, Y. & Okazaki, M. (1966), *J. Phys. Soc. Japan*, **21**, 1273   11
Onsager, L. (1952), *Phil. Mag.* **43**, 1006   119
Ornstein, L. S. & Zernike, F. (1914), *Amst. Proc.* **17**, 793   296
O'Sullivan, W. J., *see* Anderson, J. R.
Otter, M. (1961), *Z. Phys.* **5**, 539   312
Overhauser, A. W. (1962), *Phys. Rev.* **126**, 517, 1437   379
— (1964), *Phys. Rev. Lett.* **13**, 190   89
— & Rodriguez, S. (1966), *Phys. Rev.* **141**, 431   248

Paskin, A. & Weiss, R. J. (1962), *Phys. Rev. Lett.* **9**, 199   78
Patrick, L. (1954), *Phys. Rev.* **93**, 384   408
Patterson, J. (1901), *Proc. Camb. Phil. Soc.* **11**, 118   176
Paul, W., *see* Deutsch, T.
Pauling, L. (1938), *Phys. Rev.* **54**, 899   390
— (1939), *Coll. Int. Mag. CNRS Strasbourg*   342

Pauling, L. (1945), *The Nature of The Chemical Bond* (Ithaca: Cornell University Press)   318
— (1957), *Acta Cryst.* **10**, 374   318
Pearson, W. (1966), *Phase Stability in Metals and Alloys* ed. P. S. Rudman (New York: McGraw-Hill)   54
Peierls, R. E. (1933), *Z. Physik.* **81**, 186   148
— (1955), *Quantum Theory of Solids* (Oxford)   151
Penn, D. R. (1965), *Phys. Rev.* **142**, 350   379
— *see* Falicov, L. M.
Perel', V. I., *see* Konstantinov, O. V.
Peschanskii, V. G., *see* Azbel', M. Ya.
Phariseau, P. & Ziman, J. M. (1963), *Phil. Mag.* **8**, 1487   316
Phillipp, H. R., *see* Ehrenreich, H.
Phillips, D., *see* Dugdale, J. C.
Phillips, J. C. (1965), *Phys. Rev.* **140**, A1254   28
— (1967), *Teoria del magnetismo nei metalli di transizione*, Varenna Summer School (New York), XXXVII   343, 345
— & Kleinman, L. (1959), *Phys. Rev.* **116**, 287   6, 305
— (1962), *Phys. Rev.* **128**, 2098   16
— & Mueller, F. M. (1967), *Phys. Rev.* **155**, 594   26, 28
— *see* Cohen M. H.
— *see* Kleinman, L.
— *see* Mueller, F. M.
Pick, R. & Blandin, A. (1964), *J. Cond. Matter*, **3**, 1   366
— & Sarma, G. (1964), *Phys. Rev.* **135**, A1363   37
Pickart, S. J., *see* Forsyth, J. B.
— & Nathans, R. (1961), *Phys. Rev.* **123**, 1163   337
— *see* Nathans, R.
Pincus, P., *see* Grimes, C. C.
Pines, D. (1955), *Solid St. Phys.* **1**, 367   16
— (1961), *The Many-Body Problem* (Benjamin)   154
— (1963), *Elementary Excitations in Solids* (New York: Benjamin)   27
— *see* Nozieres, P.
Pippard, A. B. (1947), *Proc. Roy. Soc.* A, **191**, 385   199
— (1954), *Proc. Roy. Soc.* A, **224**, 273   199

— (1957), *Phil. Trans. Roy. Soc.* A, **250**, 325   76, 89, 96, 223
— (1960), *Rep. Prog. Phys.* **23**, 176   163, 176, 203, 224, 236
— (1962), *Proc. Roy. Soc.* A, **270**, 1   130, 135
— (1964a), *Phil. Trans. Roy. Soc.* A, **265**, 317   121, 130, 132, 136, 163, 172
— (1964b), *Proc. Roy. Soc.* A, **282**, 464   72, 163, 167, 178, 188
— (1965a), *Proc. Roy. Soc.* A, **287**, 165   144, 162, 170, 172
— (1965b), *The Dynamics of Conduction Electrons* (Gordon and Breach)   153, 163, 176, 183, 203, 236
— (1966a), *Phil. Mag.* **13**, 1143   197
— (1966b), *Optical Properties of Metals and Alloys*, ed. F. Abeles (Amsterdam and New York: North-Holland and Wiley), p. 622   202, 242
— *see* Falicov, L. M.
Placzek, G. (1952), *Phys. Rev.* **86**, 377   291
Platzman, P. M. & Tzoar, N. (1965), *Phys. Rev.* **39**, A410   77
— & Walsh, W. M. (1967), *Phys. Rev. Lett.* **19**, 514; **20**, 89   245
— *see* Walsh, W. M.
Plummer, R. D. & Gordon, W. L. (1964), *Phys. Rev. Lett.* **13**, 432   158
Pogorelov, A. V., *see* Lifshitz, I. M.
Pomeroy, A. R., *see* Campbell, I. A.
Postill, D. R., Ross, R. G. & Cusack, N. E. (1967), *Adv. Phys.* **16**, 493   305
Powell, B. M., *see* Woods, A. D. B.
Prange, R. E. & Kadanoff, L. P. (1964), *Phys. Rev.* **134**, A566   61
— & Nee, T. W. (1968), *Phys. Rev.* **168**, 779   222
— & Sachs, A. (1967), *Phys. Rev.* **158**, 672   61
Price, P. J. (1957), *IBMJ Res. Dev.* **1**, 147, 239   177
— (1958), *IBMJ Res. Dev.* **2**, 200   177
— (1960), *IBMJ Res. Dev.* **4**, 152   183
Priestley, M. G., *see* Ketterson, J. B.
— Windmiller, L. R., Ketterson, J. B. & Ekstein, Y. (1967), *Phys. Rev.* **154**, 671   2
— *see* Vuillemin, J. J.

Rajagopal, A. K. & Brooks, H.
(1967), *Phys. Rev.* **158**, 552   392
Rao, K. R., *see* Brockhouse, B. N.
Rayne, S. A. (1956), *Aust. J. Phys.*
9, 189   360
— *et al.* (1962), *Phys. Rev. Lett.* **8**,
199   358
— (1963), *Phys. Rev.* **132**, 1945   358
— (1964), *Phys. Lett.* **8**, 155   358
Raynor, G. V. (1949), *Progr.
Metal Phys.* **1**, 1   319
Reed, W. A. (1964), *Bull. Am.
Phys. Soc.* II, **9**, 633   168
— *see* Fawcett, E.
Reuter, G. E. H. & Sondheimer,
E. H. (1948), *Proc. Roy. Soc.
A,* **195**, 336   204, 243
Rhodes, P. & Wohlfarth, E. P.
(1963), *Proc. Roy. Soc. A,* **273**,
247   402
Rice, S. A., *see* Wilson, E. G.
Rice, T. M. (1965), *Ann. Phys.* **31**,
100   57, 256
Rider, J. G. & Foxon, C. T. B.
(1966), *Phil. Mag.* **13**, 289   266
Roaf, D. J. (1962), *Phil. Trans.
A,* **255**, 135   91, 94
Roberts, S. (1960), *Phys. Rev.* **118**,
1509   360
Rodriguez, S., *see* Overhauser,
A. W.
Rose, F., *see* Bowers, R.
Ross, R. G., *see* Postill, D. R.
Rostoker, N., *see* Kohn, W.
Roth, L. M. (1966), *Phys. Rev.*
145, 434   119, 149

Saada, G., *see* Blandin, A.
Sachs, A. *see* Prange, R. E.
Sarginson, K., *see* MacDonald,
D. K. C.
Sarma, G., *see* Pick, R.
Sato, H. & Toth, R. S. (1961)
Phys. Rev. **124**, 1833   53
— (1962), *Phys. Rev.* **127**, 469   322
Satz, H. G., *see* Blätt, F. J.
Sawada, K. & Fukuda, N. (1961),
Prog. Theor. Phys. **25**, 653
    358, 379, 386, 395
— *see* Iwamoto, F.
Schiff, L. I. (1955), *Quantum
Mechanics,* 2nd edition (New
York: McGraw-Hill)   5
Schirber, J. E., *see* Anderson, J. R.
Schrieffer, J. R., *see* Berk, N. F.
Schulz, L. G. (1957), *Adv. Phys.* **6**,
102   313
Sedov, V. L., *see* Kondorsky, E. I.

Seeger, A. (1962), *J. Phys. Rad.*
23, 616   259
— & Bross, H. (1960), *Z. Natur-
forsch.* **15a**, 663   266
Segall, B. (1961), *Phys. Rev.* **124**,
1797   11, 33
Seidel, G., *see* Broshar, W.
Seitz, F. (1940), *Modern Theory of
Solids* (New York)   392
Sham, L. J. (1961), *Proc. Phys.
Soc.* **78**, 895   255
— (1963), Thesis Univ. of Cam-
bridge (unpublished). Available
in microfilm or photocopy from
Micromethods, Ltd, East
Ardsley, Wakefield, Yorkshire
    16, 37, 38
— (1965), *Proc. Roy. Soc. A,* **283**,
33   16, 38
— & Ziman, J. M. (1963), *Solid
St. Phys.* **15**, 221   36, 252–7
— *see* Austin, B. J.
Sharma, K. C. & Joshi, S. K.
(1965), *Phys. Rev.* **140**, A1799   272
Sharvin, Yu. V. (1965), *Sov. Phys.
J.E.T.P.* **21**, 655   198
— & Fisher, L. M. (1965), *Sov.
Phys. J.E.T.P. Lett.* **1**, 152   198
— *see* Gantmakher, V. F.
Shimizu, M. *et al.* (1962), *J. Phys.
Soc. Japan,* **17**, 1740   393
— (1963), *J. Phys. Soc. Japan,*
18, 240   393
— (1964), *J. Phys. Soc. Japan,*
19, 1135   393
Shirane, G. (1964), *Proc. Int. Conf.
Mag., Nottingham*   386
— & Takei, W. J. (1962), *J. Phys.
Soc. Japan,* **17**, 35   386, 403
Shoemaker, C. B. & Shoemaker,
D. P. (1964), *Trans. Met. Soc.
AIME,* **230**, 486   318
Shoenberg, D. (1962), *Phil. Trans.
A,* **255**, 85   91, 154, 158
— (1965), *Low Temperature
Physics* (*LT* 9) (Plenum Press),
p. 680   63
— & Stiles, P. J. (1964), *Proc.
Roy. Soc. A,* **281**, 62   85, 86
— *see* Caplin, A. D.
Shrinivasan, T. M., Claus, H.,
Viswanathan, R., Beck, P. A.
& Bardos, D. I. (1966), *Phase
Stability in Metals and Alloys,*
ed. P. S. Rudman (New York:
McGraw Hill)   41
Shubin, S. (1934), *Phys. Z. Sowjet*
5, 81   290

Shull, C. G. (1963), *Electronic Structure and Alloy Chemistry of Transition Elements*, ed. P. A. Beck (Wiley) p. 69   335
— & Yamada, Y. (1961), *Int. Conf. Mag. and Crystallo.* (Kyoto)   391
— *see* Mook, H. A.
— *see* Wilkinson, M. K.
Sievert, P. R., *see* Falicov, L. M.
Silin, V. P. (1957), *Sov. Phys. J.E.T.P.* 6, 985   59
— (1958), *Sov. Phys. J.E.T.P.* 7, 486   58
Simmons, R. O. & Baluffi, R. W. (1962), *Phys. Rev.* 125, 862   261
Skobov, V. G., *see* Kaner, E. A.
Slater, J. C. (1930), *Phys. Rev.* 36, 57   340, 374, 376
— (1937), *J. Appl. Phys.* 8, 385   390
— (1951a), *Phys. Rev.* 81, 385   14
— (1951b), *Phys. Rev.* 82, 535   53, 392
— (1965), *Quantum Theory of Molecules and Solids*, vol. 2 (New York: McGraw-Hill)   17
— (1966), *Phys. Rev.* 145, 599   12
— & Koster, G. F. (1954), *Phys. Rev.* 94, 1498   29, 351
Smallman, R. E., *see* Edington, J. W.
Smith, D. A. (1967), *Proc. Roy. Soc.* A, 297, 205   230
Smith, N. V. (1967), *Adv. Phys.* 16, 629   313
Smithells, C. J. (1949), *Metals Reference Book* (London)   361
Snyder, T. M., *see* Bearden, J. A.
Sondheimer, E. H. (1948), *Proc. Roy. Soc.* A, 193, 484   342
— (1950), *Phys. Rev.* 80, 401   190
— (1952), *Adv. Phys.* 1, 1   189
— (1962), *Proc. Roy. Soc.* A, 268, 100   280
— *see* Hebborn, J. E.
— *see* Reuter, G. E. H.
Sparlin, O. M. & Marcus, J. A. (1966), *Phys. Rev.* 144, 484   358
Spector, H. N. (1960), *Phys. Rev.* 120, 1261   74
Spong, F. W. & Kip, A. F. (1965), *Phys. Rev.* 137, A431   228, 229
Spong, F., *see* Grimes, C. C.
Springer, B. (1964), *Phys. Rev.* A, 136, 115   311
Stachowiak, H., *see* Falicov, L. M.
Stanford, J. L., *see* McGroddy, J. C.
— & Stern, E. A. (1966), *Phys. Rev.* 144, 534   248

Stark, R. W. (1964), *Phys. Rev.* 135, A1698   128, 169
— *see* Falicov, L. M.
— *see* Ketterson, J. R.
— *see* Tsiu, D. C.
Stedman, R. & Nilsson, G. (1965), *Phys. Rev. Lett.* 15, 634   78
Stern, E. A. (1963), *Phys. Rev. Lett.* 10, 91   248
— *see* McGroddy, J. C.
— *see* Stanford, J. L.
Stevens, K. W. H., *see* Mott, N. F.
Stewart, A. T., *see* Donaghy, J. J.
Stiles, P. J., *see* Hebborn, J. E.
— *see* Shoenberg, D.
Stoelinga, J. H. M., Gersdorf, R. & de Vries, G. (1965), *Physica*, 31, 349   408
Stoltz, H. (1963), *Phys. St. Solidi*, 3, 1153, 1493, 1957   236
Stone, I. (1898), *Phys. Rev.* 6, 1   176
Stoner, E. C. (1946), *Rep. Progr. Phys.* 9, 43   383
Stradling, R. A., *see* Koch, J. F.
— *see* Grimes, C. C.
Sueoka, O., *see* Fujiwara, K.
Sundström, L. H. (1965), *Phil. Mag.* 11, 657   303, 310
Switendick, A. C., *see* Earn, V.

Tachiki, M. & Nagamiya, T. (1963), *Phys. Lett.* 3, 214   330, 358, 379, 386, 395
— & Teramoto, K. (1966), *J. Phys. Chem. Solids*, 27, 335   53
Takei, W. J., *see* Shirane, G.
Takeuchi, T. & Noguchi, S. (1966), *J. Phys. Soc. Japan*, 21, 2222   311
Taylor, M. T. (1965), *Phys. Rev.* 137, A1145   248
Taylor, P. L. (1963a) *Phys. Lett.* 3, 245   279
— (1963b), *Proc. Roy. Soc.* A, 275, 200, 209   177, 280
— (1964), *Phys. Rev.* 135, A1333   264
Taylor, W. H. (1954), *Acta Met.* 2, 684   319
— *see* Dickens, G. J.
Templeton, I. M., *see* Okumura, K.
Teramoto, K., *see* Tachiki, M.
Thalmayer, C. E., *see* Kleppa, O. J.
Thellung, A., *see* Chester, G. V.
Thomson, J. J. (1901), *Proc. Camb. Phil. Soc.* 11, 120   176
Thorsen, A. C., Joseph, A. S. & Valby, L. E. (1966), *Phys. Rev.* 150, 523   28
— *see* Joseph, A. S.

Tieche, Y. & Zareba, A. (1963),
Phys. Kond. Mat. 1, 402   282
Toombs, G. A. (1965), Proc. Phys.
Soc. 86, 273, 277   308
Toth, R. S., see Sato, H.
Toulouse, G. (1966), Solid St.
Comm. 4, 593   377, 385
Treusch, J. (1967), Physica St.
Solidi, 19, 603   11
Tsuboya, I. (1961), J. Phys. Soc.
Japan, 16, 1875   325
Tsui, D. C. & Stark, R. W. (1966),
Phys. Rev. Lett. 16, 19
107, 108, 358
Twose, W. D., see Deegan, R.
Tzoar, N., see Platzman, P. M.

Ubelacker, E. (1965), Comptes
Rendus, 261, 976   402

Valby, L. E., see Joseph, A. S.
— see Thorsen, A. C.
Van Baarle, C., see Klemens, P. G.
Van Hove, L. (1953), Phys. Rev.
89, 1189   356
Van Vleck, J. H. (1932), The
Theory of Electric and Magnetic
Susceptibilities (Oxford)
115, 150, 383
— (1953), Rev. Mod. Phys. 25, 220   373
Vasvari, B., Animalu, A. O. E. &
Heine, V. (1967), Phys. Rev.
154, 535   22
Viswanathan, R., see Shrinivasan,
T. M.
Vogt, E. (1966), Z. angenew. Phys.
21, 287   356
Volski, E. P. (1964), J.E.T.P.
(Russian), 46, 123   100
Von Rauschenplat, G., see Borne-
mann, K.
Vuillemin, J. J. (1966), Phys. Rev.
144, 396   358
— see Ketterson, J. B.
— & Priestley, M. G. (1965), Phys.
Rev. Lett. 14, 307   28

Wagenmann, G., see Bornemann,
K.
Wagner, C. N. J. & Halder, N. C.
(1967), Adv. Phys. 16, 241   309
Wagner, T. K., see Koch, J. F.
Walsh, N. H., see Fawcett, E.
Walsh, W. M. & Platzman, P. M.
(1965), Phys. Rev. Lett. 15, 784   245
— see Platzman, P. M.
Watts, B. R. (1964a), Proc. Roy.
Soc. A, 282, 521   109

— (1964b), Phys. Lett. 10, 275   358
Weaire, D. (1967), Proc. Phys.
Soc. 92, 956   43
— (1968), J. Phys. C. 1, 210   49, 50
Weaire, D., see Heine, V.
Weinberg, I. (1966), Phys. Rev.
146, 486   278
Weiss, R. J. & De Marco, J. J.
(1958), Rev. Mod. Phys. 30, 59   332
— see Cooper, M.
— see De Marco, J. J.
— see Paskin, A.
Weisz, G. (1966), Phys. Rev. 149,
504   10, 22
Welles, S. J., see Häussler, P.
Wells, A. F. (1950), Structural
Inorganic Chemistry (2nd ed.)
(Oxford: University Press)   318
White, G. K. & Woods, S. B.
(1958), Phil. Trans. A, 251, 273   402
— see Fawcett, E.
Wilkins, J. W., see Ashcroft,
N. W.
— see Heine, V.
Wilkinson, C. (1965), Dissertation
for Ph.D., Cambridge   325, 337
— see Brown, P. J.
Wilkinson, M. K. et al. (1962),
Phys. Rev. 127, 2080   385, 403
—, Gingrich, N. S. & Shull, C. G,
(1957), J. Phys. Chem. Solids,
2, 289   324
Williams, R. W., Loucks, T. L. &
Mackintosh, A. R. (1966), Phys.
Rev. Lett. 16, 168   54
Wilson, A. J. C. (1950), Research,
3, 387   267
Wilson, E. G. & Rice, S. A. (1966),
Phys. Rev. 145, 55   312
Wilson, J. R. (1965), Metall. Rev.
10, 381   284, 308
Windmiller, L. R., see Priestley,
M. G.
Wiser, N. (1966), Phys. Rev. 143,
393   272
Wohlfarth, E. P. (1953), Rev.
Mod. Phys. 25, 211   373
— see Fletcher, G. C.
— see Rhodes, P.
Wollan, E. O., see Koehler, W. C.
Woods, A. D. B., see Brockhouse,
B. N.
— & Powell, B. M. (1965), Phys.
Rev. Lett. 15, 778   78
Woods, S. B., see White, G. K.
Wyder, P. (1965), Phys. Kond.
Mat. 3, 263   188, 189
— see Lüthi, B.

Yafet, Y. (1963), *Solid St. Phys.*
14, 1   126
Yamada, Y., *see* Shull, C. G.
Young, W. H., *see* Dickey, J. M.
— *see* Meyer, A.

Zak, J. (1964), *Phys. Rev.* 134,
A 1607   132
Zareba, A., *see* Tièche, Y.
Zernike, F., *see* Ornstein, L. S.
Ziman, J. M. (1954), *Proc. Roy.*
Soc. A, 226, 436   272
— (1955), *Proc. Camb. Phil. Soc.*
51, 707   253
— (1959a), *Proc. Roy. Soc.* A,
252, 63   265
— (1959b), *Phil. Mag.* 4, 371   277
— (1960), *Electrons and Phonons*
(Oxford: Clarendon Press)   250–281
— (1961a), *Phil. Mag.* 6, 1013
279, 290, 300

— (1961b), *Adv. Phys.* 10, 1
52, 278, 279, 280, 281
— (1961c), *Phys. Rev.* 121, 1320
280, 281
— (1963), *Electrons in Metals*
(London: Taylor and Francis)   319
— (1964a), *Principles of the*
Theory of Solids (Cambridge
University Press)
2, 4, 6, 9, 10, 15, 35, 57, 59, 60, 77,
129, 175, 191, 194, 250–281, 311
— (1964b), *Adv. Phys.* 13, 89
34, 255, 263, 267
— (1965), *Proc. Phys. Soc.* 86, 337
11, 256, 272
— (1967), *Adv. Phys.* 16, 551   273
— *see* Collins, J. G.
— *see* Faber, T. E.
— *see* Nabarro, F. R. N.
— *see* Phariseau, P.
— *see* Sham, L. J.

SUBJECT INDEX